Electrochemical Energy Storage Devices

Electrochemical Energy Storage Devices

Non-Conventional Technologies and Materials

Edited by Yongbing Tang and Luojiang Zhang

WILEY-VCH

Editors

Prof. Yongbing Tang
Shenzhen Institutes of Advanced Technoloy
Chinese Academy of Sciences
1068 Xueyuan Boulevard
Shenzhen University Town
Shenzhen
China, 518055

Dr. Luojiang Zhang
Shenzhen Institutes of Advanced Technoloy
Chinese Academy of Sciences
1068 Xueyuan Boulevard
Shenzhen University Town
Shenzhen
China, 518055

Cover Image: © sesame/Getty Images

Library of Congress Card No.: applied for

British Library Cataloguing-in-Publication Data
A catalogue record for this book is available from the British Library.

Bibliographic information published by the Deutsche Nationalbibliothek The Deutsche Nationalbibliothek lists this publication in the Deutsche Nationalbibliografie; detailed bibliographic data are available on the Internet at <http://dnb.d-nb.de>.

The manufacturer's authorized representative according to the EU General Product Safety Regulation is WILEY-VCH GmbH, Boschstr. 12, 69469 Weinheim, Germany, e-mail: Product_Safety@wiley.com.

Print ISBN: 978-3-527-34969-2
ePDF ISBN: 978-3-527-83479-2
ePub ISBN: 978-3-527-83480-8
oBook ISBN: 978-3-527-83481-5

Typesetting Straive, Chennai, India
Druck und Bindung: CPI books GmbH Leck, Germany

Contents

Preface

In recent decades, the indiscriminate use of fossil fuels has resulted in severe and irreversible resource depletion and environmental challenges, including pollution. These issues pose significant barriers to achieving sustainable social development. Renewable energy sources such as sunlight, wind, and tides present immense potential to supplement electricity generation in an environmentally sustainable manner, effectively addressing the demands of modern society. Simultaneously, the seamless integration of information technology with energy transformation is revolutionizing the processes of generating, acquiring, and using renewable energy, paving the way for a low-carbon society free from fossil fuels. Electrochemical energy storage devices and associated technologies are pivotal in modern energy systems. Their ability to flexibly adjust power and energy configurations to meet diverse application needs, coupled with their quick response capabilities, is essential for ensuring the coordinated operation of energy sources, grids, loads, and storage systems.

Electrochemical energy storage technologies date back to 1836 with the invention of the first copper-zinc primary battery, known as the Daniell cell. Among the many emerging technologies, lithium-ion batteries have swiftly dominated mainstream markets, such as mobile phones, laptops, and electric vehicles, since their initial commercialization in 1991. This success can be attributed to their recharging ability and impressive electrochemical performance. In 2019, lithium-ion batteries were awarded the Nobel Prize in Chemistry, which significantly accelerated advancement in the field and inspired a surge in research and development activities related to electrochemical energy storage technologies. In this evolving context, numerous "beyond lithium-ion" technologies are emerging to address diverse application requirements, such as large-scale energy storage, high power/energy/voltage demands, flexible electronics, and operation under extreme conditions.

This book provides a comprehensive and insightful exploration of nonconventional electrochemical energy storage devices, covering metal-air batteries (Chapter 2), lithium-sulfur batteries (Chapter 3), metal-CO_2 batteries (Chapter 4), multivalent-ion batteries (Chapter 5), dual-ion batteries (Chapter 6), M/N/C catalyst-based fuel cells (Chapter 7), aqueous batteries (Chapter 8), flow batteries (Chapter 9), hybrid capacitor (Chapter 10), and flexible energy storage devices (Chapter 11). The core principles and essential materials have been explored.

Furthermore, the book outlines the current challenges and offers effective strategies to optimize the performance of these devices. The primary audience for this book includes those specializing in electrochemical energy storage, as well as those focused on electrode/electrolyte materials in material science and material chemistry. The advanced research presented in this book is accessible to a broad audience with general technical knowledge.

In conclusion, sincere gratitude is extended to the experts who dedicated their time and effort to drafting and revising the manuscripts. Heartfelt thanks are also offered to my team members and students for their contributions. Special acknowledgment goes to the Wiley Editors for their exceptional support and assistance. This book aspires to engage a wide audience in the field of electrochemical energy storage research and serve as a practical guide for newcomers embarking on related studies.

24 January 2025

Yongbing Tang and Luojiang Zhang
Advanced Energy Storage
Technology Research Center
Institute of Technology for Carbon Neutrality
Shenzhen Institutes of Advanced Technology
Chinese Academy of Sciences
Shenzhen
518055
P.R. China

1

Introduction

Qingguang Pan[1,2] *and Yongbing Tang*[1,2]

[1] *Advanced Energy Storage Technology Research Center, Shenzhen Institute of Advanced Technology, Chinese Academy of Sciences, 1068 Xueyuan Avenue, Shenzhen University Town, Shenzhen 518055, China*
[2] *University of Chinese Academy of Sciences, College of Materials Science and Opto-Electronic Technology, No.19A Yuquan Road, Shijingshan District, Beijing 100049, China*

1.1 Introduction

The past decades have witnessed significantly irreversible resource and environmental degradation issues, especially for exhaustion and pollution due to the increasing global industrialization and indiscriminate consumption of fossil fuels, which critically restrict the sustainable development of society [1–3]. Proverbially, the sun, wind, hydro, and tides as renewable energy sources provide huge prospects to supplement electricity in an environment-friendly way to satisfy our social needs [4–6]. However, these renewable energies are restrained by the geographical conditions (longitude, latitude, altitude, geomorphology, etc.) and natural environments (alternation of sunrise and sunset, seasonal and weather variations, etc.) [7, 8]. To store these random, uncontrolled, and intermittent energies, secondary batteries can be utilized to accommodate these energies in the form of chemical energy and transform chemical energy into the electric energy required, which have presented dominant application and popularization in consumer electronics, electric vehicles (EVs), and intelligent grids for electrochemical energy storage (EES), which are attributed to their rechargeability and considerable electrochemical performance [9, 10]. However, safety, cost, and service life are plaguing their applications [11, 12]. Accordingly, it is particularly urgent to develop novel EES devices and corresponding materials with low cost, high device performance, and environmental friendliness [13].

As displayed in Figure 1.1, the number of literature published per year on batteries gradually increased from 2000 to 2022, and more than 31,000 articles were published in just 2022, demonstrating the persistent investigation enthusiasm on batteries among researchers, scientists, and experts from all over the world. Noticeably, the lithium-ion batteries (LIBs) account for a larger and larger share of these publications from 2000 to 2010. As known, LIBs have become the research hotspot for portable devices since they were commercialized by Sony in 1991 [14].

Electrochemical Energy Storage Devices: Non-Conventional Technologies and Materials,
First Edition. Edited by Yongbing Tang and Luojiang Zhang.

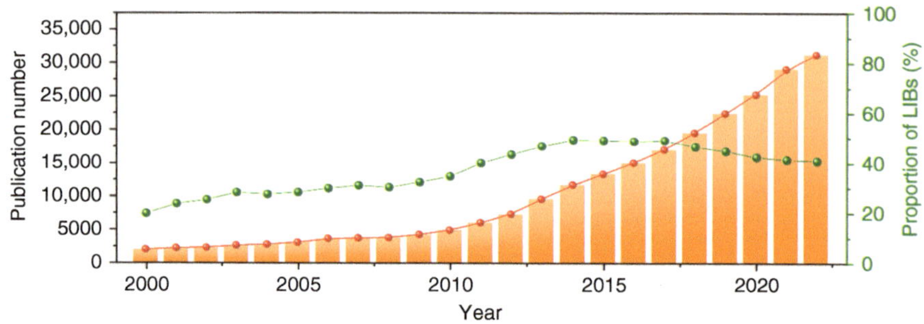

Figure 1.1 Left *y*-axis: the publication number of papers per year relevant to batteries from 2000 to 2022. Right *y*-axis: the percentage of papers published for LIBs. (Here, the results were refined from the Web of Science with "topic = Batteries" and "document types = article" for the left data and "topic = lithium-ion batteries or Li-ion batteries" and "document types = article" for the right data until 8 October 2023.)

Soon afterward, they outperformed their competition, i.e. nickel metal hydride and nickel cadmium batteries as the leading battery technology in digital cameras, laptop computers, and cell phones with an overall market share of more than 60% worldwide [15]. With the popularization of LIBs in portable consumer electronics, the next trial focuses on the application of LIBs for EVs based on the updated battery engineering in 2010s; therefore, the research attention on LIBs is a new level among batteries [16]. Although LIBs are still research hotspots in EES fields relevant to high energy density and their mature technologies on pivotal electrode materials and devices [17]. The reserves of resources such as lithium, cobalt, and nickel from cathode are cumulatively scarce and their costs are increasingly expensive, which restrict their sustainable development [18]. Therefore, novel EES devices have been investigated to complement the traditional LIBs and satisfy growing demand [19]. For example, some cost-effective cathodes can be considered, such as air, sulfur, and CO_2 [20–23]. Moreover, the construction of these batteries can be evolved into multivalent-ion or dual-ion types [24]. Perhaps, they can also be reconstituted into aqueous type, flow type, hybrid type, or flexible type [25–29]. Furthermore, fuels such as hydrogen or methanol can also be utilized as active materials to generate electric energy from chemical energy [30]. Under the circumstances, various new alternative EES devices such as metal-air batteries, metal-S batteries, metal-CO_2 batteries, multivalent-ion batteries, dual-ion batteries, fuel cells, aqueous batteries, flow batteries, hybrid capacitors, and flexible energy storage devices have been developed to achieve the aims of high energy/power density, long cycling life-time, inherently safe, cost-effective, or environmentally benign [31]. Meanwhile, numerous scientific challenges in critical electrode materials, electrolytes, and construction of these devices and research attention concerned with the system integration for energy storage and utilization are extremely vibrant [32]. First, the reaction mechanisms of these novel EES devices should be elaborated through advanced characterization techniques and theoretical simulations [33, 34]. Second, structural and electrochemical stability of device components should be enhanced to improve the device lifetime by rational electrode design [35, 36]. Third, the electrode activity and ion transport property in electrolytes should be optimized to promote their energy storage efficiency [37, 38]. The specific issues faced by different devices are briefly introduced and feasible solutions could be summarized and proposed as follows.

1.2 New Energy Storage Devices

1.2.1 Metal-Air Batteries

Metal-air batteries with an invincible theoretical specific energy and rich feedstock abundance of air have attracted much attention, especially, Li-air batteries with a theoretical specific energy of 3500 Wh kg^{-1}, whose anodes undergo the stripping/plating of metals and cathodes underpin the formation/decomposition of metal peroxides (M_2O_2) [39]. In 1996, Abraham et al. reported a rechargeable Li-O_2 battery constructed by a Li metal foil anode and a carbon composite cathode for oxygen reduction reaction with an organic polymer electrolyte membrane, which obtained a discharge capacity of 630 mAh g^{-1} [40]. This investigation is usually considered the pioneer of the aprotic Li-air batteries and inspires researchers to engage in metal-air batteries in the following years [41]. However, metal-air batteries are still in their infancy stage, and formidable challenges involving air cathodes, metal anodes, and electrolytes need to be addressed for practical applications [42].

As for the air cathode, especially for O_2 cathode, the solid and insulative discharge product M_2O_2 passivates the cathode surface and prohibits the diffusion of metal ions and O_2, which results in a high overpotential, low Coulombic efficiency, and inferior discharge capacity, even the oxidative degradation of cathode materials and electrolytes [43]. Moreover, O_2 could exacerbate the side reactions. For example, M_2CO_3, a common disreputable side reaction product, can ever-growingly deposit on the cathode/electrolyte interface because it is hard to eliminate even at a high voltage, which also degenerates the device performance and induces decomposition of the device components [44]. Noticeably, the formation of M_2O_2 might experience two different pathways of the surface-mediated and solution-mediated routes, which could influence the physical properties of M_2O_2 [45]. Usually, the solution route could alleviate the electrode passivation and contribute to a high discharge capacity by utilizing organic solvents with high solvating ability, which enhances the solvation of MO_2 intermediate [46]. However, superoxide species might lead to the solvent decomposition in the solvents with high solvating ability. In this case, designing stable electrolyte solvents without sacrificing the solvating ability can maintain the robust M_2O_2 formation property in electrolyte solution [47]. Moreover, the reactivity occupancy of superoxide intermediates is an effective strategy by incorporating the reduction mediator (RM_{disch}), which delivers electrons between the electrode and M_2O_2, and forms a soluble intermediate, i.e. MO_2–RM_{disch} complex, where the side reactions from electrode and electrolyte disintegration are restrained, increasing the M_2O_2 product upon discharge [48]. To decrease the side reactions and reduce the charging overpotential to enhance the kinetics of O_2 reduction, stable noncarbon materials such as metal oxides and nanostructured gold are employed to replace the unstable carbon-based cathodes. Meanwhile, noncarbonate electrolytes, e.g. ionic liquid and ether, are combined with carbonate electrolytes to construct a stable metal-O_2 battery [49]. In terms of metal anodes, dendritic growth of the metallic anode is a vexing puzzle, which has been extensively investigated in metal-ion batteries, causing low Coulombic efficiency, poor cycle life, and safety issues [50]. Herein, the high-quality solid electrolyte interphase (SEI) can enhance the anode stability and improve the Coulombic efficiency via the electrolyte and additive design [51]. Simultaneously, the quasi-solid-state electrolyte and solid-state electrolyte could be applied as modified separators to protect metal anodes from ambient air, i.e. CO_2 and H_2O [52]. In addition, metal anodes could be replaced by several alternatives including silicon, aluminum, tin, and their derivatives to guarantee safety [53].

1.2.2 Li-S Batteries

Li-S batteries are considered one of the promising EES devices for applications in both EVs and stationary energy storages due to the merits of high energy density based on the conversion reaction between Li and S and abundant S with low cost [54]. Notably, the Li-S batteries could achieve a theoretical energy density of ~2600 Wh kg^{-1}, which is four times higher than that of LIBs. In 1989, a rechargeable Li-S battery was established with an average discharge voltage of 2.1 V and an energy density of ~80 Wh kg^{-1} by the introduction of ether electrolyte [55]. Subsequently, the sulfur/porous carbon composition and the lithium nitrate additive were verified to be feasible in inhibiting the shuttle effect of polysulfides and enhancing the cycling robustness during the early 2000s, which ignites the researchers' enthusiasm to explore Li-S batteries in the following decades [56]. Nevertheless, formidable challenges still slow down the evolution of Li-S batteries for the practical application.

First, the S electrode faces the structural collapse after the lithiation of S to Li_2S due to the tremendous volume expansion, which is adverse to the application of high-density S electrodes [57]. Meanwhile, the electric/ionic insulating discharge products (Li_2S) and S electrode usually cause the reaction passivation [58]. Another notorious plight is the shuttle effect, where the intermediate discharge products (Li_2S_n, $4 \leq n \leq 8$) react with Li by the chemical reaction rather than by the electrochemical reaction ascribed to their dissolution, as a result of the remediless dissipation of both S and Li with the electrolyte decomposition due to the enhanced polarization [59]. Accordingly, functionalizing separators with catalytic materials are effective approaches to cope with polysulfides shuttle effect, where catalytic materials can boost sulfur redox reaction kinetics to achieve Li-S batteries with long cycle life and high energy density [60]. Besides, the multitudinous conductive matrix such as porous carbonaceous materials, conductive frameworks or polymers, and other novel materials could be introduced to alleviate the volume variation and suppress the shuttle effect by the dispersion or confinement of the active materials and to enhance the electric/ionic conductivity [61]. However, the carbonaceous hosts are inadequate to absorb Li_2S_n by physical approach. To anchor and convert

Li_2S_n to Li or S for the restraint of the shuttle effect by the catalytic methods, it is feasible to conduct the heteroatom doping or construct transition metal compounds in the carbonaceous hosts [62]. Simultaneously, the microstructure optimization of the carbonaceous hosts with aligned pores or hierarchical microstructures can decrease the porosity of S electrodes to improve the sulfur utilization and reaction dynamics [63]. Second, highly active Li anodes not only bring the safety troubles together with volatile and flammable ether-based electrolytes but also cause the depletion of the active materials derived from the shuttle effect [64]. Accordingly, it is significant to protect Li anode by introducing the protective coating or constructing effective SEI films [65]. Meanwhile, highly efficient electrolytes and solid-state electrolytes are also necessary to promote the battery performance [66].

1.2.3 Metal-CO_2 Batteries

Metal-CO_2 batteries are intriguing research hotspots in the background of global carbon neutrality due to the advantages of the real-time power input that drives CO_2 conversion and high battery energy density [67]. Compared with LIBs, metal-CO_2 batteries possess safer and higher energy density, providing cost-efficient techniques for renewable energy storage [68]. In contrast to normal CO_2 conversion and storage methods, which need energy input after fixing carbon, metal-CO_2 batteries supply electrical energy as required once recharging the battery to release the concentrated CO_2 [69]. In all, they demonstrate significant potential involving the capture of the greenhouse gas CO_2 to alleviate climate warming, the conversion of CO_2 into valuable chemicals, the storage of surplus electricity from other sustainable power systems, fossil fuels, and nuclear and carbon cycle balance [70]. Metal-CO_2 batteries are explored by supplementing CO_2 in Li/Na-O_2 batteries, and the battery capacity and energy density were greatly improved in 2011 [71, 72]. The pioneer metal-CO_2 batteries with pure CO_2 gas cathode and Li, Mg, and Al anodes in the nonaqueous electrolyte were presented in 2013 [73]. Soon afterward, the first rechargeable Li-CO_2 battery was demonstrated in organic electrolytes, which could be cycled reversibly at room temperature for several cycles [74]. Shortly after, various metal-CO_2 batteries with Li, Na, K, Mg, Zn, and Al metal anodes were dramatically developed [75]. Herein, metal-CO_2 batteries encounter several bottlenecks for commercial applications such as obscure electrochemical mechanisms, catalyst with low catalytic activity, high charge potential, poor rate capability, weak reversibility, short cycle life, and various parasitic side reactions [76].

The metal anodes in metal-CO_2 batteries contain monovalent metals (Li, Na, K, etc.), divalent metals (Mg, Zn, etc.), and trivalent metals (Al, etc.). Generally, the monovalent metal anode-based metal-CO_2 batteries select organic electrolytes due to the high metal activity [70]. In comparison, divalent and trivalent metals demonstrate weaker reaction activity; therefore, they usually utilize aqueous or ionic liquid electrolytes along with multifarious by-products [77]. Notably, various by-products involving carbonates or methanoic acid and disproportionation products such as CO or oxalate make the electrochemical mechanisms complicated; [78] therefore, it is significant to elaborate the electrochemical mechanisms via advanced characterization techniques such as *operando/in situ* gas chromatography, Fourier transform infrared spectroscopy, X-ray synchrotron radiation spectra, and neutron techniques [79]. In the aspect of catalysts, carbonaceous materials demonstrate multiple superiorities of abundant reserves, low cost, great electrical conductivity, large surface area, and glorious porosity, whose catalytic activity could be enhanced by defect engineering, nanostructure design, or surface modifications [80]. Besides, noble metal-based catalysts generally deliver high catalytic activity and great stability, which could be promoted by alloying, composition with other metals or carbonaceous materials, etc. [81]. Transition metal-based catalysts with low cost including Cu, Fe, Co, Ni, Mo, and Mn are also research focuses, especially for the single atom catalysts and coupling with carbonaceous materials, which can enhance rate capacity, cycle life, and stability by exposing abundant active sites and promoting the utilization efficiency of active sites [82]. Besides, porous framework-based catalysts such as metal-organic frameworks and covalent organic frameworks also attract extensive consideration by virtue of the large surface area, structural adjustability, and porosity, which deliver significant potential as highly stable catalysts [83]. Moreover, the exploration of novel liquid electrolytes with low volatility and great electrochemical stability is also significant for metal-CO_2 batteries along with suitable additives to promote battery reaction kinetics and rate performance [84].

1.2.4 Multivalent-Ion Batteries

Multivalent-ion batteries involving Zn^{2+}, Mg^{2+}, Ca^{2+}, and Al^{3+} have been regarded as potential candidates for grid energy storage due to low cost, good safety, and large volumetric energy density, which have the feasibility to fundamentally break through the energy density of current LIBs because they can achieve doubled or tripled capacities with the same number of reactive ions as Li^+ [85]. Besides, the reserves of multivalent-ion metal elements such as Zn, Mg, Ca, and Al are distinctly more than that of Li on the Earth's crust, displaying cost-effectiveness [86]. However, commercial or mature cathode materials in LIBs cannot be suitable for multivalent-ion batteries because the guest ions suffer from serious Coulombic interaction with the surrounding lattice due to their high charge density, which causes large energy barrier. It would block the ion diffusion process and induce the irreversible phase transition, structural collapse, and electrode element dissolution, leading to inferior rate capability and cycling robustness [87]. Therefore, the development of multivalent-ion batteries still suffers from several challenges such as sluggish ion mobility kinetics, poor ion storage reversibility, dendrite growth of metal anodes, and lack of appropriate electrode materials [88].

In the case of Zn-ion batteries, they suffer from inferior reversible capacity, low Coulombic efficiency, and poor cycle life ascribed to dendrite growth and electrochemical corrosion of Zn anode, high polarization of divalent Zn^{2+}, and unsatisfactory side reactions [89]. As for Al-ion batteries, although the graphite materials demonstrate a robust $AlCl_4^-$ intercalation stability, the low specific capacities restrict their practical application [90]. Moreover, the electrochemical properties are blocked by low discharge voltages and capacities, various by-products, unstable charge voltage platforms, the passivation of Al anode, inferior charge/discharge reversibility, severe corrosivity of chloroaluminate species for current collectors and battery assembling cans, and structural instability of cathode materials [91]. Mg-ion batteries and Ca-ion batteries are usually hindered by the incompatibility between electrolytes with metal anodes, and the generation of an impermeable passivation layer on surface of metal anodes, causing the poor reversibility of metal plating/stripping [92, 93]. Besides, Ca-ion batteries are usually operated at high temperature, which incurs high-security risks [94]. To overcome the above bottlenecks, nanostructure construction of domain size and hierarchical structure configuration are employed to boost the reaction dynamics by shortening the ion diffusion length, and the defect engineering can also promote the ion diffusion kinetics as well as enhance the electronic conductivity intrinsically [95]. Besides, the surface modification and metal alloying or compounds can effectively alleviate the metal anode surface passivation and corrosion, contributing to enhanced comprehensive performance [96]. The electrolytes also suffer from several crucial barriers, where the desolvation energy of cations increases with the rising of the charge density of the multivalent ions, causing sluggish mobility kinetics of multivalent ions across the electrode interface, especially for Mg-ion batteries and Al-ion batteries [31]. Simultaneously, the side reactions consume the electrolyte continuously, especially for aqueous system [97]. To mitigate these challenges, electrolyte design through altering electrolyte and additive composition is a feasible strategy along with the anode electrode optimization, which can improve the reversibility of ion deposition/stripping [98].

1.2.5 Dual-Ion Batteries

Dual-ion batteries exhibit a different working mechanism from the traditional LIBs with a "rocking-chair" mechanism, where anions (such as PF_6^-, FSI^-, and ClO_4^-) and cations (such as Li^+, Na^+, and K^+) are simultaneously intercalated/deintercalated into cathode and anode materials, respectively, during the charge/discharge process. This endows dual-ion battery with a competitive cutoff voltage of more than 4.5 V, contributing to a relatively high energy density [99, 100]. Besides, the energy density of dual-ion battery can be improved by regulating the concentration of electrolytes as they also serve as the active substances in this system [101]. Noteworthily, the low specific capacity, inferior rate performance, and insufficient cycling life are significant barriers for dual-ion battery to practical popularize because of the tardy ion diffusion kinetics and dissolution of electrolyte at high voltage [102].

Electrolytes as the crucial part of dual-ion battery mainly include organic liquid electrolytes, aqueous electrolytes, ionic liquid electrolytes, and solid-state electrolytes, whose energy gap depends on their highest occupied molecular orbital (HOMO) and lowest unoccupied molecular orbital (LUMO), which determines the electrochemical stability window [103]. To circumvent the electrolyte decomposition, electrolytes need to exhibit excellent compatibility with electrodes and robust electrochemical stability at high voltage [104]. Carbonates (e.g. ethylene carbonate, diethyl carbonate, ethyl methyl carbonate, and dimethyl carbonate) are one of the promising options, which can readily dissolve cations and anions due to their low viscosity and high ion conductivity, and exhibit a wide potential window [105]. Meanwhile, the compatibility between electrodes and electrolytes could be regulated by the categories and concentrations of salts, additives, and solvents [106]. Especially, fluorinated electrolytes can enhance the oxidative robustness due to their high oxidation potentials [107]. Meanwhile, highly concentrated electrolytes are promising selection for dual-ion battery to achieve the long cycle stability [108]. As for graphite electrode materials, the design strategies are significant to promote the battery performance [109]. First, an elevated graphitization degree for electrodes could contribute to the enhanced voltage efficiency and Coulombic efficiency as well as the decreased polarization and improved ion storage capability, because the graphitization degree of electrode materials could influence ion diffusion kinetics, electronic conductivity, and structural stability [110]. Besides, the nanostructure construction of electrode materials can sustain the volume variation during the charge/discharge process, thus enhancing the battery cycling stability and rate capability, because the nanostructures can improve ion transport via shortened diffusion channels and increase ion storage sites with enlarged specific surface areas [111]. Further, adequate defects or dopants in electrode materials might introduce fresh ion storage sites, boost ion diffusion kinetics, and promote the electronic conductivity, contributing high capacity and rate capability [112]. In addition to graphite, other high conductive carbon materials, organic materials, layered transitional metal disulfides, and even metals show the potential to host the active anions [113]. Amazingly, organic materials have been investigated hotspots because of their flexible designability, rich ion storage sites, and redox reversibility, which endow them with the high theoretical capacity and great cycling stability [114]. Noteworthily, the nanostructural construction and interfacial modification are extensively conducted to promote the ion reaction kinetics and strengthen the battery robustness [115]. While for the anode materials, the metallic materials, intercalation materials, alloying materials, and conversion materials have been developed, which are similar to those in rocking-chair batteries.

1.2.6 Fuel Cells

Fuel cells have reached the early stage of commercial deployment due to their outstanding energy efficiency, high energy/power densities, potential for zero/low-emission lifecycles, quick refill ability, and versatile applications involving portable devices, automotive vehicles, and power grid stations. In the fuel cells, the chemical energy stored in fuels (hydrogen, methanol, or ethanol) and oxygen is directly converted to electrical energy through electrochemical reactions with fuel oxidized reaction at the anode and oxygen reduced reaction (ORR) at the cathode, respectively [116]. However, the widespread applications of fuel cells still suffer from several technical barriers on their electrochemical performance, durability, and cost [117]. To accelerate electrochemical processes, electrocatalysts should demonstrate high activity and stability, especially for the cathodic ORR, which could be multielectron and multistep reactions with sluggish kinetics [118]. Usually, electrocatalysts are composed of metal nanoparticles as the active species and substrate materials as the support to facilitate the electron/mass transfer and enhance the stability [119]. Carbon-based materials as the common support deliver glorious conductivity, high specific surface area, and abundant porosity, which can disperse metal nanoparticles, contributing to high electrochemically active surface area [120]. Currently, carbon-supported Pt-based electrocatalysts have been regarded as the feasible choice to achieve the high catalytic activity and durability for the ORR [121]. Unfortunately, the scarce reserves and high cost of Pt-based electrocatalysts severely block the practical applications of fuel cells [122]. Therefore, more intensive efforts are still needed to promote the activity and durability of electrocatalysts under the condition of lowering the utilization of noble metal electrocatalysts.

To decrease the usage of Pt and reduce the cost, it is imperative to explore nonprecious metal electrocatalysts with great durability and high active site density. Recently, several types of nonprecious metal electrocatalysts have been reported, such as metal-free carbon materials, transition metal oxides, and M/N/C catalysts [123]. Notably, M/N/C represents metal atoms coordinated with nitrogen functionalities and encapsulated into carbonaceous matrix, where M denotes transition metals such as Fe, Co, and Ni or main group metals such as Mg, Al, and Ca have been regarded as the most promising catalysts [124]. However, the activity and stability of M/N/C catalysts need to be improved comparable to those of Pt-based electrocatalysts in practical operating environments of fuel cells by engineering the active sites. Generally, the metal-N_4 species were proved to be the active sites for the ORR; therefore, the design and increase of active sites are the research focus for M/N/C catalysts [125]. Besides, the aggregation of metal atoms should be considered. Noticeably, the metal-organic frameworks or metal macrocyclic compounds with metal-nitrogen coordination structures are usually utilized as precursors to evenly disperse metal ions [126]. Besides, metal single atom catalysts of the M/N/C are also research hotspots due to their maximized metal atom utilization efficiency and distinct electronic structure, where their catalytic properties can be regulated by adjusting the microenvironment of active sites involving the coordination number of M–N, heteroatom doping, and vacancy creation for the improved catalytic performance [127]. However, the scalable preparation of M/N/C catalysts for the practical application in fuel cells is also crucial adjective. Accordingly, the synthesis methods still face great challenges.

1.2.7 Aqueous Batteries

Aqueous batteries without utilization of flammable organic electrolytes demonstrate their superiority of high safety, low cost, and easy operation [128]. The early research on aqueous LIBs can be traced back to 1994, when the battery configuration involving $LiMn_2O_4$ as a cathode and VO_2 as an anode with 5 M $LiNO_3$ aqueous electrolyte presented a voltage of ~1.5 V and an energy density of 55 Wh kg^{-1}, outperforming than those of aqueous lead-acid batteries (~30 Wh kg^{-1}) and Ni-Cd batteries (~50 Wh kg^{-1}) [129]. Subsequently, metal oxides, polyanionic compounds, and Prussian blue analogs (PBAs) were utilized as cathodes for aqueous LIBs, where spinel $LiMn_2O_4$ is one of the research focuses owing to its availability in aqueous solutions [130]. Unfortunately, most anode materials for nonaqueous LIBs are not applicative for aqueous batteries due to their low redox potentials beyond the cathodic limit of aqueous electrolytes, where hydrogen evolution reaction (HER) would occur on the anode as a competing reaction, causing the decomposition of the aqueous electrolytes [131]. Further, various metal-based aqueous batteries (M = Li, Na, K, Zn, Mg, Ca, Al) were investigated to enhance energy density and cycle life [132]. Noticeably, the rechargeable aqueous Zn-based battery is one of the prospective battery devices due to the rich reserve of Zn, suitable redox potential (−0.763 V vs. standard hydrogen electrode [SHE]), and high theoretical specific capacity (820 mAh g^{-1}) [133]. Although Al anode possesses larger volumetric capacities of ~8046 mAh cm^{-3}, its instability in aqueous solutions results in the poor cycle performance [134]. Zn-based batteries demonstrate glorious reversibility in diversified aqueous electrolytes, which include alkaline Zn metal batteries and neutral or acid Zn-ion batteries [135]. However, the reactivity of water splitting influences the promotion of aqueous batteries, where the low potentials on the anode lead to water reduction, i.e. HER and the high potentials on the cathode cause water oxidation, i.e. oxygen evolution reaction (OER) [136]. Therefore, aqueous batteries also suffer from several obstacles such as self-discharge resulting in inferior battery storage and narrow electrochemical stability window leading to unavoidable side reactions in aqueous media, which cause the limited energy density and poor long-term stability [137].

Considering that the theoretical capacity of cathode materials depends on the electron transfer number and material mass, reducing the inactive mass of the electrodes is a theoretically feasible strategy to improve the battery capacity, but is quite challenging in the case of maintaining electrode structural stability [138]. Alternatively, it is a pivotal measure to increase the electron transfer numbers by introducing multiple redox reaction centers. Taking low-valence vanadium-based compounds as an example, their *in situ* electrochemical oxidations contribute to high capacities due to the rich active sites of high-valence compounds [139]. Meanwhile, the surface carbon coating is also necessary for these high-capacity oxides to keep their structural and chemical stability [140]. Besides, the poor intrinsic conductivity and weak selectivity to metal ions for electrode materials could lead to the deterioration of battery performance. Herein, it is crucial to perform optimizing strategies such as regulating the crystal structures and introducing conductive materials [141]. Noticeably, as for the layered electrode materials, preintercalating molecules or ions can weaken electrostatic interaction of the host lattice oxygen and metal ions and achieve the expansion of interlayer distance to promote the ion storage capacity [142]. Another focus is to achieve high-voltage aqueous batteries, which mainly depends on the electrochemical stability window of aqueous electrolytes, restricted by OER on the cathode and HER on the anode [143]. The working voltages of actual battery systems are usually less than 2.0 V due to the low theoretical water splitting potential (1.23 V vs. SHE). It is urgent and effective to broaden the electrochemical stability window by weakening the correlation between electrode materials and water molecules, e.g. water-in-salt electrolytes by adopting salt-concentrated electrolyte design can provide a wide electrochemical stability window by decreasing the electrochemical activity of water molecules [144].

1.2.8 Flow Batteries

Flow batteries are promising candidates for large-scale energy storage applications due to their high safety with aqueous recyclable electrolytes, glorious environmental friendliness, and excellent cycling stability. In the flow battery, the electrolyte in the tank circularly flows across the matched electrode with the aid of the pumps, allowing redox reactions of active components from the electrolyte to occur on the electrode surface to convert the electric energy from the chemical energy by the valence state variation of ions pairs [145]. Accordingly, the power and energy densities can be adjusted respectively through individual electrode engineering. Recently, traditional vanadium flow batteries and burgeoning zinc-based flow batteries have been popularized as the megawatt energy storage devices, demonstrating the potential commercial and industrial applications [146, 147]. Meanwhile, several novel aqueous flow battery systems can enlarge the power and energy density, lower electrolyte resistance, and enhance safety and environmental benignity, such as aqueous flow batteries including quinone/bromide system, quinone/iron system, etc. and nonaqueous flow batteries containing Li/ferrocene system, Li/cobaltocene system, TEMPO system, viologen system, etc. [148, 149]. However, the HER and OER in aqueous solutions are the focus of attention, which causes the narrow work voltage window, restricting its energy density [150]. Simultaneously, the poor power density needs to be improved for the practical industrialization. As for nonaqueous solutions, the active species are generally unstable and the solvents are usually flammable, leading to poor work stability [151].

It is effective to design and optimize pivotal materials involving the electrodes, electrolytes, membranes, and bipolar plates. To decrease the battery polarization and enhance the power density, electrode materials should possess high electron conductivity, great reaction activity to redox couples, excellent chemical and electrochemical stability, and large surface area for the sufficient active sites [152]. Herein, electrode modification, especially for the carbonaceous materials, is the common strategy, involving electrochemical oxidation, acid treatment, thermal activation, defect incorporation, functional group introduction, surface coating, or depositing electrocatalysts on the surface. Usually, the energy density of flow batteries depends on the concentration and volume of the electrolyte [153]. Accordingly, to develop high-performance and highly stable electrolytes, ideal active materials in electrolytes should have great solubility and stability. Generally, the membrane can separate the positive and negative electrolytes and transport charge carriers, whose properties also quite influence the battery performance [154]. Consequently, perfect membranes should deliver excellent ion conductivity, ion selectivity, chemical and mechanical stability, and low cost, which can effectively mitigate the self-discharge and capacity attenuation and efficiently improve the battery energy density and power density. Moreover, bipolar plates are the indispensable components to connect each cell electrically, separate neighbor cells, and guide the electrolyte flow, which should demonstrate great conductivity, stable mechanical strength, high corrosion resistance, and excellent compactness to prevent electrolytes leakage [155]. Simultaneously, the stack and assembly of flow battery are also significant to promote the industrialization and commercialization [156]. As for the non-aqueous flow battery systems, stable active materials and electrolyte optimizations are critical fields of exploration [157]. It is noteworthy that other novel systems or conceptions such as semisolid flow batteries and redox-targeting-based flow batteries also deliver the possibility to boost the battery performance.

1.2.9 Hybrid Capacitors

Hybrid capacitors deliver elevated power densities with a moderate energy density, short charge/discharge time interval, wide operating temperature range, and elongated cycle life, which are alternative technologies to decrease consumption of fossil fuels and greenhouse gas pollution [158]. Capacitors are divided into three categories involving pseudocapacitor with fast surface redox reactions, electric double layer capacitor with the ion adsorption, and hybrid capacitors [159], which can store energy via electrode materials' surface/near-surface faradaic redox reactions as well as the formation of an electrical double layer at the electrode/electrolyte interface from the purely electrostatic interaction [160]. Although their fast charge storage mechanism does not suffer from poor solid-state ion diffusion in battery electrodes ascribed to the electrode surface or near-surface reactions, their energy density and charge storage are quite inferior [161]. As for the electric double layer capacitors, they need high conductivity and large surface area to increase reaction rate at the electrode/electrolyte interface and achieve a fast charging/discharging rate [162]. However, the limited electric double layer usually brings about low energy density. In contrast, materials involving faradaic charge transfer processes such as metal oxides, sulfides, nitrides, phosphates, and oxynitrides can provide higher capacity [163]. Unfortunately, these materials often suffer from the phase/structure variation during the charge/discharge process or deliver poor electrical conductivity, causing the sluggish energy storage kinetics [164]. Therefore, there is still vast room to promote their performance.

As for the electrode materials, the reduced domain sizes of nanostructured electrodes can offer large specific surface area, increase the access of the electrode surface by electrolyte ions, shorten ion diffusion paths, and block the phase/structure variations, which tend to boost the capacitance, enhance the rate capability, and weaken cyclic degradation [165]. Simultaneously, nanomaterials provide the convenient reaction exchange on the atomic level or molecular level for the hybrid materials, which could synergistically promote electrode properties. Furthermore, it is effective in increasing the electronic conductivity, shortening the charging time, and maintaining the phase/structure stability by incorporating highly conducting carbonaceous-based materials [166]. Besides, to obtain high energy devices, selection of materials with high capacity is necessary, where hybrid materials with the combination of surface and intercalation charge storage such as battery-type oxide/hydroxide materials can contribute to higher energy density and better cyclic stability compared to carbonaceous materials and conducting polymers [167]. Moreover, hybrid configuration adhering disparate energy storage mechanisms with faradaic and non-faradaic material components (asymmetric electrode configuration) in a single device is a sensible approach to reach the synergistic performance of each component with high energy (faradaic process) and high power (non-faradaic process) storage, such as electric double layer capacitive electrode/pseudocapacitive electrode, or electric double layer capacitive electrode/battery-type electrode [168]. Besides the electrodes, the electrolytes are also significant components, where they are expected to have robust electrochemical stability, low volatility, low solvated ionic radius, low resistivity, low viscosity, low toxicity, high ionic concentration, wide voltage window, etc. [169].

1.2.10 Flexible Energy Storage Devices

Flexible energy storage devices play a significant role in portable, ultrathin, and flexible electronic and device markets due to their good mechanical stability, miniaturization, and multifunctionalities [170]. Multitudinous flexible devices were integrated into wearable devices or human body without being perceived, such as sensors, energy harvesters, antennas, and radio-frequency identification tags, which can maintain mechanical robustness under various mechanical deformation from human movements [171]. Currently, the traditional rechargeable energy storage devices mainly including batteries and supercapacitors incline to structure failure, safety hazard power, and supply interruption once they suffer from bending, folding, stretching, twisting, etc. because of the rigid structures lacking mechanical flexibility [172]. Accordingly, it is significant to comprehensively design and fabricate flexible energy conversion and storage devices. Recently, various novel materials and structures with mechanical pliability including electrodes, current collectors, and solid-state or quasi-solid-state electrolytes have been utilized for flexible energy devices, which will unveil substantial electrochemical mechanisms and challenging investigations in interdisciplinary fields such as chemistry, physics, material science, electronics, engineering, biology, and medicine [173]. Noteworthily, several fundamental and technological bottlenecks still need to be addressed for the practical applications of flexible energy conversion and storage devices. On the one hand, the initial interfaces including active materials/conductor additives, electrodes/current collectors, and electrodes/electrolytes are gradually deteriorated, which restrict the electronic transfer and/or ionic transport channels [174]. On the other hand, the weakened internal electronic conductivity from mechanical deformation results in local short circuits and serious side reactions [175].

The flexible design strategies for materials, electrodes, and architectures need to endow energy storage devices with bendable, implantable, and wearable properties. Usually, introducing redundancy in volume such as pores can achieve high flexibility by lowering the bending stiffness and releasing strain [176]. As for the electrolytes, fluidic properties of liquid electrolyte restrain the size and shape of batteries; therefore, plastic or rubbery ion-conductive medium is regarded as a feasible alternative to construct flexible batteries, where gel polymer electrolytes can be utilized as both the electrolyte and the separator [171]. Meanwhile, plastic crystal electrolytes combined with a polymer matrix as a flexible framework demonstrate great thermal and structure stability, good ionic conductivity, and mechanical strength [177]. Besides, the solid polymer electrolytes involving polyacrylonitrile, polyvinylidene fluoride, polyethylene oxide, and polymethyl methacrylate are utilized to eliminate the outflow of liquid electrolyte, where ceramic nanoparticles can be employed as fillers to avoid crystallization of polymers or ion conductors to enhance ionic conductivity, which are relatively ideal ion transport medium with variable shapes and controllable integration [178]. However, they still suffer from severe interfacial passivation and poor contact with active materials [179]. Furthermore, the current collector as another significant component is critical to construct flexibility batteries, where traditional Cu and Al foils have small tensile fracture strains and yield strains, which make it hard to maintain integrity in the condition of repetitive deformations [180]. Herein, carbon materials and conductive polymers as alternative current collectors can be employed due to highly conductive flexibility. Noticeably, the battery components and architecture innovation are feasible to promote flexibility such as wire, cable, or wave shapes [181].

1.3 Conclusion

This book provides a comprehensive review on promising energy storage devices including metal-air batteries, Li-S batteries, metal-CO_2 batteries, multivalent-ion batteries, dual-ion batteries, fuel cells, aqueous batteries, flow batteries, hybrid capacitors, and flexible energy storage devices. It also covers the fundamentals of energy storage devices and key materials (cathode, anode, and electrolyte). In addition, some advanced characterization techniques are introduced to achieve an in-depth understanding of the fundamentals inside the devices for the further improvement of their electrochemical performance. Moreover, the current challenges and effective strategies of new energy storage devices with high performance are also proposed. This book can attract more readers from various research fields on new energy storage and pivotal electrode/electrolyte materials, provide guidelines for those who are entering the related research fields, and enlighten readers to grasp the evolvement orientations in the future.

We would like to thank our colleagues who have amiably devoted their expertise, time, and enthusiasm in contributing to this book. It is their insistence and contributions that professional information can be gathered involving so many state-of-the-art processes and cutting-edge techniques in these booming fields. We would also like to thank the employees of the Wiley Group for their very kind and patient cooperation throughout the preparation of this book. We have enjoyed our communication with all the contributors and publishing editors. We hope that this book will serve readers who employ new energy storage devices and key materials.

References

1 Chen, L., Msigwa, G., Yang, M.Y. et al. (2022). Strategies to achieve a carbon neutral society: a review. *Environmental Chemistry Letters* 20 (4): 2277–2310.

2 He, M.Y., Sun, Y.H., and Han, B.X. (2022). Green carbon science: efficient carbon resource processing, utilization, and recycling towards carbon neutrality. *Angewandte Chemie International Edition* 61 (15): e202112835.

3 Li, X.Y., Wang, C., and Tang, J.W. (2022). Methane transformation by photocatalysis. *Nature Reviews Materials* 7 (8): 617–632.

4 Song, H., Luo, S.Q., Huang, H.M. et al. (2022). Solar-driven hydrogen production: recent advances, challenges, and future perspectives. *ACS Energy Letters* 7 (3): 1043–1065.

5 Fang, Y.X., Hou, Y.D., Fu, X.Z., and Wang, X.C. (2022). Semiconducting polymers for oxygen evolution reaction under light illumination. *Chemical Reviews* 122 (3): 4204–4256.

6 Macreadie, P.I., Costa, M.D.P., Atwood, T.B. et al. (2021). Blue carbon as a natural climate solution. *Nature Reviews Earth & Environment* 2 (12): 826–839.

7 Wang, Y., Zou, R.M., Liu, F. et al. (2021). A review of wind speed and wind power forecasting with deep neural networks. *Applied Energy* 304: 117766.

8 Pan, Q., Zheng, Y., Tong, Z. et al. (2021). Novel lamellar tetrapotassium pyromellitic organic for robust high-capacity potassium storage. *Angewandte Chemie International Edition* 60 (21): 11835–11840.

9 Huang, M., Wang, X.P., Liu, X., and Mai, L.Q. (2022). Fast ionic storage in aqueous rechargeable batteries: from fundamentals to applications. *Advanced Materials* 34 (9): 2105611.

10 Yuan, D., Dou, Y.H., Wu, Z.Z. et al. (2022). Atomically thin materials for next-generation rechargeable batteries. *Chemical Reviews* 122 (1): 957–999.

11 Zhang, H.L., Liu, H., Piper, L.F.J. et al. (2022). Oxygen loss in layered oxide cathodes for li-ion batteries: mechanisms, effects, and mitigation. *Chemical Reviews* 122 (6): 5641–5681.

12 Pan, Q., Tong, Z., Su, Y. et al. (2021). Energy storage mechanism, challenge and design strategies of metal sulfides for rechargeable sodium/potassium-ion batteries. *Advanced Functional Materials* 31 (37): 2103912.

13 Zhang, N., Deng, T., Zhang, S.Q. et al. (2022). Critical review on low-temperature Li-ion/metal batteries. *Advanced Materials* 34 (15): 2107899.

14 Zhang, H.M., Zhao, S.W., and Huang, F.Q. (2021). A comparative overview of carbon anodes for nonaqueous alkali metal-ion batteries. *Journal of Materials Chemistry A* 9 (48): 27140–27169.

15 Tarascon, J.M. and Armand, M. (2001). Issues and challenges facing rechargeable lithium batteries. *Nature* 414 (6861): 359–367.

16 Lai, X., Chen, Q.W., Tang, X.P. et al. (2022). Critical review of life cycle assessment of lithium-ion batteries for electric vehicles: a lifespan perspective. *Etransportation* 12: 100169.

17 Tian, Y.S., Zeng, G.B., Rutt, A. et al. (2021). Promises and challenges of next-generation "beyond Li-ion" batteries for electric vehicles and grid decarbonization. *Chemical Reviews* 121 (3): 1623–1669.

18 Kim, J., Kim, Y., Yoo, J. et al. (2023). Organic batteries for a greener rechargeable world. *Nature Reviews Materials* 8: 54–70.

19 Aziam, H., Larhrib, B., Hakim, C. et al. (2022). Solid-state electrolytes for beyond lithium-ion batteries: a review. *Renewable and Sustainable Energy Reviews* 167: 112694.

20 Duffner, F., Kronemeyer, N., Tubke, J. et al. (2021). Post-lithium-ion battery cell production and its compatibility with lithium-ion cell production infrastructure. *Nature Energy* 6 (2): 123–134.

21 Liang, Z.W., Shen, J.D., Xu, X.J. et al. (2022). Advances in the development of single-atom catalysts for high-energy-density lithium-sulfur batteries. *Advanced Materials* 34 (30): 2200102.

22 Liu, T., Vivek, J.P., Zhao, E.W. et al. (2020). Current challenges and routes forward for nonaqueous lithium-air batteries. *Chemical Reviews* 120 (14): 6558–6625.

23 Liu, B., Sun, Y., Liu, L. et al. (2019). Recent advances in understanding Li–CO_2 electrochemistry. *Energy & Environmental Science* 12 (3): 887–922.

24 Shang, W., Yu, W., Liu, Y. et al. (2020). Rechargeable alkaline zinc batteries: progress and challenges. *Energy Storage Materials* 31: 44–57.

25 Ju, Z.N., Zhao, Q., Chao, D.L. et al. (2022). Energetic aqueous batteries. *Advanced Energy Materials* 12 (27): 2201074.

26 Noack, J., Roznyatovskaya, N., Herr, T., and Fischer, P. (2015). The chemistry of redox-flow batteries. *Angewandte Chemie International Edition* 54 (34): 9775–9808.

27 Ju, J.W., Ma, J., Wang, Y.T. et al. (2019). Solid-state energy storage devices based on two-dimensional nano-materials. *Energy Storage Materials* 20: 269–290.

28 Zhao, Y.F. and Guo, J.C. (2020). Development of flexible li-ion batteries for flexible electronics. *InfoMat* 2 (5): 866–878.

29 Wang, D.G., Liang, Z.B., Gao, S. et al. (2020). Metal-organic framework-based materials for hybrid supercapacitor application. *Coordination Chemistry Reviews* 404: 213093.

30 Vichard, L., Steiner, N.Y., Zerhouni, N., and Hissel, D. (2021). Hybrid fuel cell system degradation modeling methods: a comprehensive review. *Journal of Power Sources* 506: 230071.

31 Chen, S., Zhang, M., Zou, P. et al. (2022). Historical development and novel concepts on electrolytes for aqueous rechargeable batteries. *Energy & Environmental Science* 15: 1805–1839.

32 Man, Q.Y., An, Y.L., Liu, C.K. et al. (2023). Interfacial design of silicon/carbon anodes for rechargeable batteries: a review. *Journal of Energy Chemistry* 76: 576–600.

33 Wu, J.X., Ihsan-Ul-Haq, M., Chen, Y.M., and Kim, J.K. (2021). Understanding solid electrolyte interphases: advanced characterization techniques and theoretical simulations. *Nano Energy* 89: 106489.

34 Zhou, W., Zhu, D., He, J. et al. (2020). A scalable top-down strategy toward practical metrics of Ni–Zn aqueous batteries with total energy densities of 165 Wh kg^{-1} and 506 Wh L^{-1}. *Energy & Environmental Science* 13 (11): 4157–4167.

35 Li, W.B., Song, Q.Q., Li, M. et al. (2021). Chemical heterointerface engineering on hybrid electrode materials for electrochemical energy storage. *Small Methods* 5 (8): 2100444.

36 Wang, L., Xie, X., Dinh, K.N. et al. (2019). Synthesis, characterizations, and utilization of oxygen-deficient metal oxides for lithium/sodium-ion batteries and supercapacitors. *Coordination Chemistry Reviews* 397: 138–167.

37 Zhang, Y., Tao, L., Xie, C. et al. (2020). Defect engineering on electrode materials for rechargeable batteries. *Advanced Materials* 32 (7): 1905923.

38 Xiao, K., Jiang, L., and Antonietti, M. (2019). Ion transport in nanofluidic devices for energy harvesting. *Joule* 3 (10): 2364–2380.

39 Aurbach, D., McCloskey, B.D., Nazar, L.F., and Bruce, P.G. (2016). Advances in understanding mechanisms underpinning lithium-air batteries. *Nature Energy* 1: 16128.

40 Abraham, K.M. and Jiang, Z. (1996). A polymer electrolyte-based rechargeable lithium/oxygen battery. *Journal of the Electrochemical Society* 143 (1): 1–5.

41 Cao, D., Bai, Y., Zhang, J.F. et al. (2021). Irreplaceable carbon boosts Li-O_2 batteries: from mechanism research to practical application. *Nano Energy* 89: 106464.

42 Dou, Y.Y., Xie, Z.J., Wei, Y.J. et al. (2022). Redox mediators for high-performance lithium-oxygen batteries. *National Science Review* 9 (4): nwac040.

43 Ahmed, G., Awan, Z.U., Butt, F.A. et al. (2022). The study of different redox mediators for competent Li-air batteries. *Journal of Power Sources* 538: 231379.

44 Zhao, Z., Huang, J., and Peng, Z. (2018). Achilles' heel of lithium–air batteries: lithium carbonate. *Angewandte Chemie International Edition* 57 (15): 3874–3886.

45 Liu, L.L., Liu, Y.H., Wang, C. et al. (2022). Li_2O_2 formation electrochemistry and its influence on oxygen reduction/evolution reaction kinetics in aprotic Li-O_2 batteries. *Small Methods* 6 (1): 2101280.

46 Chen, K., Yang, D.Y., Huang, G., and Zhang, X.B. (2021). Lithium-air batteries: air-electrochemistry and anode stabilization. *Accounts of Chemical Research* 54 (3): 632–641.

47 de Larramendi, I.R. and Ortiz-Vitoriano, N. (2020). Unraveling the effect of singlet oxygen on metal-O_2 batteries: strategies toward deactivation. *Frontiers in Chemistry* 8: 605.

48 Shen, X., Zhang, S., Wu, Y., and Chen, Y. (2019). Promoting Li–O_2 batteries with redox mediators. *ChemSusChem* 12 (1): 104–114.

49 Chen, L., Bao, J.L., Dong, X. et al. (2017). Aqueous Mg-ion battery based on polyimide anode and Prussian blue cathode. *ACS Energy Letters* 2 (5): 1115–1121.

50 Hosseini, S., Soltani, S.M., and Li, Y.Y. (2021). Current status and technical challenges of electrolytes in zinc-air batteries: an in-depth review. *Chemical Engineering Journal* 408: 127241.

51 Liu, Z., Huang, J., Zhang, Y. et al. (2019). Taming interfacial instability in lithium–oxygen batteries: a polymeric ionic liquid electrolyte solution. *Advanced Energy Materials* 9 (41): 1901967.

52 Qiao, Y., Wang, Q., Mu, X. et al. (2019). Advanced hybrid electrolyte Li-O_2 battery realized by dual superlyophobic membrane. *Joule* 3 (12): 2986–3001.

53 Liu, T., Cheng, X., Yu, H. et al. (2019). An overview and future perspectives of aqueous rechargeable polyvalent ion batteries. *Energy Storage Materials* 18: 68–91.

54 Wu, C., Lai, W.H., Cai, X.L. et al. (2021). Carbonaceous hosts for sulfur cathode in alkali-metal/S (alkali metal = lithium, sodium, potassium) batteries. *Small* 17 (48): 2006504.

55 Peled, E., Gorenshtein, A., Segal, M., and Sternberg, Y. (1989). Rechargeable lithium sulfur battery (extended abstract). *Journal of Power Sources* 26 (3): 269–271.

56 Ji, X., Lee, K.T., and Nazar, L.F. (2009). A highly ordered nanostructured carbon-sulphur cathode for lithium-sulphur batteries. *Nature Materials* 8: 500–506.

57 Li, F., Liu, Q.H., Hu, J.W. et al. (2019). Recent advances in cathode materials for rechargeable lithium-sulfur batteries. *Nanoscale* 11 (33): 15418–15439.

58 Wang, L., Wang, T., Peng, L.L. et al. (2022). The promises, challenges and pathways to room-temperature sodium-sulfur batteries. *National Science Review* 9 (3): nwab050.

59 Ren, W.C., Ma, W., Zhang, S.F., and Tang, B.T. (2019). Recent advances in shuttle effect inhibition for lithium sulfur batteries. *Energy Storage Materials* 23: 707–732.

60 Yuan, C., Yang, X.F., Zeng, P. et al. (2021). Recent progress of functional separators with catalytic effects for high-performance lithium-sulfur batteries. *Nano Energy* 84: 105928.

61 Wang, Y.Z., Huang, X.X., Zhang, S.Q., and Hou, Y.L. (2018). Sulfur hosts against the shuttle effect. *Small Methods* 2 (6): 1700345.

62 Wang, P., Xi, B.J., Huang, M. et al. (2021). Emerging catalysts to promote kinetics of lithium-sulfur batteries. *Advanced Energy Materials* 11 (7): 2002893.

63 Huang, Y.Z., Lin, L., Zhang, C.K. et al. (2022). Recent advances and strategies toward polysulfides shuttle inhibition for high-performance Li-S batteries. *Advanced Science* 9 (12): 2106004.

64 Angulakshmi, N., Dhanalakshmi, R.B., Sathya, S. et al. (2021). Understanding the electrolytes of lithium-sulfur batteries. *Batteries Supercaps* 4 (7): 1064–1095.

65 Ng, S.F., Lau, M.Y.L., and Ong, W.J. (2021). Lithium-sulfur battery cathode design: tailoring metal-based nanostructures for robust polysulfide adsorption and catalytic conversion. *Advanced Materials* 33 (50): 2008654.

66 Li, S.L., Zhang, W.F., Zheng, J.F. et al. (2021). Inhibition of polysulfide shuttles in Li-S batteries: modified separators and solid-state electrolytes. *Advanced Energy Materials* 11 (2): 2000779.

67 Wang, L.G., Wang, D.S., and Li, Y.D. (2022). Single-atom catalysis for carbon neutrality. *Carbon Energy* 4: 1021–1079.

68 Yu, X. and Manthiram, A. (2020). Recent advances in lithium–carbon dioxide batteries. *Small Structures* 1 (2): 2000027.

69 Mu, X., Pan, H., He, P., and Zhou, H. (2020). Li–CO_2 and Na–CO_2 batteries: toward greener and sustainable electrical energy storage. *Advanced Materials* 32 (27): 1903790.

70 Wang, F., Li, Y., Xia, X. et al. (2021). Metal–CO_2 electrochemistry: from CO_2 recycling to energy storage. *Advanced Energy Materials* 11 (25): 2100667.

71 Takechi, K., Shiga, T., and Asaoka, T. (2011). A Li–O_2/CO_2 battery. *Chemical Communications* 47 (12): 3463–3465.

72 Das, S.K., Xu, S., and Archer, L.A. (2013). Carbon dioxide assist for non-aqueous sodium–oxygen batteries. *Electrochemistry Communications* 27: 59–62.

73 Xu, S., Das, S.K., and Archer, L.A. (2013). The Li–CO_2 battery: a novel method for CO_2 capture and utilization. *RSC Advances* 3 (18): 6656–6660.

74 Liu, Y., Wang, R., Lyu, Y. et al. (2014). Rechargeable Li/CO_2–O_2 (2 :1) battery and Li/CO_2 battery. *Energy & Environmental Science* 7 (2): 677–681.

75 Xie, J. and Wang, Y. (2019). Recent development of CO_2 electrochemistry from Li–CO_2 batteries to Zn–CO_2 batteries. *Accounts of Chemical Research* 52 (6): 1721–1729.

76 Ezeigwe, R.E., Dong, L., Manjunatha, R. et al. (2022). A review of lithium-O_2/CO_2 and lithium-CO_2 batteries: advanced electrodes/materials/electrolytes and functional mechanisms. *Nano Energy* 95: 106964.

77 Wu, M.J., Zhang, G.X., Yang, H.M. et al. (2022). Aqueous Zn-based rechargeable batteries: recent progress and future perspectives. *InfoMat* 4 (5): e12265.

78 Zhang, Z., Bai, W.L., Wang, K.X., and Chen, J.S. (2020). Electrocatalyst design for aprotic Li-CO_2 batteries. *Energy & Environmental Science* 13 (12): 4717–4737.

79 Chourasia, A.K., Pathak, A.D., Bongu, C. et al. (2022). In situ/operando characterization techniques: the guiding tool for the development of Li-CO_2 battery. *Small Methods* 6: 2200930.

80 Hu, A., Shu, C., Xu, C. et al. (2019). Design strategies toward catalytic materials and cathode structures for emerging Li–CO_2 batteries. *Journal of Materials Chemistry A* 7 (38): 21605–21633.

81 Liang, F., Zhang, K.W., Zhang, L. et al. (2021). Recent development of electrocatalytic CO_2 reduction application to energy conversion. *Small* 17 (44): 2100323.

82 Li, X., Yang, S.X., Feng, N.N. et al. (2016). Progress in research on Li-CO_2 batteries: mechanism, catalyst and performance. *Chinese Journal of Catalysis* 37 (7): 1016–1024.

83 Sui, D., Chang, M.J., Wang, H.Y. et al. (2021). A brief review of catalytic cathode materials for Na-CO_2 batteries. *Catalysts* 11 (5): 603.

84 Sun, X.Y., Hou, Z.P., He, P., and Zhou, H.S. (2021). Recent advances in rechargeable Li-CO_2 batteries. *Energy Fuels* 35 (11): 9165–9186.

85 Liu, Z.X., Qin, L.P., Cao, X.X. et al. (2022). Ion migration and defect effect of electrode materials in multivalent-ion batteries. *Progress in Materials Science* 125: 100911.

86 Qu, X.L., Tang, Y., Du, A.B. et al. (2021). Polymer electrolytes—new opportunities for the development of multivalent ion batteries. *Chemistry – An Asian Journal* 16 (21): 3272–3280.

87 Wang, J., Li, H., Liu, S. et al. (2021). Turning on Zn 4s electrons in a N2-Zn-B2 configuration to stimulate remarkable ORR performance. *Angewandte Chemie International Edition* 60 (1): 181–185.

88 Ponrouch, A., Bitenc, J., Dominko, R. et al. (2019). Multivalent rechargeable batteries. *Energy Storage Materials* 20: 253–262.

89 Zhang, N., Dong, Y., Jia, M. et al. (2018). Rechargeable aqueous Zn–V_2O_5 battery with high energy density and long cycle life. *ACS Energy Letters* 3 (6): 1366–1372.

90 Zhang, Q., Wang, L., Wang, J. et al. (2018). Low-temperature synthesis of edge-rich graphene paper for high-performance aluminum batteries. *Energy Storage Materials* 15: 361–367.

91 Zhang, Y., Liu, S., Ji, Y. et al. (2018). Emerging nonaqueous aluminum-ion batteries: challenges, status, and perspectives. *Advanced Materials* 30 (38): 1706310.

92 Xie, J. and Zhang, Q. (2019). Recent progress in multivalent metal (Mg, Zn, Ca, and Al) and metal-ion rechargeable batteries with organic materials as promising electrodes. *Small* 15 (15): 1805061.

93 Lei, X., Liang, X., Yang, R. et al. (2022). Rational design strategy of novel energy storage systems: toward high-performance rechargeable magnesium batteries. *Small* 18 (22): 2200418.

94 Gummow, R.J., Vamvounis, G., Kannan, M.B., and He, Y.H. (2018). Calcium-ion batteries: current state-of-the-art and future perspectives. *Advanced Materials* 30 (39): 1801702.

95 Fu, N., Xu, Y.T., Zhang, S. et al. (2022). Electrode materials for aqueous multivalent metal-ion batteries: current status and future prospect. *Journal of Energy Chemistry* 67: 563–584.

96 Wang, M., Jiang, C.L., Zhang, S.Q. et al. (2018). Reversible calcium alloying enables a practical room-temperature rechargeable calcium-ion battery with a high discharge voltage. *Nature Chemistry* 10 (6): 667–672.

97 Liang, Y.L., Dong, H., Aurbach, D., and Yao, Y. (2020). Current status and future directions of multivalent metal-ion batteries. *Nature Energy* 5 (9): 646–656.

98 Wang, Z.Y., Li, Y.M., Wang, J.W. et al. (2022). Recent progress of flexible aqueous multivalent ion batteries. *Carbon Energy* 4 (3): 411–445.

99 Zhou, X., Liu, Q., Jiang, C. et al. (2020). Strategies towards low-cost dual-ion batteries with high performance. *Angewandte Chemie International Edition* 59 (10): 3802–3832.

100 Pan, Q., Tong, Z., Su, Y. et al. (2022). Flat–zigzag interface design of chalcogenide heterostructure toward ultralow volume expansion for high-performance potassium storage. *Advanced Materials* 34 (39): 2203485.

101 Guo, S., Qin, L., Zhang, T. et al. (2021). Fundamentals and perspectives of electrolyte additives for aqueous zinc-ion batteries. *Energy Storage Materials* 34: 545–562.

102 Ou, X., Gong, D., Han, C. et al. (2021). Advances and prospects of dual-ion batteries. *Advanced Energy Materials* 11 (46): 2102498.

103 Pan, Q., Gong, D., and Tang, Y. (2020). Recent progress and perspective on electrolytes for sodium/potassium-based devices. *Energy Storage Materials* 31: 328–343.

104 Pal, P. and Ghosh, A. (2020). Robust succinonitrile plastic crystal-based ionogel for all-solid-state Li-ion and dual-ion batteries. *ACS Applied Energy Materials* 3 (5): 4295–4304.

105 Jiang, X.Y., Luo, L.B., Zhong, F.P. et al. (2019). Electrolytes for dual-carbon batteries. *ChemElectroChem* 6 (10): 2615–2629.

106 Su, S.M., Ma, J.B., Zhao, L. et al. (2021). Progress and perspective of the cathode/electrolyte interface construction in all-solid-state lithium batteries. *Carbon Energy* 3 (6): 866–894.

107 Zhang, C. and Li, X. (2021). Perspective on organic flow batteries for large-scale energy storage. *Current Opinion in Electrochemistry* 30: 100836.

108 Xiang, L., Ou, X., Wang, X. et al. (2020). Highly concentrated electrolyte towards enhanced energy density and cycling life of dual-ion battery. *Angewandte Chemie International Edition* 59 (41): 17924–17930.

109 Kravchyk, K.V. and Kovalenko, M.V. (2019). Rechargeable dual-ion batteries with graphite as a cathode: key challenges and opportunities. *Advanced Energy Materials* 9 (35): 1901749.

110 Li, Y., Lu, Y., Adelhelm, P. et al. (2019). Intercalation chemistry of graphite: alkali metal ions and beyond. *Chemical Society Reviews* 48 (17): 4655–4687.

111 Wu, H.Z., Li, L., and Yuan, W.H. (2022). Nano-cubic alpha-Fe_2O_3 anode for Li^+/Na^+ based dual-ion full battery. *Chemical Engineering Journal* 442: 136259.

112 Zhang, M.M., Yan, D., and Li, W.C. (2022). Regulation of dual-ion batteries via the defects design in carbon electrode based on the different storage behaviors of PF_6^- and Li^+. *Journal of Power Sources* 527: 231169.

113 Wang, H.G., Wang, Y.N., Wu, Q., and Zhu, G.S. (2022). Recent developments in electrode materials for dual-ion batteries: potential alternatives to conventional batteries. *Materials Today* 52: 269–298.

114 Yu, A., Pan, Q.G., Zhang, M. et al. (2020). Fast rate and long life potassium-ion based dual-ion battery through 3D porous organic negative electrode. *Advanced Functional Materials* 30 (24): 2001440.

115 Zhang, L., Wang, H., Zhang, X., and Tang, Y. (2021). A review of emerging dual-ion batteries: fundamentals and recent advances. *Advanced Functional Materials* 31 (20): 2010958.

116 Liu, H., Zhao, J., and Li, X. (2022). Controlled synthesis of carbon-supported Pt-based electrocatalysts for proton exchange membrane fuel cells. *Electrochemical Energy Reviews* 5 (4): 13.

117 Liu, Z., Chen, H., and Zhang, T. (2022). Review on system mitigation strategies for start-stop degradation of automotive proton exchange membrane fuel cell. *Applied Energy* 327: 120058.

118 Zaman, S., Wang, M., Liu, H.J. et al. (2022). Carbon-based catalyst supports for oxygen reduction in proton-exchange membrane fuel cells. *Trends in Chemistry* 4 (10): 886–906.

119 Zhao, J., Fu, C., Ye, K. et al. (2022). Manipulating the oxygen reduction reaction pathway on Pt-coordinated motifs. *Nature Communications* 13 (1): 685.

120 Hu, C. and Dai, L. (2017). Multifunctional carbon-based metal-free electrocatalysts for simultaneous oxygen reduction, oxygen evolution, and hydrogen evolution. *Advanced Materials* 29 (9): 1604942.

121 Samad, S., Loh, K.S., Wong, W.Y. et al. (2018). Carbon and non-carbon support materials for platinum-based catalysts in fuel cells. *International Journal of Hydrogen Energy* 43 (16): 7823–7854.

122 Yang, C.-L., Wang, L.-N., Yin, P. et al. (2021). Sulfur-anchoring synthesis of platinum intermetallic nanoparticle catalysts for fuel cells. *Science* 374 (6566): 459–464.

123 Xiao, F., Wang, Y.C., Wu, Z.P. et al. (2021). Recent advances in electrocatalysts for proton exchange membrane fuel cells and alkaline membrane fuel cells. *Advanced Materials* 33 (50): 2006292.

124 Zhang, X., Truong-Phuoc, L., Asset, T. et al. (2022). Are Fe–N–C electrocatalysts an alternative to Pt-based electrocatalysts for the next generation of proton exchange membrane fuel cells? *ACS Catalysis* 12 (22): 13853–13875.

125 Kumar, A., Ibraheem, S., Nguyen, T.A. et al. (2021). Molecular-MN_4 vs atomically dispersed M-N_4-C electrocatalysts for oxygen reduction reaction. *Coordination Chemistry Reviews* 446: 214122.

126 Wu, S., Liu, H.B., Lei, G.Y. et al. (2022). Single-atomic iron-nitrogen 2D MOF-originated hierarchically porous carbon catalysts for enhanced oxygen reduction reaction. *Chemical Engineering Journal* 441: 135849.

127 Zhang, L., Meng, Q., Zheng, R. et al. (2023). Microenvironment regulation of M-N-C single-atom catalysts towards oxygen reduction reaction. *Nano Research* 16 (4): 4468–4487.

128 Zhang, H., Liu, X., Li, H.H. et al. (2021). Challenges and strategies for high-energy aqueous electrolyte rechargeable batteries. *Angewandte Chemie International Edition* 60 (2): 598–616.

129 Li, W., Dahn, J.R., and Wainwright, D.S. (1994). Rechargeable lithium batteries with aqueous electrolytes. *Science* 264 (5162): 1115–1118.

130 Bin, D., Wen, Y., Wang, Y., and Xia, Y. (2018). The development in aqueous lithium-ion batteries. *Journal of Energy Chemistry* 27 (6): 1521–1535.

131 von Wald Cresce, A. and Xu, K. (2021). Aqueous lithium-ion batteries. *Carbon Energy* 3 (5): 721–751.

132 Huang, J.H., Guo, Z.W., Ma, Y.Y. et al. (2019). Recent progress of rechargeable batteries using mild aqueous electrolytes. *Small Methods* 3 (1): 1800272.

133 Deng, Y.P., Liang, R.L., Jiang, G.P. et al. (2020). The current state of aqueous Zn-based rechargeable batteries. *ACS Energy Letters* 5 (5): 1665–1675.

134 Ru, Y., Zheng, S.S., Xue, H.G., and Pang, H. (2019). Different positive electrode materials in organic and aqueous systems for aluminium ion batteries. *Journal of Materials Chemistry A* 7 (24): 14391–14418.

135 Tang, B.Y., Shan, L.T., Liang, S.Q., and Zhou, J. (2019). Issues and opportunities facing aqueous zinc-ion batteries. *Energy & Environmental Science* 12 (11): 3288–3304.

136 Posada, J.O.G., Rennie, A.J.R., Villar, S.P. et al. (2017). Aqueous batteries as grid scale energy storage solutions. *Renewable and Sustainable Energy Reviews* 68: 1174–1182.

137 Dai, C.L., Hu, L.Y., Jin, X.T. et al. (2021). The emerging of aqueous zinc-based dual electrolytic batteries. *Small* 17 (33): 2008043.

138 Ou, L.N., Ou, H.H., Qin, M.L. et al. (2022). Recent progress on phosphate cathode materials for aqueous zinc-ion batteries. *ChemSusChem* 15 (19): e202201184.

139 Zhang, Y.R., Chen, A.B., and Sun, J. (2021). Promise and challenge of vanadium-based cathodes for aqueous zinc-ion batteries. *Journal of Energy Chemistry* 54: 655–667.

140 Chen, D., Lu, M.J., Cai, D. et al. (2021). Recent advances in energy storage mechanism of aqueous zinc-ion batteries. *Journal of Energy Chemistry* 54: 712–726.

141 Ruan, P., Liang, S., Lu, B. et al. (2022). Design strategies for high-energy-density aqueous zinc batteries. *Angewandte Chemie International Edition* 61 (17): e202200598.

142 Lewis, C.E.M., Fernando, J.F.S., Siriwardena, D.P. et al. (2022). Vanadium-containing layered materials as high-performance cathodes for aqueous zinc-ion batteries. *Advanced Materials Technologies* 7 (4): 2100505.

143 Sui, Y.M. and Ji, X.L. (2021). Anticatalytic strategies to suppress water electrolysis in aqueous batteries. *Chemical Reviews* 121 (11): 6654–6695.

144 Gao, H., Tang, K.K., Xiao, J. et al. (2022). Recent advances in "water in salt" electrolytes for aqueous rechargeable monovalent-ion (Li^+, Na^+, K^+) Batteries. *Journal of Energy Chemistry* 69: 84–99.

145 Zhang, H., Lu, W., and Li, X. (2019). Progress and perspectives of flow battery technologies. *Electrochemical Energy Reviews* 2 (3): 492–506.

146 Lourenssen, K., Williams, J., Ahmadpour, F. et al. (2019). Vanadium redox flow batteries: a comprehensive review. *Journal of Energy Storage* 25: 100844.

147 Xue, T. and Fan, H.J. (2021). From aqueous Zn-ion battery to Zn-MnO_2 flow battery: a brief story. *Journal of Energy Chemistry* 54: 194–201.

148 Wei, X., Cosimbescu, L., Xu, W. et al. (2015). Towards high-performance nonaqueous redox flow electrolyte via ionic modification of active species. *Advanced Energy Materials* 5 (1): 1400678.

149 Kwabi, D.G., Ji, Y., and Aziz, M.J. (2020). Electrolyte lifetime in aqueous organic redox flow batteries: a critical review. *Chemical Reviews* 120 (14): 6467–6489.

150 Liu, W., Lu, W., Zhang, H., and Li, X. (2019). Aqueous flow batteries: research and development. *Chemistry – A European Journal* 25 (7): 1649–1664.

151 Luo, J., Hu, B., Hu, M. et al. (2019). Status and prospects of organic redox flow batteries toward sustainable energy storage. *ACS Energy Letters* 4 (9): 2220–2240.

152 Kim, J. and Park, H. (2022). Recent advances in porous electrodes for vanadium redox flow batteries in grid-scale energy storage systems: a mass transfer perspective. *Journal of Power Sources* 545: 231904.

153 Chen, R.Y. (2023). Redox flow batteries: electrolyte chemistries unlock the thermodynamic limits. *Chemistry – An Asian Journal* 18: e202201024.

154 Shi, Y., Eze, C., Xiong, B. et al. (2019). Recent development of membrane for vanadium redox flow battery applications: a review. *Applied Energy* 238: 202–224.

155 Zhu, Z., Jiang, T., Ali, M. et al. (2022). Rechargeable batteries for grid scale energy storage. *Chemical Reviews* 122 (22): 16610–16751.

156 Zhang, H., Sun, C., and Ge, M. (2022). Review of the research status of cost-effective zinc-iron redox flow batteries. *Batteries* 8 (11): 202.

157 Krishnamurti, V., Yang, B., Murali, A. et al. (2022). Aqueous organic flow batteries for sustainable energy storage. *Current Opinion in Electrochemistry* 35: 101100.

158 Choi, C., Ashby, D.S., Butts, D.M. et al. (2020). Achieving high energy density and high power density with pseudocapacitive materials. *Nature Reviews Materials* 5 (1): 5–19.

159 Simon, P. and Gogotsi, Y. (2008). Materials for electrochemical capacitors. *Nature Materials* 7 (11): 845–854.

160 Chatterjee, D.P. and Nandi, A.K. (2021). A review on the recent advances in hybrid supercapacitors. *Journal of Materials Chemistry A* 9 (29): 15880–15918.

161 Zhang, Y., Mei, H.X., Cao, Y. et al. (2021). Recent advances and challenges of electrode materials for flexible supercapacitors. *Coordination Chemistry Reviews* 438: 213910.

162 Simon, P. and Gogotsi, Y. (2020). Perspectives for electrochemical capacitors and related devices. *Nature Materials* 19 (11): 1151–1163.

163 Wu, F., Liu, M.Q., Li, Y. et al. (2021). High-mass-loading electrodes for advanced secondary batteries and supercapacitors. *Electrochemical Energy Reviews* 4 (2): 382–446.

164 Tafete, G.A., Abera, M.K., and Thothadri, G. (2022). Review on nanocellulose-based materials for supercapacitors applications. *Journal of Energy Storage* 48: 103938.

165 Wang, G., Zhang, L., and Zhang, J. (2012). A review of electrode materials for electrochemical supercapacitors. *Chemical Society Reviews* 41 (2): 797–828.

166 Augustyn, V., Simon, P., and Dunn, B. (2014). Pseudocapacitive oxide materials for high-rate electrochemical energy storage. *Energy & Environmental Science* 7 (5): 1597.

167 Kumar, S., Saeed, G., Zhu, L. et al. (2021). 0D to 3D carbon-based networks combined with pseudocapacitive electrode material for high energy density supercapacitor: a review. *Chemical Engineering Journal* 403: 126352.

168 Iqbal, M.Z. and Aziz, U. (2022). Supercapattery: merging of battery-supercapacitor electrodes for hybrid energy storage devices. *Journal of Energy Storage* 46: 103823.

169 Keum, K., Kim, J.W., Hong, S.Y. et al. (2020). Flexible/stretchable supercapacitors with novel functionality for wearable electronics. *Advanced Materials* 32 (51): 2002180.

170 Wang, X.F., Lu, X.H., Liu, B. et al. (2014). Flexible energy-storage devices: design consideration and recent progress. *Advanced Materials* 26 (28): 4763–4782.

171 Li, L., Wu, Z., Yuan, S., and Zhang, X.-B. (2014). Advances and challenges for flexible energy storage and conversion devices and systems. *Energy & Environmental Science* 7 (7): 2101–2122.

172 Xu, T., Du, H.S., Liu, H.Y. et al. (2021). Advanced nanocellulose-based composites for flexible functional energy storage devices. *Advanced Materials* 33 (48): 2101368.

173 Mackanic, D.G., Chang, T.H., Huang, Z.J. et al. (2020). Stretchable electrochemical energy storage devices. *Chemical Society Reviews* 49 (13): 4466–4495.

174 Zhao, D.W., Zhu, Y., Cheng, W.K. et al. (2021). Cellulose-based flexible functional materials for emerging intelligent electronics. *Advanced Materials* 33 (28): 2000619.

175 Qian, G., Liao, X., Zhu, Y. et al. (2019). Designing flexible lithium-ion batteries by structural engineering. *ACS Energy Letters* 4 (3): 690–701.

176 Pomerantseva, E., Bonaccorso, F., Feng, X. et al. (2019). Energy storage: the future enabled by nanomaterials. *Science* 366 (6468): 969–981.

177 Zhou, D., Shanmukaraj, D., Tkacheva, A. et al. (2019). Polymer electrolytes for lithium-based batteries: advances and prospects. *Chem* 5 (9): 2326–2352.

178 Zhang, Q., Liu, K., Ding, F., and Liu, X. (2017). Recent advances in solid polymer electrolytes for lithium batteries. *Nano Research* 10 (12): 4139–4174.

179 Zhao, Y., Wang, L., Zhou, Y.A. et al. (2021). Solid polymer electrolytes with high conductivity and transference number of Li ions for Li-based rechargeable batteries. *Advanced Science* 8 (7): 2003675.

180 Gwon, H., Hong, J., Kim, H. et al. (2014). Recent progress on flexible lithium rechargeable batteries. *Energy & Environmental Science* 7 (2): 538–551.

181 Liu, Y.Q., He, K., Chen, G. et al. (2017). Nature-inspired structural materials for flexible electronic devices. *Chemical Reviews* 117 (20): 12893–12941.

2

Mechanisms and Promising Cathode Catalysts for Metal-Air Batteries

Tao Zhang[1,2] *and Zhiqian Hou*[1,2]

[1] *Chinese Academy of Sciences, Shanghai Institute of Ceramics, State Key Lab of High-Performance Ceramics and Superfine Microstructure, 1295 Dingxi Road, Shanghai 200050, P.R. China*
[2] *University of Chinese Academy of Sciences, Center of Materials Science and Optoelectronics Engineering, 19 Yuquan Road, Beijing 100049, P.R. China*

2.1 Introduction

In recent decades, energy shortages and environmental pollution have pushed the development and utilization of cleaner and renewable energy sources [1]. Meanwhile, the high consumption of portable electronic devices and electric vehicles has boosted the rapid progress of energy storage and conversion technologies [2]. With the actual energy density of conventional lithium-ion batteries (LIBs) gradually approaching the theoretical value, the search for rechargeable cells with application prospects has become the current consensus [3]. Among the existing energy storage systems, metal-air batteries have attracted considerable attention from scientists due to their ultrahigh theoretical energy density and optimal feedstock abundance (air) [4]. Specifically, a metal-air battery is an electrochemical energy storage device that uses lithium, sodium, or zinc metal as the anode and oxygen from the air as the cathode active material [5]. The current research focuses on nonaqueous lithium-air batteries (LABs), sodium-oxygen batteries (NOBs), and aqueous zinc-air batteries (ZABs). In addition, besides using oxygen as the active species, carbon dioxide (CO_2) and nitrogen (N_2) have also been screened as cathode reactants to design new electrochemical technologies, such as Li-CO_2 batteries and Li-N_2 batteries [6, 7].

2.2 Overview of Metal-Air Batteries

To visualize the merits of metal-air cells, Figure 2.1a compares the energy density in current energy storage systems, from which it is observed that Li/Zn-air batteries display much higher energy density than LIBs and that LAB demonstrates the theoretical value that can compete with gasoline [8]. In terms of configuration, the metal-air battery combines a hybrid structure of battery and fuel cell characteristics.

Electrochemical Energy Storage Devices: Non-Conventional Technologies and Materials, First Edition. Edited by Yongbing Tang and Luojiang Zhang.

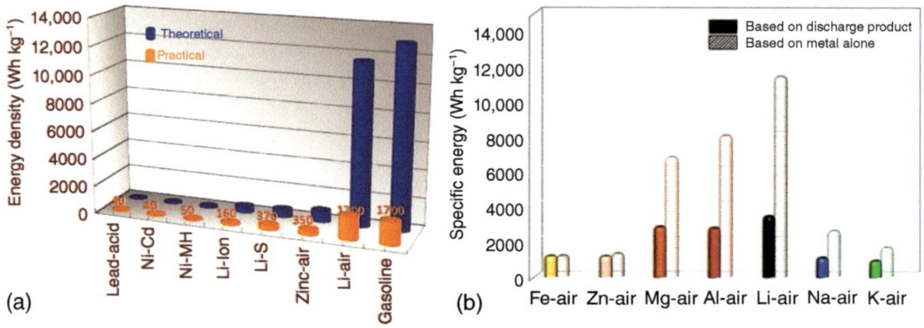

Figure 2.1 (a) Gravimetric energy densities (Wh kg^{-1}) for various types of rechargeable batteries compared to gasoline. The theoretical density is based strictly on thermodynamics and is shown as blue bars, while the practical achievable density is indicated by orange bars and numerical values. Source: Reproduced with permission [8]. Copyright 2014, American Chemical Society. (b) Specific energies of metal-air batteries. Solid-colored bars represent specific energy densities based on the discharge product and one equivalent of the metal anode, whereas patterned bars are based on the formal specific energies taking only metal into account. Source: Reproduced with permission [9]. Copyright 2020, American Chemical Society.

Figure 2.1b lists several reported energy storage systems based on alkali metals (Li, Na, and K), alkaline earth metals (Mg), and first-row transition metals (Fe and Zn) or Al as anodes and compares their specific energies for different standards [9]. It is important to note that there is ambiguity in the recorded reports of metal-O_2 batteries and metal-air batteries, and therefore all discussions, in this case, are annotated with metal-air cells for ease of understanding. However, despite the bright prospects, the development of metal-air batteries can still face many serious challenges, such as low energy efficiency, high overpotential, and poor cycle durability, which severely hinder their widespread applications [9]. Studies have shown that these issues are closely related to the sluggish kinetics of the oxygen reduction reaction (ORR) and oxygen evolution reaction (OER) during the discharge/charge [10]. Therefore, a deeper insight into the fundamental electrochemical reaction mechanism is of great scientific importance to realize the commercialization of metal-air batteries.

2.2.1 Reaction Mechanism of Metal-Air Batteries

In metal-air systems, Li, Na, and K are overreactive in aqueous solutions requiring ion-conductive membranes (sodium super ionic conductor, NASICON-type glass-ceramics) for isolation and protection for stable operation; however, the manufacturing complexity and production costs are rarely favored by researchers [11]. The fundamental mechanism of a nonaqueous Li/Na/K-air cell involves a multielectron transfer redox reaction in which oxygen is adsorbed at the catalytic site and suffers a single-electron reduction to produce the oxygen anion O_2^-, which then combines with an alkali metal cation to form peroxide MO_2 (M = Li, Na, K). The difficulty of stabilizing O_2^- due to the small size of Li^+ leads to the disproportionation of LiO_2, forming Li_2O_2 as the discharge product, whereas larger cations Na^+ and K^+ can build stable structures of Na_2O_2, NaO_2, and KO_2 [12, 13]. The anode/cathode reaction can be summarized as follows:

$$\text{Anode: } M \rightarrow M^+ + e^- \tag{2.1}$$

$$\text{Cathode: } xM^+ + O_2 + xe^- \leftrightarrow M_xO_2 \tag{2.2}$$

Superoxide/peroxide accumulates on the cathode surface as the discharge progresses, gradually blocking the porous cathode and blunting the catalytic activity, resulting in the actual capacity being much lower than the theoretical value [14].

The aqueous metal-air cells can be represented by Zn, Fe, Mg, and Al [15–18]. The thermodynamic instability in the aqueous medium drives the passivation of oxides and hydroxides, which completes the discharge/charge behavior. The corresponding reaction equations are as follows:

$$\text{Anode: } M \rightarrow M^{n+} + ne^- \tag{2.3}$$

$$\text{Cathode: } O_2 + ne^- + 2H_2O \leftrightarrow 4OH^- \tag{2.4}$$

where M presents the metal and n is the oxidation number of the metal ion. This chapter focuses on the author's understanding and knowledge of the cathode reaction mechanism of metal-air batteries and uses nonaqueous LABs and aqueous ZABs as the main discussion cases.

For LABs, the constituent components mainly include lithium metal, separator, electrolyte, porous cathode, and catalyst [19]. Unlike the sealed structure of LIBs, the unique open-ended nature of LABs ensures rapid oxygen transfer. The LAB is accompanied by the reversible generation and decomposition of lithium peroxide (Li_2O_2) during discharge/charge (Eq. 2.5) [20–22].

$$2Li + O_2 \leftrightarrow Li_2O_2 \tag{2.5}$$

Briefly, during discharge, the anode lithium metal loses electrons to Li^+, the cathode active material oxygen is reduced to O_2^- and combines with Li^+ to form lithium superoxide (LiO_2) intermediates, which then transform to Li_2O_2 through electrochemical reduction or disproportionation [23]. The insulating insolubility of Li_2O_2 in the electrolyte leads to the continuous deposition of Li_2O_2 onto the cathode surface during the discharge process, corresponding to the ORR [24]. The recharge behavior corresponds to the OER, where Li_2O_2 decomposes to release Li^+ and O_2.

Generally, the discharge product with different morphologies depends on the growth pathway of Li_2O_2 [25]. The formation of Li_2O_2 involves multistep reactions including electrochemical and chemical steps [26–28]. First, O_2 dissolves in the electrolyte and is adsorbed on the active site (*) on the surface of the cathode to create adsorbed oxygen (O_2^*) (Eq. 2.6) [18].

$$O_2 +^* \rightarrow O_2^* \tag{2.6}$$

Second, O_2^* captures an electron to combine with Li^+ coming across the electrolyte to form LiO_2^* (Eq. 2.7).

$$O_2^* + e^- + Li^+ \rightarrow LiO_2^* \tag{2.7}$$

Third, LiO_2^* undergoes electrochemical reduction (Eq. 2.8) or disproportionation (Eq. 2.9) at the surface to form $Li_2O_2^*$.

$$LiO_2^* + e^- + Li^+ \rightarrow Li_2O_2^* \tag{2.8}$$

$$LiO_2^* + LiO_2^* \rightarrow Li_2O_2^* + O_2^* \tag{2.9}$$

Indeed, the lower kinetic potential barrier and free energy endow the electrochemical reduction pathway with higher precedence than disproportionation [29, 30]. On the other hand, adsorbed LiO_2^* can also diffuse into the solution by conditioning the electrolyte [31]. Thus, the solution growth model implies that soluble LiO_2 grows into Li_2O_2 by nucleation in the electrolyte via disproportionation (Eq. 2.10) [32].

$$LiO_{2(sol)} + LiO_{2(sol)} \rightarrow Li_2O_2 + O_2 \tag{2.10}$$

When charging is carried out, the abovementioned electrochemical reactions described are reversed, with the metal plating at the anode and O_2 evolving at the cathode. The discharge product of Li_2O_2 required a high disintegration potential, which leads to a high charging plateau [33]. Furthermore, the deposition behavior of Li_2O_2 is heavily influenced by the ORR kinetics. Therefore, the design of efficient ORR/OER bifunctional electrocatalysts is essential for low overpotential and large-capacity LABs.

In terms of constituent parts, ZABs use zinc as the anode and aqueous electrolyte, and the ORR/OER reactions in a ZAB are based on the following equation [15]:

$$\text{Anode: } Zn \rightarrow Zn^{2+} + 2e^- \tag{2.11}$$

$$Zn^{2+} + 4OH^- \rightarrow Zn(OH)_4{}^{2-} \tag{2.12}$$

$$Zn(OH)_4{}^{2-} \rightarrow ZnO + H_2O + 2OH^- \tag{2.13}$$

$$\text{Cathode: } O_2 + H_2O + 4e^- \rightarrow 4OH^- \tag{2.14}$$

The overall reaction can be summarized as follows:

$$2Zn + O_2 \rightarrow 2ZnO \tag{2.15}$$

Not surprisingly, the fast ORR/OER kinetics for the ZAB system also dominate the total performance of the battery.

2.2.2 Design of the Cathode Catalysts

High-efficient electrocatalysts are considered one of the main strategies for achieving fast redox reaction kinetics at the cathode, where the adsorption/desorption of reactant intermediates at the catalytic site can be defined by the descriptor *d*-band center [34]. In nonaqueous LABs, the capture and conversion of O_2^-, LiO_2, Li_2O_2, and O_2 are mainly implicated, whereas in aqueous ZABs OH^*, O^*, OOH^*, and O_2 are represented [35, 36]. Theoretically, the activity of either catalyst may be limited by the abovementioned transfer steps. For now, there are two strategies

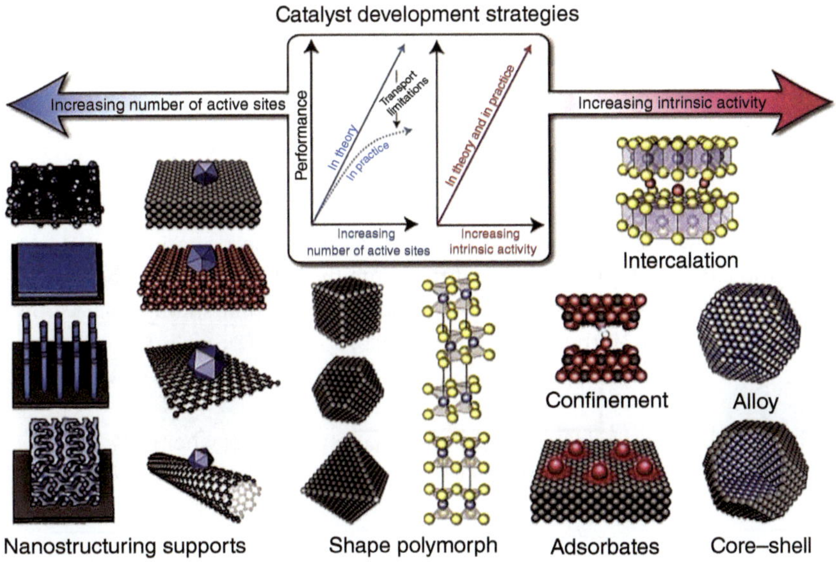

Figure 2.2 Schematic of various catalyst development strategies, which aim to increase the number of active sites and/or increase the intrinsic activity of each active site. Source: Reproduced with permission [37]. Copyright 2017, American Association for the Advancement of Science.

to systematically increase the electrocatalytic activity, as shown in Figure 2.2: (i) increasing the number of active sites per unit reaction area and (ii) enhancing the intrinsic activity of the reaction sites [37]. The enhancement of catalytic performance can be achieved via heterogeneous element doping, construction of heterostructure, metal-support interactions, confinement, and core–shell structures to strengthen/weaken the surface adsorption/desorption energy of reaction sites for intermediate species and to enable fast dynamics of each transfer pathway.

2.2.2.1 Carbon-Based Catalysts

The low cost and large specific surface area of carbon materials have stimulated extensive research into carbon-based catalysts. Currently, scientists are devoted to porous carbon, graphene, carbon nanotubes (NTs), and carbon nanofibers as ORR/OER bifunctional electrocatalysts [38, 39]. Additionally, the excellent electrical conductivity and gas diffusion of carbon materials play an integral role in the transport of electrons and reactants [40]. Therefore, the anchoring of efficient catalysts on the surface of carbon substrates retains the advantages of carbon materials while reducing the overpotential during charge/discharge cycles.

Carbon Catalysts Xia and coworkers [41] synthesized the first three-dimensional ordered mesoporous/macroporous carbon sphere arrays (MMCSAs) and used them as cathode electrocatalysts for LAB. The scanning electron microscope (SEM), transmission electron microscope (TEM) images and functioning mechanism of MMCSAs are shown in Figure 2.3a–c. The ordered mesoporous channels and

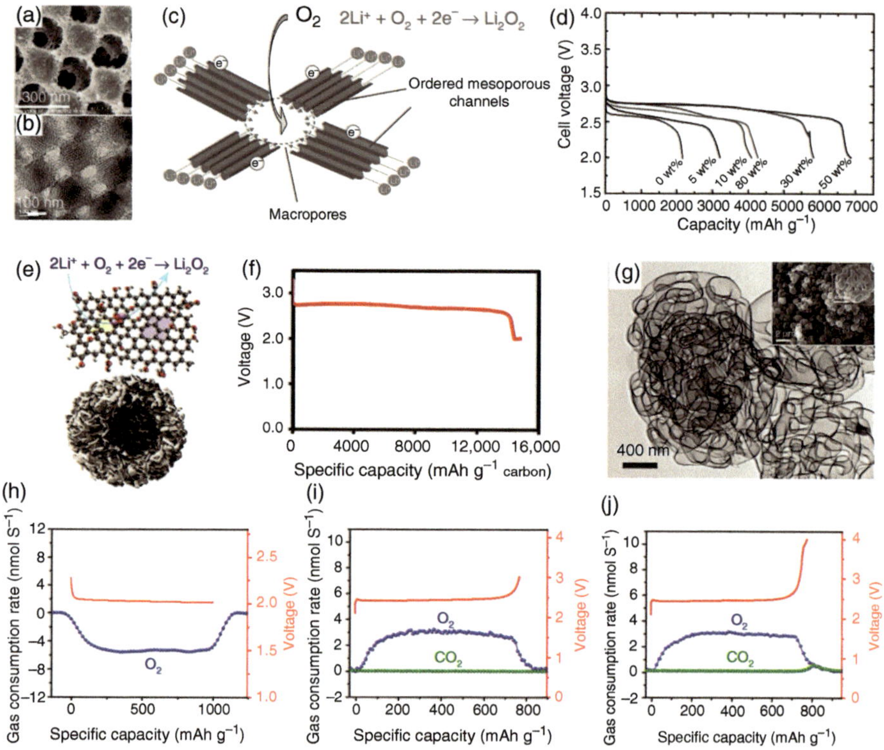

Figure 2.3 Three-dimensional ordered MMCSAs: (a) SEM image, (b) TEM image, (c) schematic illustration of O_2/Li_2O_2 conversion, and (d) discharge curves of Li-O_2 batteries at a current density of 50 mA g^{-1} with different wt% MMCSAs in porous catalytic electrodes. Source: Reproduced with permission of Guo et al. [41]. © 2013/John Wiley & Sons. (e) Schematic structure of an FGS with an ideal bimodal porous structure. (f) The discharge curve of the battery with FGS as the air electrode. Source: Reproduced with permission of Xiao et al. [42]. © 2011/American Chemical Society. Nanostructure-rich layered porous carbon spheres (PCS). (g) TEM image (inset: SEM image), *in situ* DEMS analysis of the gas consumption and evolution during Na-O_2 cell operation; discharge at 1 A g^{-1} to 1000 mAh g^{-1} (h); charge at 500 mA g^{-1} to 3.0 V (i) and 4.0 V (j). Source: Reproduced with permission of Sun et al. [51]. © 2017/John Wiley & Sons.

hierarchical mesoporous/macroporous structure of MMCSA facilitated electrolyte infiltration and Li^+ transport and provided an efficient space for O_2 diffusion and O_2/Li_2O_2 conversion. The experimental results showed that the LAB containing 50 wt% MMCSA delivers a discharge capacity of 7000 mAh g^{-1} (Figure 2.3d) and an average operating voltage of 2.75 V.

Zhang and coworkers designed layered porous air electrodes constructed from lattice defects and hydroxyl, epoxy, and carboxyl groups according to the colloidal microemulsion method [42, 43]. As illustrated in Figure 2.3e, the functional groups and lattice defects on the functionalized graphene sheet (FGS) showed epoxy and hydroxyl groups on both sides of the graphene plane, carboxyl and hydroxyl groups on the edges, 5-8-5 defects (yellow), and 5-7-7-5 (Stone-Wells) defects (blue). (The carbon atoms were grey, the oxygen atoms were red, and the hydrogen atoms were white.) [43] Lattice-defect sites like 5-8-5 were energetically favorable sites for nucleation and pegging of reaction products. In addition, the interconnected pores on the microscale and nanoscale characteristic of 3D electrodes helped to boost the rapid diffusion of O_2 and stacking of Li_2O_2. Density functional theory (DFT) calculations and electron microscope characterization confirmed that lattice defect sites on functionalized graphene facilitated the formation of nanoscale Li_2O_2 (2–50 nm) and that the synergistic effect of layered pore structure and defect sites ensured an ultrahigh capacity of 15,000 mAh g^{-1} (Figure 2.3f).

Although graphene nanosheet-assembled air cathodes can significantly increase the discharge capacity of the LAB, the inferior ORR/OER activity is not conducive to the rapid decomposition of Li_2O_2 [44]. Pyridine nitrogen in nitrogen-doped carbon materials can promote the nucleation potential reduction of the discharge products and improve the adsorption behavior of the intermediates in the rate-determining step (RDS), thereby improving the reversible cycling performance of the battery [45]. The initial nucleation pathway of Li_2O_2 on different carbon-based surfaces was investigated by theoretical calculations from Jing and Zhou [46]. The free energy step diagram of the discharge product at open circuit potential ($U = 0$ V) demonstrated that the O_2 adsorption step was always endothermic, while the two reactions containing Li were downhill in the free energy curve. Therefore, the RDS for the nucleation kinetics of Li_2O_2 at the op circuit potential depended on the O_2 adsorption behavior. In contrast, at equilibrium potential ($U = 2.92$ V), all intermediate steps in the initial nucleation of Li_2O_2 became endothermic. Interestingly, the RDS at this point was no longer reliant on O_2 adsorption but limited by the lithium-containing conversion reaction instead. The overpotential (η) for Li_2O_2 nucleation on different surfaces can be obtained from the free energy change (ΔG_c) of the control step, which is defined as $\eta = -\Delta G_c/e$. The lower the η to be overcome, the easier the reaction step is to complete.

Wei and coworkers revealed the underlying cause of the improved discharge/charge potential plateau of B and N codoped Stone–Wales defective graphene (SWG) by DFT [47]. The adsorption energy of the intermediates (E_{ads}) was studied to gain further insight into the potential mechanism of action before the overpotential and oxygen-containing species. η_{ORR}/η_{OER} was plotted as a function of the adsorption energy of intermediates in RDS, with all E_{ads} being negative implying that the adsorption of intermediates was an exothermic process. For ORR, η_{ORR} decreased from SWG to BN-SWG as the adsorption energy increased, reaching a minimum at BN-SWG. Further increases in adsorption energy led to a rapid increase in overpotential, resulting in a V-shaped curve. This phenomenon implies that larger adsorption energies limit the involvement of Li_2O_2 in subsequent reactions, leading to high discharge overpotentials. The lower adsorption energy is not beneficial for the anchoring of Li_2O_2 at the catalytic site, still forming a high overpotential. In the case of the OER pathway, the adsorption energy of LiO_2 and $(Li_2O_2)_2$ was the determining factor for η_{OER}. As the adsorption energy increased, η_{OER} increased from BN-SWG to B-SWG and had a minimum value at BN-SWG. Conversely, a further decrease in adsorption energy resulted in a rapid increase in overpotential compared to the trough (BN-SWG). Thus, the intermediate E_{ads} in RDS have a positive effect on the reduction of η_{ORR}/η_{OER} by influencing its subsequent reaction.

Similarly, Dai and coworkers [48] prepared flexible nanoporous carbon nanofiber films (NCNFs) for wearable electronic devices by the desorption of electrospun polyimide, which exhibited superior bifunctional electrocatalytic activities for ORR/OER. When NCNFs directly serve as the air-cathode in a liquid ZAB, excellent performance with a high open-circuit voltage (1.48 V), maximum power density (185 mW cm^{-2}), and energy density (776 Wh kg^{-1}) can be achieved. Flexible rechargeable ZABs based on the NCNFs showed a high round-trip efficiency (~62%) and mechanical stability. Wu and coworkers have fabricated graphene-like carbon (NPS-G) doped with heteroatoms N, P, and S simultaneously by a simple one-step pyrolysis method [49]. Specifically, the resulting metal-free NPS-G catalyst with optimized N, P, and S contents exhibited an encouraging half-wave potential ($E_{1/2}$ = 0.857 V) toward the ORR in alkaline media, relative to any single doping. The origin of high activity associated with various heteroatom dopings was elucidated through X-ray photoelectron spectroscopy analysis and DFT studies. The enhanced chemisorption of oxygen species onto the dopants of the NPS-G catalysts reduced charge transfer resistance and facilitated the ORR. The porous 2D structure contributed to the increase in active site density and facile mass transport.

In addition to LAB and ZAB, carbon materials have been applied extensively in Na-O_2 batteries. Zhou and coworkers [50] employed carbon NT paper with a binder-free direct as the cathode catalyst. A high discharge capacity of 7530 mAh g^{-1} was obtained and an improved charge/discharge voltage gap (200 mV) was achieved in the limit of 1000 mAh g^{-1}. Wang reported a promising strategy of nanostructure-rich layered PCS (Figure 2.3g) in tailoring the porosity, pore size distribution, and discharge product morphology of oxygen electrodes [51]. Macropores formed between the PCS electrodes and nanopores inside the separate PCS not only promoted electrolyte impregnation and oxygen diffusion but also modulated the deposition of products on the PCS surface in a conformal film morphology. *In situ* differential electrochemical mass spectrometry (DEMS) monitored the gas consumption and evolution during the discharge/charge (Figure 2.3h–j) process, confirming that the discharge products were generated via single transfer to NaO_2 while revealing that a charging potential below 3.5 V favors the suppression of side reactions. Besides conventional carbon-based materials, graphene aerogels prepared by biomolecule-assisted electrochemical exfoliation have been explored as high-performance cathode catalysts. Ortiz-Vitoriano and coworkers utilized natural nucleotide adenosine monophosphate (AMP)-assisted synthesis of a colloidal suspension of graphene nanosheets [52]. It was found that the synergistic effect between the nitrogen element and the phosphate from the AMP molecule significantly enhanced the catalytic activity of the electrode.

Although carbon-based materials present favorable application prospects, numerous studies have proven that carbon materials as cathode materials will react with the discharge product Li_2O_2 to form Li_2CO_3, which decomposes at voltages higher than 4 V [53–55]. As the number of charges and discharges increases, the irreversibly decomposed Li_2CO_3 gradually accumulates in the cathode and clogs the porous structure, while the electrolyte decomposes at high potential, eventually leading to a sharp degradation of the battery performance [56–58]. Therefore, the design of ideal catalyst/carbon substrate composites with both high activity and conductivity is seen as a hopeful pathway.

Catalyst/Carbon Composite Catalysts Based on the earlier discussion of pure carbon-based catalysts, we further summarize the progress of research on catalysts/carbon-based air cathodes. First, innovative work on composite carbon substrates of noble metals and their compounds is outlined [59, 60]. Amine's and coworkers has proposed a concept for a new LAB cathode structure, as shown in Figure 2.4a [61]. An atomic layer deposition (ALD) technique was used to anchor palladium (Pd) nanoparticles to the porous carbon surface, while an aluminum oxide (Al_2O_3) coating was applied to passivate the carbon defects. Microscale Pd nanoparticles acted as efficient electrocatalysts to promote the formation of Li_2O_2 nanocrystals, while the Al_2O_3 protective layer hindered the decomposition of the electrolyte at the carbon substrate contact interface, resulting in an ultralow overpotential of 0.2 V (inset).

Jeong et al. systematically investigated Pt, Pd, and Ru nanoparticles loaded on reduced graphene oxide (rGO) as electrocatalysts [62]. The results showed that all noble metals can decrease the overpotential, with the Ru-rGO hybrids exhibiting the most robust cycling performance and the lowest overpotential. Furthermore, the action behavior of Ru nanoparticles differed from that of conventional electrocatalysts that reduced the activation potential through electron transfer, because the main contribution of Ru nanoparticles in reducing the overpotential was to modulate the morphology of the discharge products. Ru nanoparticles promoted the formation of thin film-like or nanoparticle-like Li_2O_2 in the ORR, which had a lower decomposition potential during charging (Figure 2.4b,c). In contrast, Pt and Pd-rGO hybrids showed a significantly fluctuating potential distribution during cycling. Although Pt- and Pd-rGO electrolytes decomposed after electrochemical cycling, no instability of the electrolyte was observed for Ru-rGO hybrids.

Chou and coworkers [63] designed porous AgPd-Pd composite NTs as an efficient bifunctional catalyst for the ORR/OER. The porous NT structure promoted rapid diffusion of O_2 and electrolyte through the NTs and provided an abundance of catalytic sites, forming a continuous conductive network with good cycling performance throughout the energy conversion process. Xu reported the synthesis of nitrogen-doped porous carbon-anchored Ru single-atom electrocatalytic materials using ion replacement at the Ru^{3+} to Zn^{2+} node during high-temperature pyrolysis and a domain-limited dual strategy of microporous metal-organic frameworks (MOFs) (Ru SAs-NC, including $Ru_{0.1}$ SAs-NC and $Ru_{0.3}$ SAs-NC) [64]. The experimental results showed that the lowest overpotential of 0.55 V was achieved at a current density of 0.02 mA cm^{-2} for LAB using the optimized $Ru_{0.3}$ SAs-NC as an electrocatalyst. *In situ* DEMS results showed that the battery had a full cycle e^-/O_2 value of only 2.14, providing outstanding electrocatalytic performance in metal-air battery applications. Theoretical calculations revealed that the size of

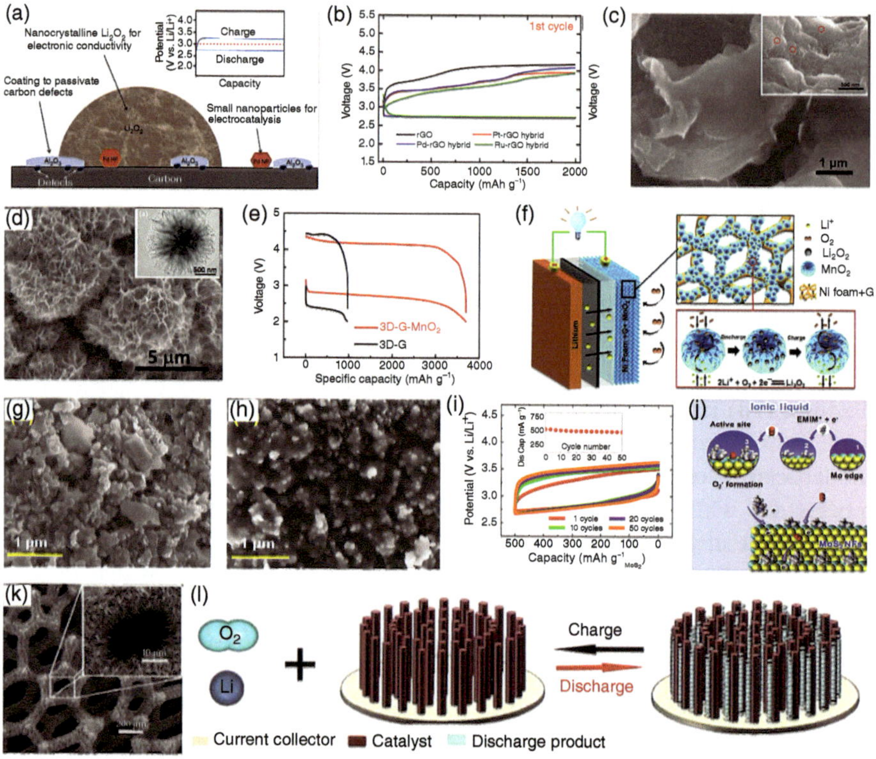

Figure 2.4 (a) The Al_2O_3 coating, the Pd nanoparticles, and the nanocrystalline Li_2O_2, all of which contribute to lowering the overpotential. The inset shows a hypothetical charge/discharge voltage profile vs. capacity. Source: Reproduced with permission of Lu et al. [61]. © 2013/Springer Nature. (b) The first discharge/charge voltage curves of catalyst-free and catalyst-loaded rGO electrodes at 200 mA g^{-1}, (c) SEM images of Ru-rGO hybrid electrode at the charge capacity of 10,000 mAh g^{-1}. Source: Reproduced with permission of Jeong et al. [62]. © 2015/American Chemical Society. (d) SEM and TEM images of δ-MnO_2 on 3D graphene (3D-G-MnO_2), (e) discharge–charge curves at a current density of 0.083 mA cm^{-2} of Li-O_2 batteries with 3D-G electrode and 3D-G-MnO_2 electrode at 2.0–4.5 V, (f) structure and working mechanism of the air electrode based on 3D-G-MnO_2. Source: Reproduced with permission of Liu et al. [71]. © 2014/John Wiley & Sons. SEM images of (g) a discharged and (h) a pristine MoS_2 cathode, (i) charge/discharge voltage profiles of the battery, (j) illustration of the ORR on MoS_2 nanoflakes in ionic liquid electrolyte. Source: Reproduced with permission of Asadi et al. [73]. © 2016/American Chemical Society. (k) SEM images of Co_3O_4@Ni, (l) the schematic diagram of the Co_3O_4@Ni-based electrode during cycling in the Li-O_2 battery. Source: Reproduced with permission of Cui et al. [81]. © 2011/Royal Society of Chemistry.

the configuration of Ru-N_4 as the driving active center significantly affected the internal affinity of the intermediate. Furthermore, the RDS of the ORR/OER on the catalyst surface was the occurrence of $2e^-$ reactions to generate Li_2O_2 and the oxidation of Li_2O_2, respectively. Although precious metals and their compounds have substantially improved the battery performance as cathode catalysts, their scarcity and high cost significantly limit their practical application.

Transition metals such as Fe, Co, Ni, Mn, Cu, Zn, Mo, W, and Ti oxides or compounds such as MnO_2, Co_3O_4, ternary spinel MCo_2O_4 (M for Mn, Fe, Ni, Zn, Cu), Mo_2C, Ni_3S_2, MoS_2, and WS_2 have shown catalytic potential for ORR/OER [65–70]. They are abundant and inexpensive compared to noble metals, so they have been extensively investigated by researchers as efficient catalysts for metal-air batteries. Liu et al. prepared a flower-like δ-MnO_2 grown directly on 3D graphene (encapsulated in nickel foam) to be used as a free-standing cathode (Figure 2.4d–f) [71].

The results showed that the cell can operate stably for 110 cycles at a high current density of 0.333 mA cm^{-2} while exhibiting an energy density of up to 1350 Wh kg^{-1}. This electrode had the following merits: (i) the high conductivity nickel foam can effectively promote electron transport, while the porous network structure facilitated O_2 diffusion and electrolyte wetting; (ii) graphene encapsulated on the nickel foam skeleton further expanded the electron transport channels; (iii) the unique flower-like structure of δ-MnO_2 exhibited a highly specific surface area and suitable pore distribution, which both greatly increased the discharge capacity and provided sufficient active sites to accelerate the electrochemical reaction rate; and (iv) the excellent catalytic activity of δ-MnO_2 contributed to the efficient deposition/decomposition of Li_2O_2. Liu and coworkers [72] reported that CoO/C composites with oxygen vacancies exhibit superior electrocatalytic performance compared to commercial CoO and vacancy-free CoO. The analysis showed that the high activity of the oxygen vacancy-rich CoO/C electrode was attributable to the oxygen vacancy enhancing the electron and Li^+ mobility, which acted as an active site for the reaction between O_2 and Li_2O_2, improving the reactivity of the ORR/OER. This provides a good reference for subsequent investigations.

Asadi et al. [73] presented a system based on MoS_2 nanosheets as an electrocatalyst combined with ionic liquids, as shown in Figure 2.4g–j. Cyclic voltammetric results showed that MoS_2 exhibited promising electrocatalytic performance in terms of ORR/OER compared to noble metal gold and platinum catalysts, and the cell with MoS_2 as a catalyst exhibited a round-trip efficiency of 85% and up to 50 reversible cycles. On the other hand, Dong and coworkers [74] successfully fabricated an open-air cathode with a sisal architecture by directly growing Co_9S_8 nanorods on porous carbon foil (PCF). The study discovered that the open structure not only provided sufficient storage space for the reaction products but also effectively avoided the blockage of the oxygen electrode by insoluble Li_2O_2. Moreover, the special open structure facilitated the capture and release of oxygen, guaranteeing an efficient and fast electrochemical reaction; second, the excellent electrocatalytic activity of Co_9S_8 significantly improved the kinetics of the redox reaction and substantially increased the electrode reaction speed; finally, the favorable oxygen affinity of Co_9S_8 can induce oxygen to react on the electrode surface to form Li_2O_2, forming a well-contacted Li_2O_2/cathode contact interface, thus facilitating the complete decomposition of Li_2O_2.

Additionally, the formation of stable M-N-C (M = Fe, Co, Ni) sites by transition metals with nitrogen-doped carbon materials also shows satisfactory catalytic activity, especially in aqueous ZAB systems [75, 76]. Mu and coworkers [77] designed and constructed a one-dimensional Co-N_x/C nanorod array electrocatalyst by preparing three-dimensional (3D) zeolitic-imidazolate framework (ZIF) materials. The one-dimensional nanorod structure can effectively avoid the reduction of N content and porosity of the material due to over-temperature carbonization and effectively prevent the agglomeration of Co during carbonization to form a large number of Co-N active sites. The test results showed that the one-dimensional Co-N_x/C nanorod arrays had excellent ORR electrocatalytic activity and stability. As a highly efficient bifunctional catalyst, it achieved an efficiency of 65.7% at 5 mA cm^{-2} current density when used as a cathode catalyst in a rechargeable ZAB. Tan and his co-workers have improved the electrocatalytic performance of M-N-C by modulating the coordination environment of the active sites and increasing the pore structure of the catalyst [78]. The finely tuned coordination environment and pore structure showed better ORR performance and durability in alkaline media than commercial Pt/C and other similar catalysts. The aqueous ZABs assembled with Fe-N/S-C as the air cathode can achieve an ultrahigh power density of 315.4 mW cm^{-2} and a long-term stable discharge at 20 mA cm^{-2}. This work opens up new perspectives for the design of catalysts with atomic-level dispersion.

2.2.2.2 Noncarbon Catalysts

Inspired by the tendency of carbon-based materials to cause undesirable side effects in redox reactions, researchers have attempted to design carbon-free catalysts while screening suitable porous substrates to circumvent the formation of carbonate and carboxylate by-products [79, 80]. To date, research on efficient noncarbon cathode catalysts for metal-air batteries has diversified, and breakthrough performance improvements have been achieved. Wen and coworkers [81] prepared a Co_3O_4@Ni composite for use as a self-supporting air cathode (Figure 2.4k,l). The battery delivered the discharge/charge plateau at 2.95 and 3.44 V, respectively, with a specific capacity of 4000 mAh g^{-1}, while the cycling durability was significantly improved. The excellent electrochemical performance was attributed to the structural advantages of the carbon-free cathode and the highly active catalyst, which both promoted a good contact interface between the discharge products and the catalyst and effectively controlled the volume expansion during the deposition/decomposition of the discharge products. Similarly, Long's and coworkers prepared a shape-controlled flower-shaped $ZnCo_2S_4$@Ni free-standing cathode by hydrothermal method, which substantially improved the cycle life of the battery, effectively reduced the overpotential (0.78 V), and enhanced the multiplicative performance of the battery [22]. It was found that a reasonable combination of catalyst and cathode material, avoiding the use of conductive and adhesive agents, not only prevented the occurrence of side reactions but also effectively improved the cycling stability of the electrode and slowed down the polarization effect during the charging and discharging of the battery which was benefited from the three-dimensional open structure. Analysis of the formation path of Li_2O_2 revealed that the special flower-like structure of $ZnCo_2S_4$ can induce the discharge products to deposit uniformly on the cathode surface in a solution-model growth manner, thus effectively reducing the overpotential of the battery. Furthermore, the team prepared high-reactivity $CuCo_2S_4$ nanosheets vertically anchored to a nickel foam backbone, effectively reducing the cell overpotential (0.82 V) and lifetime (83 cycles) [82]. Soft-pack batteries based on the free-standing cathode can achieve an energy density of 536 Wh kg^{-1}, which is much higher than that of conventional LIBs. The soft pack cells showed good reversibility and excellent flexibility in an air atmosphere and can operate under different bending and twisting conditions, exhibiting high energy densities of 428 mAh (~8000 mAh g^{-1}) and 403 mAh (~7800 mAh g^{-1}), respectively. This work provides new perspectives for the development of carbon- and a binder-free air cathode.

2.2.3 Li/Na/Zn-CO_2 Batteries

Encouraged by the side reactions of CO_2 with Li_2O_2, researchers have carried out a series of studies on metal-CO_2 batteries [83–85]. Németh and Srajer [86] proposed that dissolved CO_2 molecules captured e^- from the cathode and reduced it to oxalate via a single electron (Eq. 2.16). In analogy to the disproportionation of lithium superoxide in an LAB (Eq. 2.17), Chen and coworkers hypothesized that a similar disproportionation reaction would occur during CO_2 reduction [87]. The first step was consistent with Eq. (2.16), and subsequently, the unstable $C_2O_4^{2-}$ underwent a two-step disproportionation reaction to form CO_3^{2-} and C, as shown in Eqs. (2.18) and (2.19). Finally, the formed CO_3^{2-} was coupled with Li^+ to form crystalline Li_2CO_3 (Eq. 2.20). Thus, the electrochemical reduction mechanism of the nonprotonic Li-CO_2 cell can be described in Eq. (2.21).

$$2CO_2 + 2e^- \rightarrow C_2O_4^{2-} \tag{2.16}$$

$$2LiO_2 \rightarrow Li_2O_2 + O_2 \tag{2.17}$$

$$C_2O_4^{2-} \rightarrow CO_2^{2-} + CO_2 \tag{2.18}$$

$$C_2O_4^{2-} + CO_2^{2-} \rightarrow 2CO_3^{2-} + C \tag{2.19}$$

$$2CO_3^{2-} + Li^+ \rightarrow Li_2CO_3 \tag{2.20}$$

$$4Li^+ + 4e^- + 3CO_2 \rightarrow 2Li_2CO_3 + C \tag{2.21}$$

In addition, Peng and coworkers revealed that in dimethyl sulfoxide (DMSO, a high donor number solvent) there is a primary solution-phase "electrochemical" mechanism (—OC(=O)OC(=O)OO— as the reaction intermediate) and a secondary solution-phase surface mechanism (—OC(=O)OOC(=O)O— as the reaction intermediate), whereas in CH_3CN (a low donor number solvent) the main surface phase "chemical" reaction mechanism occurs, i.e. the ORR first produces Li_2O_2 at the electrode surface, followed by a chemical reaction between Li_2O_2 and CO_2 from solution to form Li_2CO_3, as shown in Figure 2.5a [88]. In CH_3CN, in this case, trace amounts of O_2 act as a "quasi-catalyst" to immobilize CO_2, especially in high donor number solvents, and the solution phase "electrochemical" mechanism can greatly improve the capacity of the Li-CO_2 battery discharge. It is worth noting that the final detection of Li_2CO_3 as the only product can be attributed to the key role of trace O_2 in the cell, while CO_2 itself is not undergoing electrochemical reduction.

On the other hand, Zhou and coworkers found that the Ru catalyst facilitated the interaction between Li_2CO_3 and carbon, by tracing the detailed products of the discharge–charge process by *in situ* surface-enhanced Raman spectroscopy (SERS) [89]. The high electrocatalytic activity of specific crystalline surfaces on Ru thermodynamically lowered the reaction barrier, thus facilitating a reversible process during charging. At the same time, the charge/discharge curves in Figure 2.5b confirmed that Li_2CO_3 had a lower decomposition potential in the Li-CO_2 system containing Ru elements, while the cathode without Ru catalyst exhibited typical polarization effects [90]. Even if a reversible reaction path is achieved at lower charging voltages with the assistance of a suitable catalyst, the exact catalytic mechanism still needs to be further explored. More importantly, overpotentials typically exceeding 1.2 V are still too large to be used in practical Li-CO_2 cells and more efficient and robust catalysts are required.

Recently, Yuan and coworkers [91] have theoretically demonstrated that $Li_2C_2O_4$ can be stabilized as the final discharge product by systematically investigating the Gibbs-free energy changes of different intermediates during the nucleation of $Li_2C_2O_4$ and Li_2CO_3, thus preventing the further formation of Li_2CO_3. DFT calculations showed that $Li_2C_2O_4$ was thermodynamically favored by using Mo_2C as the final discharge product for the cathode catalyst of the Li-CO_2 cell, whereas β-Mo_2C (001) and (101) catalysts had the lowest Gibbs free energy change on the surface to facilitate $Li_2C_2O_4$ nucleation. The discharge and charging processes of the catalysts were investigated in depth by calculating electrochemical free energy diagrams to identify the overpotentials. Cheng and coworkers [92] proposed a design guideline for efficient bidirectional catalysts for Li-CO_2 cells: the construction of high levels of electrophilic and nucleophilic dual centers. Theoretical studies demonstrated that for transition metal dichalcogenides (TMDs e.g. MoS_2, WS_2, and ReS_2) basal plane engineering, it is hard to simultaneously achieve efficient bidirectional activity with S vacancies or N doping engineering alone, while nucleophilic N doping and electrophilic S vacancies can improve the adsorption with Li and C/O atoms, respectively, with favorable complementary effects. In the case of ReS_2, during the operation of the Li-CO_2 cell, the nucleophilic N doping and the electrophilic S hole double centers at the basal plane presented suitable adsorption interactions with intermediates containing C/O/Li atoms, thus reducing the energy potential barrier of the decisive step. Based on the above theoretical guidance, the researchers used a facile hydrothermal method to enable the controlled synthesis of different contents of N-doped atoms and S vacancies in the ReS_2 substrate on the carbon paper surface (NS_V-ReS_2(x)/CP). Test results revealed that the NS_V-ReS_2(5)/CP catalysts had substantial amounts of highly active sites, good electrical conductivity, and excellent electrochemical stability and exhibited an exceptionally small voltage gap (0.66 V) and ultrahigh energy efficiency (81.1%) at 20 $\mu A\,cm^{-2}$. This work shows that the engineering of electrophilic and nucleophilic dual centers provides important implications for excellent bifunctional catalysts for metal-CO_2 batteries.

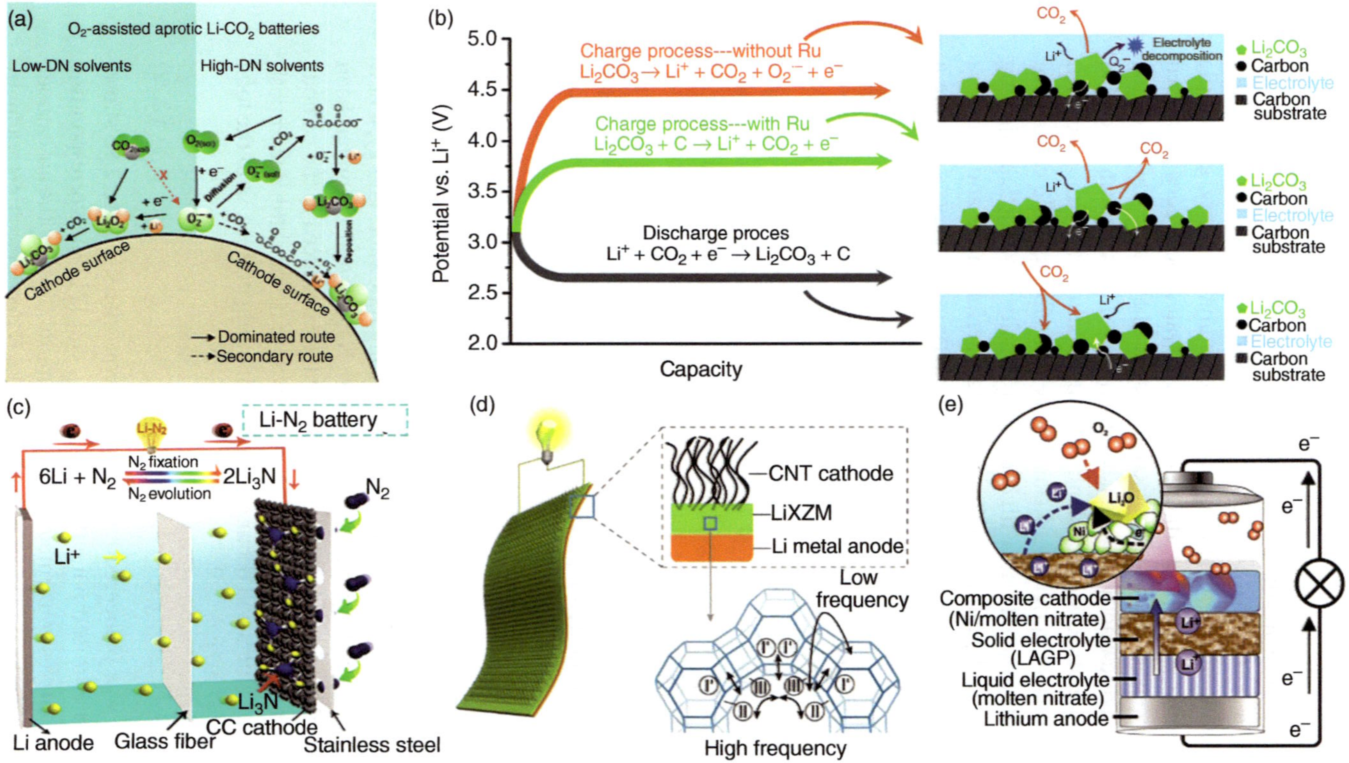

Figure 2.5 (a) Scheme diagram of possible reaction routes for the O_2-assisted Li-CO_2 batteries with electrolytes of different donor number values. Source: Reproduced with permission [88]. Copyright 2019, American Chemical Society. (b) Schematic diagram of the reaction mechanism of the charging process in Li-CO_2 battery using $LiCF_3SO_3$-TEGDME (mole ratio of 1 : 4) electrolyte with and without Ru catalyst, respectively. Source: Reproduced with permission [90]. Copyright 2017, Royal Society of Chemistry. (c) Structure of a Li-N_2 battery with a Li-foil anode, ether-based electrolyte, and CC cathode. Source: Reproduced with permission [7]. Copyright 2017, Elsevier. (d) Schematic of the integrated solid-state LAB with lithium-ion-exchanged zeolite X (LiX) zeolite membrane (LiXZM) and the conduction mechanism of Li ions in LiX. Source: Reproduced with permission [112]. Copyright 2021, Springer Nature. (e) Configuration of the inorganic electrolyte Li-O_2 cell and schematic illustration of Li_2O formation during discharge. Source: Reproduced with permission Xia et al. [123] Copyright 2018, American Association for the Advancement of Science.

To further seek a cost-effective energy storage system, Chen and coworkers constructed a Na-CO_2 cell with excellent performance using sodium metal sheets as the negative electrode and tetraethylene glycol dimethyl ether (TEGDME) treated multiwalled carbon nanotubes (MWCNTs) as the positive electrode [93]. For the anode, the gas electrode with a unique three-dimensional structure using a chemically modified porous three-dimensional network structure had extremely high ion/electron conduction, efficient and selective catalytic activity, and ample storage space for discharge products. In addition, the treated electrodes possessed excellent electrolyte wettability, which effectively reduced the polarization of the cell, thus exhibiting high-capacity storage capability. The cell system can be cycled 200 times at room temperature without significant degradation, exhibiting excellent reversible charge/discharge activity and stability. A reversible specific capacity of 4000 $mAh\,g^{-1}$ was still available at a high current density of 4 $A\,g^{-1}$, indicating that this battery had good high multiplier discharge performance and could be rapidly charged and discharged. These excellent battery performances provide a new idea for the conversion of CO_2 absorption into clean electrical energy at room temperature.

Xin and coworkers [94] designed and synthesized ultrathin MXene coupled SnO_2 quantum dots to exploit their unique and stable structure for the electrochemical reduction of CO_2. The Zn-CO_2 cell based on the SnO_2/MXene cathode assembly exhibited a maximum power density of 4.28 $mW\,cm^{-2}$ and an open-circuit voltage of 0.83 V as well as excellent rechargeability. Similarly, Wang and coworkers [95] successfully implemented a designed aqueous reversible metal-CO_2 cell by designing and synthesizing a bifunctional 3D porous palladium catalyst electrode material, untying the product limitations of CO_2 electrochemistry in Zn-CO_2 cells. This work broadens the pathway for the development of highly active and selective CO_2RR catalysts for electrochemical CO_2 reduction energy conversion and device applications.

2.2.4 Li-N_2 Batteries

In recent years, inspired by the rapidly developing metal-air batteries and the experimental fact that nitrogen can also be activated for storing renewable energy, a new Li-N_2 battery combining an energy storage system with N_2 immobilization has been proposed [96]. According to previous reports, Li_3N will be produced at the cathode during the discharge of the Li-N_2 battery (Eq. 2.22), and, in turn, the decomposition of the discharge products will occur during the charging process, with the total reaction shown below [7, 97]:

$$6Li + N_2 \leftrightarrow 2Li_3N \tag{2.22}$$

A proof of concept was pioneered by Zhang and coworkers. Figure 2.5c shows the configuration and core components of the Li-N_2 cell [7]. The corresponding charge and discharge curves confirmed that N_2 was immobilized at the air cathode as a Li_3N product during the discharge process, with Li_3N decomposition gradually releasing N_2 as the charging reaction progresses. In addition, an integrated electrode was successfully constructed by chemically loading the catalysts based on their better adsorption activity for nitrogen molecules by growing them uniformly on a carbon cloth substrate. Cyclic voltammetry (CV) results showed that Li-N_2 cells using Ru-CC cathodes achieved a more positive cathodic potential and a higher current density than ZrO_2-CC or pure CC cathodes. Under the same test conditions, Li-N_2 cells using Ru-CC (72th) or ZrO_2-CC (80th) cathodes had a longer cycle life than those using pure CC cathodes (65th), demonstrating the efficiency of the catalyst in improving the stability of the cells. In an attempt to achieve superior stability, Zhou and coworkers [98] introduced graphene as a cathode material into Li-N_2 batteries. A series of tests indicated that Li_3N and LiOH generated *in situ* at the Li anode could effectively modulate the deposition behavior of Li and inhibit Li dendrite growth, enabling a quicker and feasible migration of Li^+ during the de-embedding process, thus improving the rechargeability. This work provides a new strategy for the development of nitrogen fixation technology, which has implications for the subsequent development of Li-N_2 batteries and the expansion of other metal-air battery applications.

2.2.5 Solid Li/Zn-Air Batteries

Notwithstanding the impressive progress made in the study of nonaqueous LABs [99–103], a series of safety challenges such as the flammable and volatile nature of organic liquid electrolytes and the decomposition of electrolytes have greatly hindered their practical application [104–107]. Solid electrolytes, with their high mechanical strength, high chemical stability in an open environment, wide electrochemical window, and low flammability, offer a viable strategy to address the safety issues faced and achieve a stable and practicable LAB system [108–111].

Yu and coworkers [112] presented a new solid-state electrolyte material based on molecular sieve films, which exhibited an ionic conductivity as high as $2.7\times10^{-4}\,S\,cm^{-1}$, an electronic conductivity as low as $1.5\times10^{-10}\,S\,cm^{-1}$, and high stability, to effectively solve the problems of difficult interface construction, internal lithium dendrites, and poor stability of conventional solid-state electrolyte materials, and enabled the construction of an integrated flexible solid-state LAB through an *in situ* growth strategy (Figure 2.5d). Benefiting from a good "electrolyte–electrode" low-impedance contact interface, the battery exhibited an ultrahigh capacity of 12,020 mAh g^{-1} and a long cycle life of 149 cycles (500 mA g^{-1} and 1000 mAh g^{-1}) in a real-world air environment, far better than solid-state LABs based on the most stable NASICON-type $Li_{1.5}Al_{0.5}Ge_{1.5}(PO_4)_3$ (LAGP) solid-state electrolyte available (12 cycles) and even better than LABs using organic electrolytes under equivalent conditions (102 cycles). Meanwhile, the battery exhibited excellent flexibility, high safety, and good environmental suitability and took into account the need for environment-friendly, cost-effective, and simple production processes.

Recently, Asadi and coworkers [113] reported an LAB using a $Li_{10}GeP_2S_{12}$ (LGPS) nanoparticle – poly (ethylene oxide) – a based composite solid electrolyte that can achieve a reversible four-electron Li_2O reaction for 1000 cycles at room temperature and operate at high rates and low polarization gaps in air. This composite electrolyte combined the advantages of organic and inorganic solid electrolytes and has a conductivity (0.52 mS cm^{-1}) that was 10 times higher than that of a solid electrolyte without nanoparticles. The four-electron Li_2O reaction based on the LGPS solid electrolyte and catalyst Mo_3P was achieved with 1000 reversible cycles at 25 °C. It was found that the products consisted mainly of LiO_2 and Li_2O_2 during the first 15 minutes. Subsequently, Li_2O became the main discharge product. Presumably, the initially formed LiO_2/Li_2O_2 interface is homogeneous at the anode, which could provide the required mixed electron/ion conduction properties for further electrochemical reduction. After reaching a steady state, Li_2O_2 was isolated below LiO_2 without O_2 and was in direct contact with the anode where electron conduction was most favorable. These conditions promoted a two-electron reduction of Li_2O_2 to Li_2O. Thus, the combination of two two-electron reactions (O_2/Li_2O_2 and Li_2O_2/Li_2O) yielded a highly desirable four-electron Li-air/O_2 chemistry. This result demonstrates that it is possible to construct a Li-air battery system based on a solid electrolyte that facilitates the four-electron Li_2O reaction and obtains an expected specific energy of >1000 Wh kg^{-1} (bulk energy density of 1000 Wh l^{-1}), which would provide a much higher battery energy density than current LIB.

Solid-state ZABs are typically limited by a relatively low multiplicative capacity (<10 mA cm^{-2}) due to slow oxygen electrocatalysis and an unstable electrochemical interface [114, 115]. Lei and coworkers [116] demonstrated a high-speed and robust quasi-solid-state ZAB consisting of atomically dispersed cobalt sites anchored on nitrogen-doped graphene (Co-SA-NDGs) as the air cathode and a polyacrylamide organo-hydrogel electrolyte with a hydrogen bonding network enabled by the addition of dimethyl sulfoxide. The catalyst of Co-SA-NDGs enabled efficient ORR kinetics while the ZABs were charged/discharged at current densities of up to 100 mA cm^{-2} and stable cycling for 50 hours. In addition, more than 300 hours of low-temperature cycling stability (at 0.5 mA cm^{-2}) at −60 °C was achieved with over 90% capacity retention and wide temperature adaptation (−60 to 60 °C). Yan and coworkers [117] presented an advanced oxygen electrode catalyst with carbon nanosheets having active Fe-N$_x$ sites. It was shown that the improved ORR/OER kinetics benefited from the abundance of Fe-N$_x$ active sites and large specific surface area and that the high electronic conductivity and highly graphitized substrate provided a highway for electron transfer. Notably, this catalyst exhibited superior electrochemical performance in flexible all-solid-state batteries using gel electrolytes and can be used in illuminated light-emitting diode (LED) bracelets.

2.2.6 Sealed Li/Zn-O_2 Batteries

Typically, the reduction of O_2 at the anode produces Li_2O_2 through the transfer of two electrons; however, the high activity of Li_2O_2 can lead to undesired side reactions with the organic electrolyte and the carbon substrate [118, 119]. When the product of O_2 reduction is Li_2O, it not only provides a higher energy density but also avoids reactions with the electrolyte [120–122]. However, thermodynamic studies have established that oxygen reduction at room temperature is more inclined to produce Li_2O_2. Interestingly, higher temperatures (>150 °C) are more inclined to produce Li_2O [123]. Based on this, Nazar and coworkers [123] designed a closed Li-O_2 cell based on an inorganic electrolyte, as shown in Figure 2.5e. The $LiNO_3/KNO_3$ eutectic molten salt and a solid electrolyte LAGP were used as the electrolyte to avoid degradation of the organic electrolyte. A noncarbon composite anode consisting of Ni nanoparticles coated *in situ* to form Li_xNiO_2 was used as a bifunctional ORR/OER catalyst. The carefully tailored cell operated at high temperatures to achieve a reversible four-electron redox reaction while delivering a high capacity of 11 mAh cm^{-2} and a very low overpotential. An in-depth analysis of the underlying reaction mechanism revealed that the battery overcame thermodynamic and kinetic barriers, allowing the electrochemical reversible formation of Li_2O rather than Li_2O_2, and improved the reversibility of Li_2O, resulting in a Li-O_2 cell with high capacity, low overpotential when transferring $4\,e^-/O_2$, and excellent cycling performance.

On a similar note, Zhou and coworkers [118] successfully realized a closed Li_2O cell based on the reversible conversion of Li_2O/Li_2O_2 and elucidated the role and formation principles of metal Ir and its associated stabilizing protective layer. This work revealed the importance of controlling and effectively utilizing the reversible Li_2O/Li_2O_2 conversion. In addition, the researchers have successfully calibrated the reversible redox reaction interval and demonstrated the reversible conversion of the reaction products using a full range and multiple perspectives of *in situ/non-in situ* spectroscopy combined with precise quantitative analysis. Ultimately, the sealed Li-O_2 battery system delivered high specific energy (1090 Wh kg^{-1}, based on full cathode loading mass), high reversibility (stable coulomb efficiency of ~99.5%), high energy efficiency (overpotential of only ~0.12 V), and high stability (>2000 cycles).

Winter and coworkers [124] have shown that a highly reversible two-electron Zn-O_2/ZnO_2 chemistry ($Zn + O_2 \leftrightarrow ZnO_2$) can be achieved by screening for an optimal nonbasic electrolyte. This ZnO_2 chemistry was caused by the formation of an inner Helmholtz layer lacking water and rich in Zn^{2+} by the hydrophobic trifluoromethane sulfonate anions on the air cathode. By making the electrolyte hydrophobic, water was excluded from the near surface of the cathode, thereby preventing four-electron reduction. Nonalkaline-sealed ZABs can operate stably in ambient air and exhibit better reversibility than alkaline batteries. Finally, stable cycling was achieved at a current density of 0.1/1 mA cm^{-2} for 1600/160 hours, respectively. These works provide important lessons for the development of highly reversible sealed energy storage systems and provide theoretical guidance for the development of high-energy density devices.

2.3 Summary and Outlook

It is well known that the development of practicable metal-air batteries still faces many challenges. Among them, the customization of the ideal cathode catalyst and the still unknown potential reaction mechanism severely limit the development of high-energy-density and high-power-density metal-air batteries. In this section, we focus on summarizing efficient cathode catalysts for nonaqueous LAB and aqueous

ZAB, while briefly outlining other alkali metal-air cell developments. It is worth noting that the main gas components of air include O_2, CO_2, and N_2 and to further screen promising energy storage systems, we also address metal-CO_2/N_2 cell technology in detail. In addition, we discuss in depth the prospects for the development of solid electrolytes in metal-air batteries. Finally, based on the current demand for energy storage devices, we try to present closed Li-O_2 cells and ZABs that are similar to conventional LIBs.

References

1 Daehn, K., Basuhi, R., Gregory, J. et al. (2022). Innovations to decarbonize materials industries. *Nature Reviews Materials* 7 (4): 275–294.

2 Jiao, K., Xuan, J., Du, Q. et al. (2021). Designing the next generation of proton-exchange membrane fuel cells. *Nature* 595 (7867): 361–369.

3 Rahman, M.A., Wang, X.J., and Wen, C.E. (2013). High energy density metal-air batteries: a review. *Journal of the Electrochemical Society* 160 (10): A1759–A1771.

4 Lee, J.S., Kim, S.T., Cao, R. et al. (2011). Metal-air batteries with high energy density: Li-air versus Zn-air. *Advanced Energy Materials* 1 (1): 34–50.

5 Li, Y.G. and Lu, J. (2017). Metal air batteries: will they be the future electrochemical energy storage device of choice? *ACS Energy Letters* 2 (6): 1370–1377.

6 Zhang, Z., Zhang, Q., Chen, Y.A. et al. (2015). The first introduction of graphene to rechargeable Li-CO_2 batteries. *Angewandte Chemie International Edition* 54 (22): 6550–6553.

7 Ma, J.L., Bao, D., Shi, M.M. et al. (2017). Reversible nitrogen fixation based on a rechargeable lithium-nitrogen battery for energy storage. *Chem* 2 (4): 525–532.

8 Lu, J., Li, L., Park, J.-B. et al. (2014). Aprotic and aqueous Li–O_2 batteries. *Chemical Reviews* 114 (11): 5611–5640.

9 Kwak, W.J., Rosy, Sharon, D. et al. (2020). Lithium-oxygen batteries and related systems: potential, status, and future. *Chemical Reviews* 120 (14): 6626–6683.

10 Liu, T., Vivek, J.P., Zhao, E.W. et al. (2020). Current challenges and routes forward for nonaqueous lithium-air batteries. *Chemical Reviews* 120 (14): 6558–6625.

11 Grande, L., Paillard, E., Hassoun, J. et al. (2015). The lithium/air battery: still an emerging system or a practical reality? *Advanced Materials* 27 (5): 784–800.

12 Hassoun, J., Croce, F., Armand, M., and Scrosati, B. (2011). Investigation of the O_2 electrochemistry in a polymer electrolyte solid-state cell. *Angewandte Chemie International Edition* 50 (13): 2999–3002.

13 Johnson, L., Li, C., Liu, Z. et al. (2014). The role of LiO_2 solubility in O_2 reduction in aprotic solvents and its consequences for Li–O_2 batteries. *Nature Chemistry* 6 (12): 1091–1099.

14 Lu, J., Cheng, L., Lau, K.C. et al. (2014). Effect of the size-selective silver clusters on lithium peroxide morphology in lithium-oxygen batteries. *Nature Communications* 5: 4895.

15 Pan, J., Xu, Y.Y., Yang, H. et al. (2018). Advanced architectures and relatives of air electrodes in Zn-air batteries. *Advanced Science* 5 (4): 1700691.

16 Alegre, C., Stassi, A., Modica, E. et al. (2015). Investigation of the activity and stability of Pd-based catalysts towards the oxygen reduction (ORR) and evolution reactions (OER) in iron-air batteries. *RSC Advances* 5 (32): 25424–25427.

17 Li, C.S., Sun, Y., Gebert, F., and Chou, S.L. (2017). Current progress on rechargeable magnesium-air battery. *Advanced Energy Materials* 7 (24): 1700869.

18 Yang, S.H. and Knickle, H. (2002). Design and analysis of aluminum/air battery system for electric vehicles. *Journal of Power Sources* 112 (1): 162–173.

19 Wen, Z.Y., Shen, C., and Lu, Y. (2015). Air electrode for the lithium-air batteries: materials and structure designs. *Chempluschem* 80 (2): 270–287.

20 Abraham, K.M. and Jiang, Z. (1996). A polymer electrolyte-based rechargeable lithium/oxygen battery. *Journal of the Electrochemical Society* 143 (1): 1–5.

21 Peng, Z.Q., Freunberger, S.A., Chen, Y.H., and Bruce, P.G. (2012). A reversible and higher-rate Li-O_2 battery. *Science* 337 (6094): 563–566.

22 Hou, Z.Q., Feng, S., Hei, P. et al. (2019). Morphology regulation of Li_2O_2 by flower-like $ZnCo_2S_4$ enabling high performance Li-O_2 battery. *Journal of Power Sources* 441: 227168.

23 Ren, X.D. and Wu, Y.Y. (2013). A low-overpotential potassium-oxygen battery based on potassium superoxide. *Journal of the American Chemical Society* 135 (8): 2923–2926.

24 Burke, C.M., Pande, V., Khetan, A. et al. (2015). Enhancing electrochemical intermediate solvation through electrolyte anion selection to increase nonaqueous Li-O_2 battery capacity. *Proceedings of the National Academy of Sciences of the United States of America* 112 (30): 9293–9298.

25 Lyu, Z.Y., Zhou, Y., Dai, W.R. et al. (2017). Recent advances in understanding of the mechanism and control of Li_2O_2 formation in aprotic Li-O_2 batteries. *Chemical Society Reviews* 46 (19): 6046–6072.

26 Shao, Y.Y., Park, S., Xiao, J. et al. (2012). Electrocatalysts for nonaqueous lithium-air batteries: status, challenges, and perspective. *ACS Catalysis* 2 (5): 844–857.

27 Luntz, A.C. and McCloskey, B.D. (2014). Nonaqueous Li-air batteries: a status report. *Chemical Reviews* 114 (23): 11721–11750.

28 Safari, M., Adams, B.D., and Nazar, L.F. (2014). Kinetics of oxygen reduction in aprotic Li-O_2 cells: a model-based study. *Journal of Physical Chemistry Letters* 5 (20): 3486–3491.

29 Lim, H.K., Lim, H.D., Park, K.Y. et al. (2013). Toward a lithium-"Air" battery: the effect of CO_2 on the chemistry of a lithium-oxygen cell. *Journal of the American Chemical Society* 135 (26): 9733–9742.

30 Sharon, D., Etacheri, V., Garsuch, A. et al. (2013). On the challenge of electrolyte solutions for Li–air batteries: monitoring oxygen reduction and related reactions in polyether solutions by spectroscopy and EQCM. *The Journal of Physical Chemistry Letters* 4 (1): 127–131.

31 Aurbach, D., McCloskey, B.D., Nazar, L.F., and Bruce, P.G. (2016). Advances in understanding mechanisms underpinning lithium-air batteries. *Nature Energy* 1: 16128.

32 Zhai, D.Y., Wang, H.H., Lau, K.C. et al. (2014). Raman evidence for late stage disproportionation in a Li-O_2 battery. *Journal of Physical Chemistry Letters* 5 (15): 2705–2710.

33 Kwabi, D.G., Tulodziecki, M., Pour, N. et al. (2016). Controlling solution-mediated reaction mechanisms of oxygen reduction using potential and solvent for aprotic lithium-oxygen batteries. *Journal of Physical Chemistry Letters* 7 (7): 1204–1212.

34 Man, I.C., Su, H.Y., Calle-Vallejo, F. et al. (2011). Universality in oxygen evolution electrocatalysis on oxide surfaces. *Chemcatchem* 3 (7): 1159–1165.

35 McCloskey, B.D., Scheffler, R., Speidel, A. et al. (2011). On the efficacy of electrocatalysis in nonaqueous Li-O_2 batteries. *Journal of the American Chemical Society* 133 (45): 18038–18041.

36 Wang, H.F., Tang, C., and Zhang, Q. (2018). A review of precious-metal-free bifunctional oxygen electrocatalysts: rational design and applications in Zn-air batteries. *Advanced Functional Materials* 28 (46): 1803329.

37 Seh, Z.W., Kibsgaard, J., Dickens, C.F. et al. (2017). Combining theory and experiment in electrocatalysis: insights into materials design. *Science* 355 (6321): eaad4998.

38 Xu, W., Hu, J.Z., Engelhard, M.H. et al. (2012). The stability of organic solvents and carbon electrode in nonaqueous Li-O_2 batteries. *Journal of Power Sources* 215: 240–247.

39 Kim, D.Y., Kim, M., Kim, D.W. et al. (2015). Flexible binder-free graphene paper cathodes for high-performance Li-O_2 batteries. *Carbon* 93: 625–635.

40 Cao, D., Bai, Y., Zhang, J.F. et al. (2021). Irreplaceable carbon boosts Li-O_2 batteries: from mechanism research to practical application. *Nano Energy* 89: 106464.

41 Guo, Z.Y., Zhou, D.D., Dong, X.L. et al. (2013). Ordered hierarchical mesoporous/macroporous carbon: a high-performance catalyst for rechargeable Li-O_2 batteries. *Advanced Materials* 25 (39): 5668–5672.

42 Xiao, J., Mei, D.H., Li, X.L. et al. (2011). Hierarchically porous graphene as a lithium-air battery electrode. *Nano Letters* 11 (11): 5071–5078.

43 Punckt, C., Pope, M.A., Liu, J. et al. (2010). Electrochemical performance of graphene as effected by electrode porosity and graphene functionalization. *Electroanalysis* 22 (23): 2834–2841.

44 Zhao, C.T., Yu, C., Liu, S.H. et al. (2015). 3D porous N-doped graphene frameworks made of interconnected nanocages for ultrahigh-rate and long-life Li-O_2 batteries. *Advanced Functional Materials* 25 (44): 6913–6920.

45 Shui, J.L., Lin, Y., Connell, J.W. et al. (2016). Nitrogen-doped holey graphene for high-performance rechargeable Li-O_2 batteries. *ACS Energy Letters* 1 (1): 260–265.

46 Jing, Y. and Zhou, Z. (2015). Computational insights into oxygen reduction reaction and initial Li_2O_2 nucleation on pristine and N-doped graphene in Li-O_2 batteries. *ACS Catalysis* 5 (7): 4309–4317.

47 Yao, Y.S., Cao, J.X., Yin, W.J. et al. (2021). Modified graphene sheets as promising cathode catalysts for Li-O_2 batteries: a first-principles study. *Journal of Physical Chemistry C* 125 (8): 4363–4370.

48 Liu, Q., Wang, Y.B., Dai, L.M., and Yao, J.N. (2016). Scalable fabrication of nanoporous carbon fiber films as bifunctional catalytic electrodes for flexible Zn-air batteries. *Advanced Materials* 28 (15): 3000–3006.

49 Zheng, X.J., Wu, J., Cao, X.C. et al. (2019). N-, P-, and S-doped graphene-like carbon catalysts derived from onium salts with enhanced oxygen chemisorption for Zn-Air battery cathodes. *Applied Catalysis B-Environmental* 241: 442–451.

50 Jian, Z.L., Chen, Y., Li, F.J. et al. (2014). High capacity Na-O_2 batteries with carbon nanotube paper as binder-free air cathode. *Journal of Power Sources* 251: 466–469.

51 Sun, B., Kretschmer, K., Xie, X.Q. et al. (2017). Hierarchical porous carbon spheres for high-performance Na-O_2 batteries. *Advanced Materials* 29 (48): 1606816.

52 Enterría, M., Gómez-Urbano, J.L., Munuera, J.M. et al. (2021). Boosting the performance of graphene cathodes in Na–O_2 batteries by exploiting the multifunctional character of small biomolecules. *Small* 17 (2): 2005034.

53 Ottakam Thotiyl, M.M., Freunberger, S.A., Peng, Z., and Bruce, P.G. (2013). The carbon electrode in nonaqueous Li–O_2 cells. *Journal of the American Chemical Society* 135 (1): 494–500.

54 Zhang, Y., Zhang, X., Wang, J. et al. (2016). Potential-dependent generation of O_2^- and LiO_2 and their critical roles in O_2 reduction to Li_2O_2 in aprotic Li–O_2 batteries. *The Journal of Physical Chemistry C* 120 (7): 3690–3698.

55 Peng, Z., Chen, Y., Bruce, P.G., and Xu, Y. (2015). Direct detection of the superoxide anion as a stable intermediate in the electroreduction of oxygen in a non-aqueous electrolyte containing phenol as a proton source. *Angewandte Chemie International Edition* 54 (28): 8165–8168.

56 Wang, J., Zhang, Y., Guo, L. et al. (2016). Identifying reactive sites and transport limitations of oxygen reactions in aprotic lithium-O_2 batteries at the stage of sudden death. *Angewandte Chemie International Edition* 55 (17): 5201–5205.

57 Zhang, T. and Zhou, H. (2012). From Li–O_2 to Li–air batteries: carbon nanotubes/ionic liquid gels with a tricontinuous passage of electrons, ions, and oxygen. *Angewandte Chemie International Edition* 51 (44): 11062–11067.

58 McCloskey, B.D., Scheffler, R., Speidel, A. et al. (2012). On the mechanism of nonaqueous Li–O_2 electrochemistry on C and its kinetic overpotentials: some implications for Li–air batteries. *The Journal of Physical Chemistry C* 116 (45): 23897–23905.

59 Jung, H.-G., Jeong, Y.S., Park, J.-B. et al. (2013). Ruthenium-based electrocatalysts supported on reduced graphene oxide for lithium-air batteries. *ACS Nano* 7 (4): 3532–3539.

60 Lee, Y.J., Park, S.H., Kim, S.H. et al. (2018). High-rate and high-areal-capacity air cathodes with enhanced cycle life based on RuO_2/MnO_2 bifunctional electrocatalysts supported on CNT for pragmatic Li–O_2 batteries. *ACS Catalysis* 8 (4): 2923–2934.

61 Lu, J., Lei, Y., Lau, K.C. et al. (2013). A nanostructured cathode architecture for low charge overpotential in lithium-oxygen batteries. *Nature Communications* 4 (1): 2383.

62 Jeong, Y.S., Park, J.-B., Jung, H.-G. et al. (2015). Study on the catalytic activity of noble metal nanoparticles on reduced graphene oxide for oxygen evolution reactions in lithium–air batteries. *Nano Letters* 15 (7): 4261–4268.

63 Luo, W.-B., Gao, X.-W., Chou, S.-L. et al. (2015). Porous AgPd–Pd composite nanotubes as highly efficient electrocatalysts for lithium–oxygen batteries. *Advanced Materials* 27 (43): 6862–6869.

64 Hu, X., Luo, G., Zhao, Q. et al. (2020). Ru single atoms on N-doped carbon by spatial confinement and ionic substitution strategies for high-performance Li–O_2 batteries. *Journal of the American Chemical Society* 142 (39): 16776–16786.

65 Yang, Y., Qin, Y., Xue, X. et al. (2018). Intrinsic properties affecting the catalytic activity of 3d transition-metal carbides in Li–O_2 battery. *The Journal of Physical Chemistry C* 122 (31): 17812–17819.

66 Lee, Y.J., Kim, D.H., Kang, T.-G. et al. (2017). Bifunctional MnO_2-coated Co_3O_4 hetero-structured catalysts for reversible Li-O_2 batteries. *Chemistry of Materials* 29 (24): 10542–10550.

67 Deng, Y.-P., Jiang, Y., Luo, D. et al. (2017). Hierarchical porous double-shelled electrocatalyst with tailored lattice alkalinity toward bifunctional oxygen reactions for metal–air batteries. *ACS Energy Letters* 2 (12): 2706–2712.

68 Wang, X., Li, Y., Jin, T. et al. (2017). Electrospun thin-walled $CuCo_2O_4$@C nanotubes as bifunctional oxygen electrocatalysts for rechargeable Zn–air batteries. *Nano Letters* 17 (12): 7989–7994.

69 Yoon, K.R., Shin, K., Park, J. et al. (2018). Brush-like cobalt nitride anchored carbon nanofiber membrane: current collector-catalyst integrated cathode for long cycle Li–O_2 batteries. *ACS Nano* 12 (1): 128–139.

70 Weinrich, H., Come, J., Tempel, H. et al. (2017). Understanding the nanoscale redox-behavior of iron-anodes for rechargeable iron-air batteries. *Nano Energy* 41: 706–716.

71 Liu, S., Zhu, Y., Xie, J. et al. (2014). Direct growth of flower-like δ-MnO_2 on three-dimensional graphene for high-performance rechargeable Li-O_2 batteries. *Advanced Energy Materials* 4 (9): 1301960.

72 Gao, R., Li, Z., Zhang, X. et al. (2016). Carbon-dotted defective CoO with oxygen vacancies: a synergetic design of bifunctional cathode catalyst for Li–O_2 batteries. *ACS Catalysis* 6 (1): 400–406.

73 Asadi, M., Kumar, B., Liu, C. et al. (2016). Cathode based on molybdenum disulfide nanoflakes for lithium–oxygen batteries. *ACS Nano* 10 (2): 2167–2175.

74 Lin, X., Yuan, R., Cai, S. et al. (2018). An open-structured matrix as oxygen cathode with high catalytic activity and large Li_2O_2 accommodations for lithium–oxygen batteries. *Advanced Energy Materials* 8 (18): 1800089.

75 Wang, J., Wu, H., Gao, D. et al. (2015). High-density iron nanoparticles encapsulated within nitrogen-doped carbon nanoshell as efficient oxygen electrocatalyst for zinc–air battery. *Nano Energy* 13: 387–396.

76 Zhao, Y., Kamiya, K., Hashimoto, K., and Nakanishi, S. (2015). Efficient bifunctional Fe/C/N electrocatalysts for oxygen reduction and evolution reaction. *The Journal of Physical Chemistry C* 119 (5): 2583–2588.

77 Amiinu, I.S., Liu, X., Pu, Z. et al. (2018). From 3D ZIF nanocrystals to Co–N_x/C nanorod array electrocatalysts for ORR, OER, and Zn–air batteries. *Advanced Functional Materials* 28 (5): 1704638.

78 He, Y., Jia, Y., Yu, B. et al. (2023). Heteroatom coordination regulates iron single-atom-catalyst with superior oxygen reduction reaction performance for aqueous Zn-air battery. *Small* 19 (8): 2206478.

79 Oh, S.H., Black, R., Pomerantseva, E. et al. (2012). Synthesis of a metallic mesoporous pyrochlore as a catalyst for lithium–O_2 batteries. *Nature Chemistry* 4 (12): 1004–1010.

80 Zhao, G., Mo, R., Wang, B. et al. (2014). Enhanced cyclability of Li–O_2 batteries based on TiO_2 supported cathodes with no carbon or binder. *Chemistry of Materials* 26 (8): 2551–2556.

81 Cui, Y., Wen, Z., and Liu, Y. (2011). A free-standing-type design for cathodes of rechargeable Li–O_2 batteries. *Energy & Environmental Science* 4 (11): 4727–4734.

82 Long, J., Hou, Z., Shu, C. et al. (2019). Free-standing three-dimensional $CuCo_2S_4$ nanosheet array with high catalytic activity as an efficient oxygen electrode for lithium–oxygen batteries. *ACS Applied Materials & Interfaces* 11 (4): 3834–3842.

83 Ahmadiparidari, A., Warburton, R.E., Majidi, L. et al. (2019). A long-cycle-life lithium–CO_2 battery with carbon neutrality. *Advanced Materials* 31 (40): 1902518.

84 Xie, Z., Zhang, X., Zhang, Z., and Zhou, Z. (2017). Metal–CO_2 batteries on the road: CO_2 from contamination gas to energy source. *Advanced Materials* 29 (15): 1605891.

85 Gao, S., Liu, Y., Xie, Z. et al. (2021). Metal-free bifunctional ordered mesoporous carbon for reversible Zn-CO_2 batteries. *Small Methods* 5 (4): 2001039.

86 Németh, K. and Srajer, G. (2014). CO_2/oxalate cathodes as safe and efficient alternatives in high energy density metal–air type rechargeable batteries. *RSC Advances* 4 (4): 1879–1885.

87 Hou, Y., Wang, J., Liu, L. et al. (2017). Mo_2C/CNT: an efficient catalyst for rechargeable Li–CO_2 batteries. *Advanced Functional Materials* 27 (27): 1700564.

88 Zhao, Z., Su, Y., and Peng, Z. (2019). Probing lithium carbonate formation in trace-O_2-assisted aprotic Li-CO_2 batteries using in situ surface-enhanced raman spectroscopy. *The Journal of Physical Chemistry Letters* 10 (3): 322–328.

89 Qiao, Y., Yi, J., Wu, S. et al. (2017). Li-CO_2 electrochemistry: a new strategy for CO_2 fixation and energy storage. *Joule* 1 (2): 359–370.

90 Yang, S., Qiao, Y., He, P. et al. (2017). A reversible lithium–CO_2 battery with Ru nanoparticles as a cathode catalyst. *Energy & Environmental Science* 10 (4): 972–978.

91 Yang, C., Guo, K., Yuan, D. et al. (2020). Unraveling reaction mechanisms of Mo_2C as cathode catalyst in a Li-CO_2 battery. *Journal of the American Chemical Society* 142 (15): 6983–6990.

92 Chen, B., Wang, D., Tan, J. et al. (2022). Designing electrophilic and nucleophilic dual centers in the ReS_2 plane toward efficient bifunctional catalysts for Li-CO_2 batteries. *Journal of the American Chemical Society* 144 (7): 3106–3116.

93 Hu, X., Sun, J., Li, Z. et al. (2016). Rechargeable room-temperature Na–CO_2 batteries. *Angewandte Chemie International Edition* 55 (22): 6482–6486.

94 Han, L., Peng, X., Wang, H.-T. et al. (2022). Chemically coupling SnO_2 quantum dots and MXene for efficient CO_2 electroreduction to formate and Zn-CO_2 battery. *Proceedings of the National Academy of Sciences of the United States of America* 119 (42): e2207326119.

95 Xie, J., Wang, X., Lv, J. et al. (2018). Reversible aqueous zinc–CO_2 batteries based on CO_2–HCOOH interconversion. *Angewandte Chemie International Edition* 57 (52): 16996–17001.

96 Wang, X.-G., Zhang, Q., Zhang, X. et al. (2019). Promoting nitrogen electroreduction on Mo_2C nanoparticles highly dispersed on N-doped carbon nanosheets toward rechargeable Li–N_2 batteries. *Small Methods* 3 (6): 1800334.

97 Yi, S., Liu, G., Liu, Z. et al. (2019). Theoretical insights into nitrogen fixation on Ti_2C and Ti_2CO_2 in a lithium–nitrogen battery. *Journal of Materials Chemistry A* 7 (34): 19950–19960.

98 Zhang, Z., Wu, S., Yang, C. et al. (2019). Li-N_2 batteries: a reversible energy storage system? *Angewandte Chemie International Edition* 58 (49): 17782–17787.

99 Hou, Z., Shu, C., and Long, J. (2018). Honeycomb-like Ni_3S_2 supported on Ni foam as high performance free-standing cathode for lithium oxygen batteries. *Electrochimica Acta* 290: 657–665.

100 Bruce, P.G., Freunberger, S.A., Hardwick, L.J., and Tarascon, J.-M. (2012). Li–O_2 and Li–S batteries with high energy storage. *Nature Materials* 11: 19–29.

101 Hou, Z., Long, J., Shu, C. et al. (2019). Two-dimensional spinel $CuCo_2S_4$ nanosheets as high efficiency cathode catalyst for lithium-oxygen batteries. *Journal of Alloys and Compounds* 798: 560–567.

102 Black, R., Adams, B., and Nazar, L.F. (2012). Non-aqueous and hybrid Li-O_2 batteries. *Advanced Energy Materials* 2 (7): 801–815.

103 Hou, Z., Shu, C., Zheng, R. et al. (2020). Anionic vacancy-dependent activity of the $CoSe_2$ with a tunable interfacial electronic structure on the N-doped carbon cloth for advanced Li–O_2 batteries. *Journal of Materials Chemistry A* 8 (32): 16636–16648.

104 Hou, Z., Shu, C., Hei, P. et al. (2020). Configuration of gradient-porous ultrathin $FeCo_2S_4$ nanosheets vertically aligned on Ni foam as a noncarbonaceous freestanding oxygen electrode for lithium–oxygen batteries. *Nanoscale* 12 (3): 1864–1874.

105 Zhang, X.-P., Sun, Y.-Y., Sun, Z. et al. (2019). Anode interfacial layer formation via reductive ethyl detaching of organic iodide in lithium–oxygen batteries. *Nature Communications* 10 (1): 3543.

106 Hou, Z., Shu, C., Hei, P. et al. (2020). A 3D free-standing Co doped Ni_2P nanowire oxygen electrode for stable and long-life lithium–oxygen batteries. *Nanoscale* 12 (12): 6785–6794.

107 Hou, Z., Hei, P., Shu, C. et al. (2020). Multifunctional selenium vacancy coupling with interface engineering enables high-stability Li–O_2 battery. *ACS Sustainable Chemistry & Engineering* 8 (17): 6667–6674.

108 Yang, Y.-N., Li, Y.-X., Li, Y.-Q., and Zhang, T. (2020). On-surface lithium donor reaction enables decarbonated lithium garnets and compatible interfaces within cathodes. *Nature Communications* 11 (1): 5519.

109 Takada, K. (2018). Progress in solid electrolytes toward realizing solid-state lithium batteries. *Journal of Power Sources* 394: 74–85.

110 Wang, H.-F., Wang, X.-X., Li, F., and Xu, J.-J. (2022). Fundamental understanding and construction of solid-state Li–air batteries. *Small Science* 2 (5): 2200005.

111 Manthiram, A., Yu, X., and Wang, S. (2017). Lithium battery chemistries enabled by solid-state electrolytes. *Nature Reviews Materials* 2 (4): 16103.

112 Chi, X., Li, M., Di, J. et al. (2021). A highly stable and flexible zeolite electrolyte solid-state Li–air battery. *Nature* 592 (7855): 551–557.

113 Kondori, A., Esmaeilirad, M., Harzandi, A.M. et al. (2023). A room temperature rechargeable Li_2O-based lithium-air battery enabled by a solid electrolyte. *Science* 379 (6631): 499–505.

114 Wang, Y., Cao, Q., Guan, C., and Cheng, C. (2020). Recent advances on self-supported arrayed bifunctional oxygen electrocatalysts for flexible solid-state Zn–air batteries. *Small* 16 (33): 2002902.

115 Yang, L., Shi, L., Wang, D. et al. (2018). Single-atom cobalt electrocatalysts for foldable solid-state Zn-air battery. *Nano Energy* 50: 691–698.

116 Wang, Q., Feng, Q., Lei, Y. et al. (2022). Quasi-solid-state Zn-air batteries with an atomically dispersed cobalt electrocatalyst and organohydrogel electrolyte. *Nature Communications* 13 (1): 3689.

117 Wang, M., Liu, S., Xu, N. et al. (2017). Active Fe-N_x sites in carbon nanosheets as oxygen reduction electrocatalyst for flexible all-solid-state zinc–air batteries. *Advanced Sustainable Systems* 1 (10): 1700085.

118 Qiao, Y., Jiang, K., Deng, H., and Zhou, H. (2019). A high-energy-density and long-life lithium-ion battery via reversible oxide–peroxide conversion. *Nature Catalysis* 2 (11): 1035–1044.

119 Qiao, Y., Deng, H., He, P., and Zhou, H. (2020). A 500 Wh/kg lithium-metal cell based on anionic redox. *Joule* 4 (7): 1445–1458.

120 Zhu, Z., Kushima, A., Yin, Z. et al. (2016). Anion-redox nanolithia cathodes for Li-ion batteries. *Nature Energy* 1 (8): 16111.

121 Qiao, Y., Yang, H., Chang, Z. et al. (2021). A high-energy-density and long-life initial-anode-free lithium battery enabled by a Li_2O sacrificial agent. *Nature Energy* 6 (6): 653–662.

122 Yang, H., Qiao, Y., Chang, Z. et al. (2021). A safe and sustainable lithium-ion–oxygen battery based on a low-cost dual-carbon electrodes architecture. *Advanced Materials* 33 (24): 2100827.

123 Xia, C., Kwok, C.Y., and Nazar, L.F. (2018). A high-energy-density lithium-oxygen battery based on a reversible four-electron conversion to lithium oxide. *Science* 361 (6404): 777–781.

124 Sun, W., Wang, F., Zhang, B. et al. (2021). A rechargeable zinc-air battery based on zinc peroxide chemistry. *Science* 371 (6524): 46–51.

3

Rechargeable Lithium-Sulfur Batteries

Girum Girma Bizuneh[1], Fang Li[1], Abdullah N. Alodhayb[2], and Jianmin Ma[1]

[1]*Hunan University, School of Physics and Electronics, Lushan Road, Yuelu District, Changsha 410082, P.R. China*
[2]*King Saud University, Department of Physics and Astronomy, College of Science, Riyadh 11451, Saudi Arabia*

3.1 Background

The fast-changing lifestyles and economic and technological advancements of modern society are largely dependent on the energy generated from various sources to meet the insatiable demands for tremendous applications of day-to-day activities. However, satisfying the wider need for energy is largely based on irreplaceable, costly, and environmentally polluting sources such as fossil fuels. In the view of environmental protection, shifting the energy source dependencies from fossil fuels toward safe, clean, and inexhaustible sources such as solar power, hydropower, wind, and geothermal energies is crucial. However, these alternative energy sources are intermittent and fail to generate uniform energy due to seasonal inconsistency, and still the intense demand for powering various utilities is not satisfied. Therefore, leveling the fluctuating energy generation from these sources is the primary importance with electrochemical energy storage and conversion devices (e.g., batteries, capacitors, and fuel cells). Developing reliable energy storage devices under thorough consideration of economic feasibility and high-performance efficiency is an immense need [1]. In this regard, the earlier rechargeable batteries such as metal hydride, nickel-cadmium, and lead-acid, as well as conventional Li-ion batteries, have been widely used for a long time to power enormous electronic devices [2]. Among various rechargeable batteries, the Li-ion battery has been serving the market demand of electronic devices for over three decades preferably due to its high energy density and performance stability. For instance, the energy density of 387 Wh kg^{-1} [3] can be harnessed from the battery based on the coupling of the $LiCoO_2$ cathode and a graphite anode. However, this sort of energy density from Li-ion battery is not sufficient for the unlimited demand for electromobility, grid storage, and other high-energy-demanding applications. Furthermore, the practical and theoretical energy densities of Li-ion batteries are nearly identical at the moment, implying that the technology's future growth is limited [4]. Hence,

Electrochemical Energy Storage Devices: Non-Conventional Technologies and Materials, First Edition. Edited by Yongbing Tang and Luojiang Zhang.

the alternative new-generation high-energy (Li-S) battery coupling sulfur cathode and Li metal anode operating with redox chemistry transferring $2e^-$ needs to be introduced instead of the intercalation-based conventional Li-ion battery [5, 6]. Thus, the Li-S battery is expected to meet high energy demands while still being environmental friendly and cost-effective.

3.2 Components and Mechanism of Lithium-Sulfur Batteries

In contrast to the Li-ion battery, where the cathode and the anode materials are the transition metal oxides (TMOs) and the graphite framework, respectively, in the Li-S battery, the positive electrode elemental sulfur coupled with the negative electrode Li metal are separated by the porous polymeric membrane for efficient ionic and electrolyte transportation between the electrodes while avoiding direct contact of the electrodes Figure 3.1. During operation, electrons running along the external wire complete the circuit, while the ions are supported by the organic electrolytes and shuttling between the electrodes via the porous membrane separator, connecting the cathode and the anode. In the discharging phase of the Li-S battery, the Li metal anode simultaneously releases electrons and Li^+ ions: the electrons move to the cathode (sulfur) along the external wire connecting both electrodes, and the Li^+ ions travel internally through the electrolyte, which triggers sulfur to undergo reduction to form lithiated sulfides of various chain lengths (high-order soluble polysulfides (Li_2S_x, $2 < x \leq 8$) and short-chain insoluble sulfides (Li_2S_2/Li_2S)), thereby releasing electric current to power various appliances. While in the charging phase, the electrons and Li^+ ions move in the opposite direction (i.e., from cathode to anode) along the external wire and through the separator, respectively, being forced by the charging current applied from the external source. The process is accompanied by the delithiation of the discharge products (lithium sulfides) to form the oxidized product (elemental sulfur) and the Li^+ ions being plated on the Li metal anode.

Relying on the basic working mechanism and building a battery by bringing the two high-specific-capacity electrodes (Li metal ~3860 mAh g^{-1} and sulfur ~1675 mAh g^{-1}) together, a very huge theoretical energy density of 2600 Wh kg^{-1}

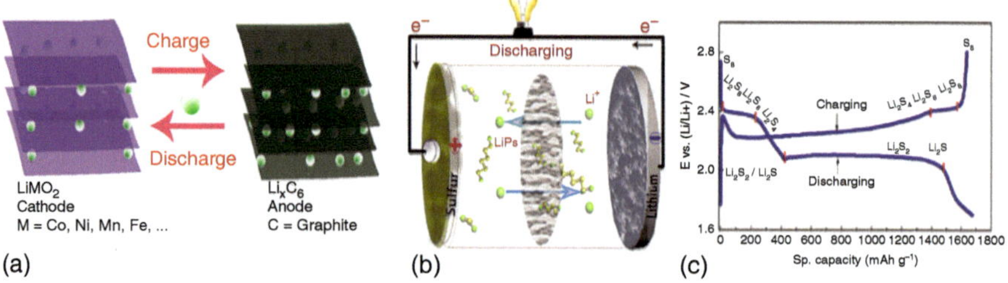

Figure 3.1 Schematics of the working mechanisms of (a) Li-ion battery and (b) Li-S battery. (c) Charge/discharge voltage profile of typical Li-S battery.

can be extracted, promising to satisfy the long-awaited demand for power-intensive appliances [7–9]. It can also offer a pretty high open-circuit cell potential equivalent to the voltage difference between the sulfur cathode and the Li metal anode [9, 10]. However, as shown in Eqs. (3.1–3.5) [3], the path to complete electrochemical reduction of sulfur passes through several steps, resulting in chain products of lithiated sulfur at various voltage levels. Apparently, there are two voltage plateaus when the Li-S battery is discharging: (i) in the region from 2.4 to 2.1 V (vs. Li/Li^+), the cyclic sulfur (S_8) proceeds to form long-chain soluble lithium polysulfides accompanied with the solid/liquid phase transition (Eqs. 3.1, 3.3). Through the course of discharging, the chain length of the intermediate products continues to shorten. (ii) At a plateau potential of 2.1 V (vs. Li/Li^+), further reduction of the polysulfides leads to the generation of insoluble products Li_2S_2 or Li_2S, where the phase transition from liquid to solid undergoes (Eqs. 3.4 and 3.5). The discharging phase voltage plateaus also indicate the existing variation in the reaction kinetics, where the conversion from S_8 (solid) → Li_2S_4 (liquid) proceeds with fast kinetics while the process from Li_2S_4 (liquid) → S_8 (solid) undergoes relatively slower kinetics associated with a much energy-demanding nucleation process to initiate solid-phase growth. The transition from Li_2S_2 (solid) → Li_2S (solid) is typically very sluggish due to the hindered solid-state diffusion [9, 11]. On the other hand, the reverse process occurs when the battery is charged, forced by the externally applied potential, whereby oxidizing the pre-formed reduction products to sequentially generate polysulfides and ultimately cyclic octa sulfur (Li_2S or Li_2S_2 (solid) → Li_2S_x, $2 < x \leq 8$ (liquid) → S_8 (solid)). The charging process is accompanied by the delithiation of the reduced products and simultaneous plating of the Li ions on the Li metal anode.

$$S_8 + 2Li^+ + 2e^- \rightarrow Li_2S_8 \tag{3.1}$$

$$3Li_2S_8 + 2Li^+ + 2e^- \rightarrow 4Li_2S_6 \tag{3.2}$$

$$2Li_2S_6 + 2Li^+ + 2e^- \rightarrow 3Li_2S_4 \tag{3.3}$$

$$Li_2S_4 + 2Li^+ + 2e^- \rightarrow 2Li_2S_2 \tag{3.4}$$

$$Li_2S_2 + 2Li^+ + 2e^- \rightarrow 2Li_2S \tag{3.5}$$

3.3 The Existing Challenges of Li-S battery

Despite the many advantages of Li-S battery technology, such as high energy density, environmental safety, and low cost, the system continues to face significant bottlenecks that limit its commercial availability [12–14]. Generally, the challenges can be tracked to emerge from either of the electrodes (sulfur cathode or lithium anode). The sulfur cathode contrasting to its attractive advantages also brings several challenges influencing the Li-S battery, as poor electronic conductivities

(~10^{-30} and 10^{-14} S cm^{-1}, respectively, for sulfur and Li_2S) [15], which induce capacity tradeoff due to insufficient active material utilization [13, 16], the huge volume change ≈80% between sulfur and Li_2S (considering ρ_{Sulfur} = 2.07 g cm^{-3} and ρ_{Li_2S} = 1.66 g cm^{-3}) when the battery is working along with phase transition of sulfur (solid) ↔ polysulfide (liquid) ↔ Li_2S (solid), which leads to loss of integrity, electronic disconnection, and the collapse of the cathode structure ultimately resulting in an irreversible capacity loss [13, 16]. The other and most prominent issue that emerges from the sulfur cathode is the evolution of polysulfide intermediates that shuttle between the electrodes, which could cause anode corrosion and depletion of the sulfur cathode, consequently leading to poor coulombic efficiency, fast capacity decay, and battery failure [17, 18].

On the other hand, the lithium metal anode is also not free from challenges as it can react with the vast majority of chemical species based on its extremely reactive nature, leading to the loss of electrolytes and thus causing poor cycle efficiency. Moreover, the nonhomogeneous stripping and plating of the Li^+ ion on the Li metal anode during cycling induce deleterious mossy dendritic features that can pierce the separator and cause internal short circuit, fire, and safety concerns with consequential battery failure [19–26]. Attributed to these and other concerns evolving from the sulfur cathode, Li metal anode, and also from electrolytes, the Li-S battery technology is hindered, and its commercialization is not yet realized [26].

3.4 Sulfur Cathode

The octa-atomic cyclic sulfur (S_8) has become an attractive potential cathode material in the prospective high-energy-density Li-S battery. In the pursuit of clean, cheaper, and high-energy-density rechargeable energy storage devices, the Li-S battery technology alternative is striving to overtake the stage from the current state-of-the-art Li-ion batteries. Due to the primary advantages of sulfur such as lightweight, environmental benignity, high abundance, and a high specific theoretical capacity (1675 mAh g^{-1}) [27], there is a tremendous rush to produce sulfur-based cathodes and replace transition metal-based cathodes in traditional Li-ion batteries, to increase energy density. For instance, the energy density of the sulfur cathode can reach more than sixfold that of the $LiCoO_2$ cathode [3]. Hence, implementing the sulfur cathode in Li-S battery has the potential to provide high volumetric as well as gravimetric energy densities of 2500 Wh L^{-1} and 2500 Wh kg^{-1}, respectively, with an average operating potential of ≈2.1 V vs. Li/Li^+ [28, 29]. However, the practical implementation down to the ground of those theoretical merits of the Li-S battery technology is not an easy task due to the challenges discussed in Section 3.3. Therefore, several strategic designs and efforts have been made recently to overcome the issues from various perspectives (cathode designs, electrolyte manipulation, and anode engineering). Especially, when considering the cathode design, due to the intrinsic limitations such as poor electronic/ionic conductivity and lack of defined crystal structure and enough space to accommodate the Li^+ during reduction, the fabrication of sulfur cathode mostly necessitates relying on host materials, which can fill the gaps [30]. For example, various cathode designs engaging enormous creative strategies have been implemented, such as carbon-based cathode architectures [31], polymer-based cathode [32], inorganic-based cathode architectures [33–35], hybrid cathodes [36], and lithium sulfide-based cathodes [37].

3.4.1 Carbon Materials for Sulfur Cathode

Carbonaceous materials with vast architectural varieties ranging from porous carbons with different subdivisions (microporous, mesoporous, hierarchical porosity, hollow carbon), to one-dimensional carbons (nanofibers and nanotubes), and two-dimensional carbon frameworks (graphene oxide and graphene) have been intensively studied to improve the sulfur cathode with various designs and configurations targeting the advantages of the low cost of carbon materials and the structural (morphological) feasibility to lodge the active sulfur in the composite. The various features in carbon materials such as high electrical conductivity and high porosity and surface area to promote the interfacial electrochemical processes and lightweights make it preferable to resolve the critical challenges of the Li-S batteries such as polysulfide shuttle effect, electrical insulation, and volume change, thus resulting in efficient sulfur utilization and high electrochemical performance [38, 39].

3.4.1.1 Porous Carbons as a Sulfur Host

Porous carbons have long been used in various applications such as sorbents for gas storage and separation, catalyst support, and electrode materials for fuel cells, supercapacitors, and batteries [40]. In the latter case, to exploit the structural merits of carbonaceous materials for improving the electrochemical performance of Li-S batteries, immense attention has been given to employ porous carbons in hosting the sulfur cathodes. The porous carbon in sulfur cathode provides the attribute of enhanced electrical conductivity, arrests the diffusing polysulfides within the micro/mesopores, and promotes both mass and charge transfer processes to improve the rate performance of the Li-S battery [41, 42]. Especially, the mesoporous carbons (with hierarchical porosity) possessing large pore volume and high surface area [43] are suitable for facilitated mass transport phenomena due to the dependency on pore size and diffusion path length inside the pores [44, 45]. Employing the mesoporous carbon as a potential cathode host for Li-S battery has been receiving much attention since an interesting work reported by Nazar's research group using an ordered nanostructured carbon [46]. The ordered nanostructured carbon possessed a pore size of 3–4 nm and was covered with the hydrophilic polymer to efficiently arrest sulfur and polysulfides while unveiling a high reversible specific capacity of $1320\,mAh\,g^{-1}$. In addition to this strategy, Dudney and coworkers [47] demonstrated high surface area hierarchically featured carbon nanostructures for sulfur encapsulation. Assisted by the soft-templating method, they effectively synthesized a mesoporous carbon (7.3 nm) where the pores were distributed uniformly. After activation with potassium hydroxide, a bimodal carbon with mesoporous and microporous features (~2 nm) can be further created. Also, in the preparation of the S/C cathode composite, the micropores were infiltrated with elemental sulfur via a solution approach, which was beneficial for sulfur utilization, resulting in improved performance of cathode.

Archer and coauthors [48] also used a template-based strategy to synthesize mesoporous hollow carbon capsules to effectively anchor sulfur in the interior and porous shell of the carbon structures. The intentionally designed mesoporous shell structure with the void room can effectively achieve high sulfur encapsulation, which minimizes polysulfide dissolution, facilitates Li^+ ion transport to the entrapped sulfur with good electrolyte percolation, and enhances electron mobility. Accordingly, the mesoporous hollow carbon–sulfur cathode (C@S) exhibited an improved capacity of 850 mAh g^{-1} at 0.5C and persisted up to 100 cycles. Similarly, based on the hard templating approach with anodized aluminum oxide (AAO) as the substrate, Wang and coworkers [49] fabricated 1D hollow porous carbon fibers (HPCFs) and loaded them with sulfur to obtain the composite cathode of HPCF/S containing 40 wt% of sulfur through low-temperature sulfur infiltration followed by high-temperature sulfur intercalation into the porous shell of the material. The structural features of the 1D HPCF can facilitate the reaction kinetics in the HPCF/S cathode owing to the conductive channels for both electrons and ions. Later, Cui and coworkers [50] came across a clever strategy for site-specific deposition of sulfur selectively into the inner space of the 1D HPCF via the heat treatment of the AAO@HPCF–sulfur mixture, as shown in Figure 3.2a. The as-obtained S@HPCF cathode after removal of the AAO substrate with sulfur mass loading of 75 wt% offered 706 mAh g^{-1} reversible capacity at 0.2C and sustained up to 150 cycles (Figure 3.2b).

Despite the fact that porous carbon architectures have broad specific surface areas and enriched porous features, manipulating and controlling the porosity, pore size, and surface area is a difficult task that still lacks the efficiency to suppress the polysulfide shuttle hurdles [54]. Therefore, further works need to be done to fill the gap created by using porous carbon materials that mitigate the existing problems and enable efficient Li-S battery performance with high sulfur utilization and polysulfide shuttle suppression.

3.4.1.2 Graphene-Supported Sulfur Cathodes

Due to the unique physicochemical properties such as exceptional electrical conductivity, mechanical stability, chemical processability, and high specific surface areas, two-dimensional materials such as graphene have received great emphasis in electrochemical energy storage research [54]. Integrating sulfur nanoparticles with a graphene nanosheet framework can exhibit outstanding benefits for Li-S battery performance. Attributed to the extraordinary electronic conductive channels, employing graphene as the cathode structure skeleton in Li-S battery can boost the active sulfur utilization while the battery is operating [55–57].

Furthermore, functionalizing graphene with different functional groups and heteroatoms will disrupt the uniformity of charge distribution in the structure, resulting in specific behavior and application of Li-S batteries. The ability of graphene to form a stacking sheet as well as the edge-attached dangling atoms and

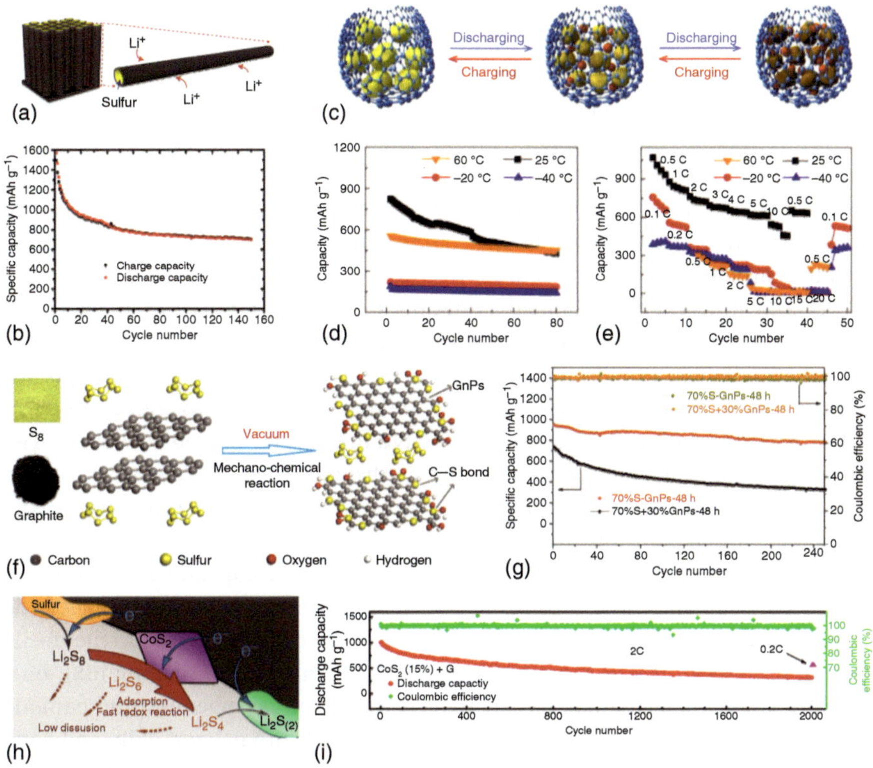

Figure 3.2 (a) The design concept demonstrating the high aspect ratio of hollow carbon nanofibers for efficient polysulfide trapping and (b) capacities of S@HPCF for charge/discharge cycling at C/5. Source: Reproduced with permission Zheng et al. [50]. Copyright 2011, American Chemical Society. (c) Scheme showing sulfur infiltration in graphene support for Li-S batteries during the charging–discharging phase, (d) cycling performance sulfur entrapped hierarchically porous graphene (PGS-1000) at a current density of 1672 mA g^{-1}, (e) rate performance of PGS-1000 at −40, −20, 25, 60 °C. Source: Reproduced with permission Huang et al. [51] Copyright 2013, Elsevier. (f) Schematic illustration of the vacuum mechano-chemical reaction used to prepare 70% S-GnPs. (g) Cycling performance of 70% S-GnPs at 0.5C in the voltage range of 1.7–2.8 V. Source: Reproduced with permission Yan et al. [52] Copyright 2017, Elsevier. (h) Scheme illustrating the discharge phase in a CoS_2-integrated carbon/sulfur cathode, (i) cycling stability of a CoS_2 (15%) + graphene-based sulfur cathode cycled for 2000 cycles at 2C rate and for 10 cycles at 0.2C. Source: Reproduced with permission of Yuan et al. [53]. © 2016/American Chemical Society.

functional groups provide an ideal environment for chemical contact with sulfur species while also effectively maintaining the electroactive material in the cathode for long-term battery performance [58].

In such systems, the structure and functional groups of graphene greatly aid in regulating the electrochemical processes [30, 59–62]. Zhang and coworkers [63] reported a chemical strategy to trap sulfur and polysulfides by depositing reactive functional groups on graphene oxide nanosheets. The approach ensures thin sulfur with a thickness of about tens of nanometers was coated uniformly on the graphene oxide sheet structures. The apparent strong interaction between sulfur or polysulfide and graphene oxide enabled the Li-S battery to deliver a pretty high specific capacity of 950–1400 mAh g^{-1} at 0.1C. The study revealed that after heat treatment of the graphene oxide-sulfur composite, a highly conjugated layer-resembling nanofeature with rich pore structures can be formed. The as-synthesized graphene–sulfur composite possessed advantages such as accommodating the huge volume change during battery cycling and offering intimate electrical contact within the sulfur cathode, which created a suitable environment to arrest the polysulfide avoiding the shuttle problem.

Wang and coworkers developed a freestanding mesoporous graphene–sulfur nanocomposite via a sulfur vapor treatment approach [64]. The strategy enabled uniform percolation and immobilization of amorphous sulfur into the porous graphene paper. Besides, the material can serve as free-standing electrode without the utilization of extra binder, conductive agent, and current collector. The flexible, porous, and conductive features of the graphene framework provided electron transfer channels and electrolyte diffusion pathways. Consequently, the mesoporous graphene–sulfur free-standing cathode exhibited a specific capacity of 1393 mAh g^{-1} and sustained up to 50 cycles with a retained capacity of 689 mAh g^{-1} at 0.1C. Further work has been reported to increase the rate capability of the free-standing graphene–sulfur matrix by Huang and coworkers via the facile synthesis strategy of freeze-drying followed by low-temperature heating [65]. As a result, the as-assembled battery based on the macroporous graphene–sulfur cathode offered a capacity of 800 mAh g^{-1} after 200 cycles at 300 mA g^{-1} and delivered good rate performance with the current density up to 1500 mA g^{-1}. The authors claimed that the enhanced battery performance was attributed to the stable macroporous structure of the graphene framework and the tight interaction between sulfur and graphene.

Further, the nitrogen-doped graphene paper free-standing electrode was fabricated by Kung and coworkers [66] via pyrolysis of poly diallyl dimethylammonium chloride and employed in a Li-S battery without a binder. The as-fabricated nitrogen-doped electrode demonstrated an outstanding specific capacity of 1000 mAh g^{-1} with a high coulombic efficiency of 98% over 100 cycles, outperforming the undoped graphene electrode. The spectroscopic study revealed that the N-doped graphene could trigger stronger binding of sulfide when compared with the undoped graphene structure. Besides, the density functional theory calculation also confirmed that the pyrrolic- and pyridinic-N underwent stronger interaction with polysulfides than the quaternary-N species, which created a suitable partition of the polysulfides between the electrode and the electrolyte, and ultimately impacted the passivation layer morphology of the electrode.

Considering the relatively complicated steps of aforementioned sulfur cathode preparation, Meng and co-investigators [52] designed a facile vacuum mechanochemical method to fabricate an edge sulfurized graphene nanoplatelet featured with 3D porous structures as the cathode for Li-S battery. The optimal sample with 70 wt% sulfur loading was achieved by ball milling for 48 hours, out of which 13.2 wt% of sulfur being chemically bonded at the edge of graphene matrix

(Figure 3.2f,g). The as-constructed battery demonstrated capacities of 1089 and 950 mAh g^{-1} at 0.1C and 0.5C, respectively, and sustained at 0.5C for 250 cycles with a retained specific capacity of 776 mAh g^{-1}. It was claimed that the enhanced performance was attributed to the combined mechano-chemical interaction occurring between sulfur and graphene frameworks. Also, Zhang and coworkers demonstrated that sulfur can be entrapped in a hierarchically porous graphene structure, where graphene served as a mini-electrochemical reaction chamber in addition to facilitating the transfer of electrons within the encapsulated sulfur particles [51] (Figure 3.2c–e). Owing to the sp^2 conjugation, the graphene network ensured a lower resistance electronic transport path, and the hierarchically porous structure provided sufficient room to hold the dynamic change of volume of sulfur and Li^+ ion diffusion channels. Moreover, the hierarchically porous graphene was decorated by the epoxy and hydroxyl functional groups, which was beneficial for the binding of sulfur. Consequently, the Li-S battery with the entrapped sulfur exhibited extraordinary performance at extreme conditions from −40 to 60 °C.

Overall, graphene plays a key role in enhancing the Li-S battery's electrochemical efficiency. It provides a confining function to sulfur and polysulfides with networked channels for electron and Li^+ ion transport as well as enough room for electrolyte percolation. Especially, graphene featured with suitable porous structures and functionalized heteroatoms (for example, nitrogen, oxygen, and other elements) offers extraordinary performance [67–69] due to the break of the charge distribution, which results in suitable interactive binding sites for grafting sulfur and alleviating the polysulfide shuttle and enhancement of electron and Li^+ ion transport, thus improving the battery performance.

3.4.2 Inorganic-Based Structures for Hosting Sulfur

The exploration of suitable materials to boost the electrochemical performance of the Li-S batteries has gone through miles, considering to provide ultimate solutions to the existing scientific and practical obstacles that hinder the applicability of the battery. The material's nature, size, morphology, and shape also impose a significant influence on the performance of the battery. Carbonaceous materials with different morphological features have been employed to host and immobilize sulfur and its reduced species to enhance the electrochemical performance of the Li-S battery. However, the carbon-based materials lack certain polarity to effectively arrest the polysulfides and hold sulfur for long time [70]. On the contrary, inorganic materials possess the characteristic chemical affinity toward the polysulfide species and are capable of restricting the polysulfide diffusion. Nanostructured inorganic materials with large specific surface areas and polarities such as oxides [71], sulfides [53], nitrides [72], borides [73], carbides [74, 75], and phosphides [76–78] have recently been studied to alleviate the challenges of polysulfides via different interaction mechanisms (physical, chemical, or electrocatalytic) [74]. In subsections, we will cover some representative work from oxides, sulfides, and nitrides.

3.4.2.1 Inorganic Sulfides

In solar cells and energy storage applications, it is usual to encounter metal sulfides [79]. Recently, an interest to exploit the catalytic roles of metal sulfides in the Li-S battery application is gaining attention.

In the view of extracting the merits of metal sulfides such as stability toward sulfur chemistries, low lithiation voltage vs. Li/Li^+ (prevents coinciding with the Li-S battery working voltage), metallic/semimetallic nature (fast electron transport), and strong affinity toward polysulfides, tendencies toward applying metal sulfides in Li-S batteries have been promoted [80]. Zhang and coworkers reported sulphiphilic cobalt disulfide (CoS_2) (Figure 3.2h,i) combined with the carbon–sulfur to serve as cathode material that can interact strongly with polysulfide [53]. The polysulfide conversion reaction was accelerated owing to the interfaces between the CoS_2 and the electrolyte, which served as the sites for polysulfide adsorption and activation. The highly reactive affinity of polysulfides toward CoS_2 contributed to lower polarization with an energy efficiency improvement of 10%, decreased capacity decay rate, high specific capacity, and ultrastable cycle performance up to 2000 cycles.

In another work, a cathode with sulfur loaded on multifunctional nanotube Co_3S_4 synthesized via a hydrothermal approach was reported [81]. The Co_3S_4 nanotube displayed good electronic conductivity and high chemical affinity with the polysulfides, which endowed Co_3S_4@S a superb specific discharge capacity of 1267 mAh g^{-1} at 0.05C and 517 mAh g^{-1} at 5C, respectively. The cathode can also sustain over 1000 cycles with a very low attenuation rate of 0.041% per cycle.

Lee and coworkers demonstrated the catalytic activity of sulfur-deficient MoS_2 nanoflakes combined with reduced graphene oxide (MoS_{2-x}/rGO) on the conversion of polysulfide [82]. It was shown that the sulfur defects on the surface induced the catalytic role of the nanoflakes and promoted the conversion kinetics. By expediting the conversion of the polysulfides, MoS_{2-x}/rGO can effectively avoid its agglomeration on the cathode surface as well as being eroding by diffusion. With the incorporation of only 4 wt% MoS_{2-x}/rGO in the sulfur cathode, an improved rate performance up to 8C and long cycle stability at 0.5C were achieved. Considering the benefit of strong chemical coupling in hybrid materials, the uniformly decorated NiS nanoparticles on 3D hollow carbon spheres (NiS@C-HS) were synthesized by *in situ* thermal reduction and sulfidation procedure [83]. The uniformly distributed NiS nanoparticles offered enhanced adsorption capability for polysulfides, while the C-HS provided advantages of high sulfur loading, accommodating volume change, suitable electron transport channel, and reducing the active material loss. Significantly, owing to the effective chemical binding between NiS and the carbon structure, the redox kinetics and the charge transfer of the sulfur electrode can be enhanced. Hence, the NiS@C-HS exhibited very stable cycle performance for 300 cycles, delivering a capacity of 695 mAh g^{-1} with a fading rate of only 0.013% per cycle.

With the intent of mitigating the challenges and pushing forward the development of the robust high-energy Li-S battery, many other inorganic sulfide-based materials have also been identified including Co_9S_8 [35], WS_2 [84], ZnS [85], SnS_2 [86], $NiCo_2S_4$ [87], and $FeCo_2S_4$ [88].

3.4.2.2 Inorganic Oxides

Inorganic oxides (typically TMOs), possessing an oxide anion with an O^{2-} oxidation state, are known to have polar surfaces, which are suitable to interact with the polar polysulfide species [71]. Attributed to the high surface area and strong surface affinity toward the polysulfides, the oxides can enhance the interaction between electrode and electrolyte while inhibiting the polysulfide shuttle-induced active material loss as well as facilitating the redox conversion between polysulfides and sulfides (Li_2S_2/Li_2S). Thus, employing the nanostructured oxides to modify the sulfur cathodes can lead to efficient sulfur utilization and performance enhancement of the Li-S battery. Due to robust metal-oxygen bonding, the metal oxides are almost insoluble in organic solvents that are commonly used in Li-S batteries [89, 90]. Hence, they are used to modify the nanostructures of the carbon/sulfur composite. The $Mg_{0.6}Ni_{0.4}O$ additive was introduced in sulfur cathodes, which exhibited an enhanced capacity due to the promising polysulfide adsorbing role of the oxide [91]. Afterward, numerous outstanding reports involving inorganic oxides in sulfur cathodes were studied, such as porous SiO_2 [92], Ti_4O_7 [33], TiO_2 [93], MnO_2 [94], Fe_2O_3 [95], VO_2 [96], V_2O_5 [97], MoO_3 [98], WO_{3-x} [99], CeO_2 [100], CoO [101], Al_2O_3 [102], MgO [103], ZrO_2 [104], and Nd_2O_3 [105].

Inspired by the design principle of drug delivery, Nazar and coworkers incorporated porous silica (10 wt% SBA-15) into mesoporous carbon/sulfur composite (SCM/S). It was found that the material can absorb the polysulfides via weak binding and allow reversible desorption and release. More importantly, it acted as an internal reservoir for polysulfides, enabling reversible electrochemical cycling and high coulombic efficiency [92]. Also in another report, by hosting sulfur in sulphiphilic Ti_4O_7, a Ti_4O_7/S composite was developed to prevent the dissolution of polysulfides and form an excellent interface with Li_2S through surface-mediated redox processes. The Ti_4O_7/S cathode delivered a high capacity of 1070 mAh g^{-1} at C/5 and good capacity retention with stable cycling at 2C for more than 500 cycles [33] (Figure 3.3a,b).

Moreover, the catalytic role of oxygen-deficient metal oxides was also demonstrated in Li-S batteries using WO_{3-x} as an example [99]. The WO_{3-x} can bidirectionally promote the conversion reaction of the sulfur and its reduction products. As a result, it prohibited the accumulation of polysulfide on the cathode, the loss of active material, and the shuttle-induced corrosion of Li anode. Hence, the as-assembled Li-S battery exhibited a capacity of 693.2 mAh g^{-1} at 4C and sustained stably over 300 cycles with a decay rate of 0.13%.

To sum up, nanostructured inorganic oxides play prominent roles in Li-S batteries in terms of suppressing the polysulfides via surface physical adsorption or chemisorption on the polar surfaces as well as catalyzing the transformation reactions to-and-from sulfur, polysulfides, and Li_2S_2/Li_2S. Therefore, employing properly tuned custom-made nanostructured oxides can improve the Li-S battery performance and lead to the realization of practical high-energy Li-S batteries.

3.4.2.3 Inorganic Nitrides

Different metal nitrides such as TiN [72], VN [108], Mo_2N [109, 110], Ni_3N [111], and NbN [112] have long been utilized in LIBs and supercapacitors due to reversible

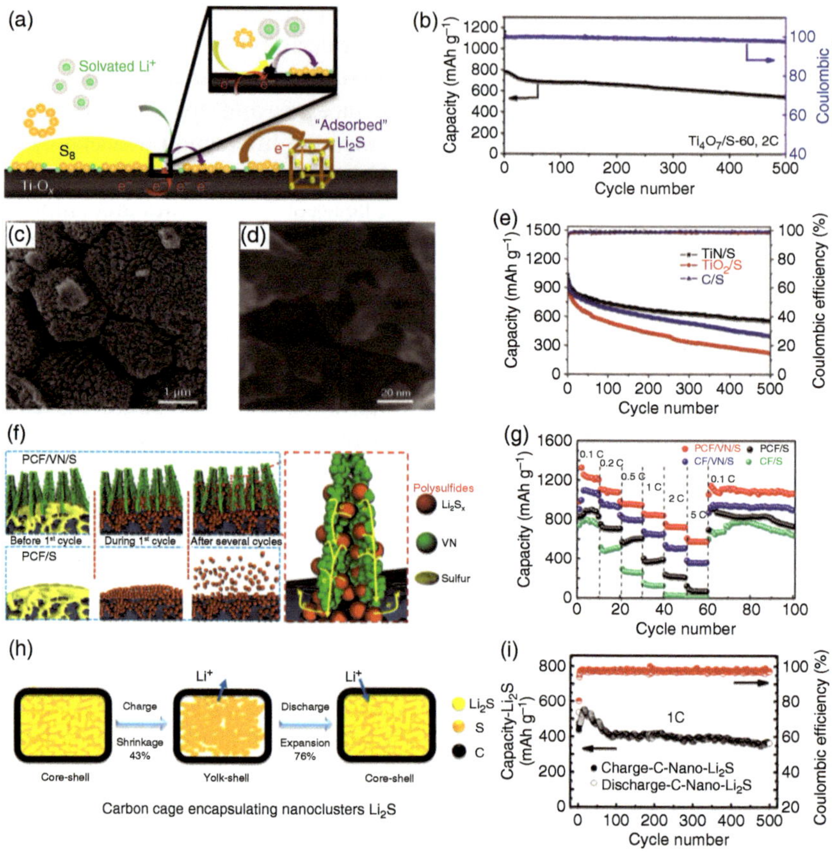

Figure 3.3 (a) Diagram illustrating surface-mediated reduction of Li_2S from LiPSs on Ti_4O_7 and (b) high-rate cycle performance of Ti_4O_7/S-60 at 2C over 500 cycles with corresponding coulombic efficiency. Source: Reproduced with permission Pang et al. [33]. Copyright 2014, Springer Nature. (c,d) SEM images of porous TiN, (e) cycle stability of TiN/S, TiO_2/S, and C/S cathodes. Source: Reproduced with permission of Cui et al. [34]. © 2016/John Wiley & Sons. (f) Schematic illustration of dual blocking effects associated with "physical block and chemical absorption" for polysulfides in the PCF/VN/S electrode. (g) The rate capability of the CF/S, PCF/S, PCF/VN/S, and CF/VN/S electrodes at varying C rates. Source: Reproduced with permission Zhong et al. [106] Copyright 2018, Wiley-VCH. (h) Schematic of the delithiation and lithiation process in carbon-coated Nano-Li_2S with cage-like structure, (i) long cycling life and coulombic efficiency of C-Nano-Li_2S at a high rate (1C) after the first cycle at 0.1C. Source: Reproduced with permission Suo et al. [107]. Copyright 2015, Elsevier.

insertion/deinsertion reactions via intercalation mechanism [113, 114]. The characteristic attributes of metal nitrides such as chemical stability, excellent electronic conductivity, and good affinity for polysulfides are the impulsive reasons that enable them to be applied in Li-S batteries. However, complicated and costly synthetic procedures and a lack of understanding of electrochemical reactions pose certain challenges in the applicability of metal nitrides [89]. The polysulfides are trapped on the surface of metal nitrides via a chemisorption mechanism whereby the polysulfides' "sulfur" and the "metal" of the metal nitride form a chemical bond on the metal nitride surface. Also, the "nitrogen atom" participates interactively with the polysulfides and boosts binding on the metal nitride surface [115].

Goodenough and coworkers developed a mesoporous tin nitride hosting sulfur as cathode for Li-S batteries [34]. Using a solid/solid phase separation approach under ammonia atmosphere, mesoporous TiN with a high surface area of 69.689 $m^2\ g^{-1}$ was produced. TiN was then loaded with sulfur through a melt-diffusion heating to obtain the TiN-S composite. Owing to the high electrical conductivity, good chemical affinity with polysulfides, and mesoporous structure, the TiN-S cathode exhibited outstanding electrochemical performance over 500 charging/discharging cycles with a decay rate of 0.07% (Figure 3.3c–e).

Besides, the electrochemical performance of other metal nitrides (WN, Mo_2N, and VN) supported sulfur cathode was also reported. Among them, WN showed good electrochemical performance owing to the effective arresting of the polysulfides via the formation of strong S—W—N bonding on the nitride surface [116], while VN showed increased sulfur utilization and effective polysulfide trapping with the help of porous carbon fiber scaffold (PCF/VN), as reported by Zhong et al. (Figure 3.3f,g) [106]. The as-obtained PCF/VN-S cathode could provide a high specific capacity of 1310.8 $mAh\ g^{-1}$ at 0.1C and retained 1052.5 $mAh\ g^{-1}$ after 250 cycles. Despite the challenges, employing metal nitrides for improving the Li-S battery performances shows a promising future, provided that the issues are effectively solved. Hence, lots of research is expected to be done.

3.4.2.4 Lithium Sulfide

Li_2S, the reduced and fully lithiated form of sulfur with a high theoretical specific capacity (1166 $mAh\ g^{-1}$), has hopeful potential to play instead of the pure sulfur cathode, enabling safer and high energy battery in joint with anodes other than Li metal (such as graphite, silicon, and tin) [117].

Li_2S cathode can bring dendrite-free battery without safety concerns and offer an intermediate energy density between the conventional Li-ion battery (400 $Wh\ kg^{-1}$) and the Li-S battery (2600 $Wh\ kg^{-1}$ based on pure S cathode) [37, 118]. Moreover, the comparatively higher melting point of Li_2S (938 °C) than that of pure sulfur (112.8 °C) ensures the processability of Li_2S cathode at an elevated temperature. In addition, Li_2S has a lower density (1.66 $g\ cm^{-3}$) than sulfur (2.07 $g\ cm^{-3}$), which would lead to less volume change during battery operation [12].

Nevertheless, Li_2S cathode is severely challenged by poor ionic and electronic conductivity as well as polysulfide dissolution in organic electrolytes, leading to feeble battery performance. Beyond this, Li_2S, especially the bulk Li_2S, suffers from a high activation energy barrier during the initial charging step. Therefore, intelligent and creative strategies are important for Li_2S to achieve high performance. In this regard, coating a conductive agent on nanoscaled Li_2S to create an electrically conductive composite is the most fascinating way, such as mixing carbon precursor and Li_2S in ethanol [119], thermally decomposing Li_2S_3-polyacrylonitrile composite [120], reacting organolithium with sulfur [121], electrospinning of PVP-Li_2S_3 mixture [122], and ball milling carbon additives and Li_2S mixture [123].

Recently, a composite of Li_2S nanoparticles (~8.5 nm) uniformly anchored thermally exfoliated graphene (TG-Li_2S) was synthesized via chemical reduction of sulfur/TG by lithium triethyl borohydride [124]. The ultrasmall Li_2S nanoparticles coupled with highly conductive TG nanosheets enabled TG-Li_2S cathode to display an initial capacity of 1609 mAh g s^{-1} at 0.1C with a negligible charging potential barrier. Moreover, it could retain the discharge capacity of 1137 mAh g s^{-1} after 100 cycles at 0.1C and 805 mAh g s^{-1} at 2C, respectively.

Wang and coworkers [107] also synthesized a carbon cage enclosed Li_2S nanoparticles (C-Nano-Li_2S) using ionic liquid as the carbon precursor (Figure 3.3h,i). Owing to the short Li^+ ion diffusion path and enhanced electron conductivity, the C-Nano-Li_2S cathode exhibited a capacity of 826 mAh g^{-1} at 0.1C and good cycling stability with a retention rate of 82% after 500 cycles at 1C. Ascribed to the effectively coated conductive cage, C-Nano-Li_2S offered a coulombic efficiency of ~100%. The polymer-derived conductive carbon encapsulation is also a feasible approach as the polymer allows to uniformly mix with the Li_2S particles due to their viscosity and also serves as the promising carbon precursor [125]. This phenomenon was also observed in another study, in which Li_2S-C_{PAN} with Li_2S particles homogeneously dispersed on polyacrylonitrile (PAN)-derived carbon exhibited an improved cycling performance up to 50 cycles and a high coulombic efficiency of 95.5% [126]. In addition to carbon materials, conducting polymers also display the potential for combining with Li_2S to achieve high electrochemical performance. Cui and coworkers demonstrated an *in situ* synthesized Li_2S-polypyrrole composite cathode delivering a capacity of 1126 mAh g^{-1} and enhanced cycle performance over 400 cycles, which was attributed to the strong chemical interaction between N atoms in polypyrrole and Li atoms in Li_2S [127]. The interaction enabled the polypyrrole to effectively cover and bind with the Li_2S particles.

The investigation for exploring the proper strategy to enable the practical application of Li-S battery is ongoing, despite that much progress has been achieved from the perspectives of materials frontiers and promising results were reported in the scientific discipline. However, more miles remain to reach for Li-S battery to appear in the market.

3.5 Lithium Anode

The performance of the battery depends on the contributory roles of every battery component (cathode, anode, separator, and electrolytes). Therefore, these components need to be designed to be safe, stable, and highly efficient for promoting the electrochemical process of batteries. Besides, the role of the anode is inevitable in enhancing battery performance. The intrinsic properties of Li metal such as the low electrochemical potential (−3.04 V vs. SHE), low density (0.59 g cm^{-3}), and ultrahigh theoretical specific capacity (3860 mAh g^{-1}) drive its candidacy in anode material for the prospective high-energy-density batteries [26]. Li metal has been applied in primary Li metal batteries since the 1970s [128]. Recently, the Li metal anode was coupled with high-voltage TMO cathodes [129] and new generation high-specific-capacity cathodes (sulfur and oxygen) to assemble rechargeable batteries to increase the energy density and satisfy the high energy demand for the electrifying various applications. Accordingly, the energy density can reach 500, 2600, and 3500 Wh kg^{-1} respectively for Li-TMO, Li-S, and Li-O_2. However, the extremely reactive nature of Li metal brings a severe challenge to its practicability in those high-energy-density batteries, due to the safety concern and poor cycling stability [130].

3.5.1 Challenges with Li Metal Anode

The high reactivity of Li induced complex challenges of Li metal-based batteries that are safety issues, low coulombic efficiency, and poor cycle stability. Particularly, undesired side reactions between Li metal and the electrolytes generate uncontrolled dendrite growth on the anode surface while the battery is operating. The sharp tip of the evolving dendrite can easily drill through the polymeric separator and connect the cathode, resulting in a short circuit that would ultimately lead to thermal runaway, fire, and explosion [131, 132]. Also, nonuniform Li deposition and stripping during the battery cycling aggravates breakage and fracture of the solid electrolyte interface and the dendrite, causing electrical disconnection "dead lithium." This fracture exposes more fresh Li to the electrolyte and fosters further degradation and consumption of both the electrolyte and Li metal, presumably leading to an early-stage loss of reversible capacity and cell failure [26]. In the case of Li-S batteries, the Li-anode undergoes a side reaction with shuttled polysulfides and results in the deposition of insoluble lithium sulfides (Li_2S_2/Li_2S) on its surface [133]. The insoluble sulfides, in turn, cannot be reutilized during subsequent cycling and cause active material loss and fast capacity decay [134, 135].

3.5.2 Strategies Enabling Li Metal Anode

For effective performance and stability of Li metal anode, the existing challenges must be addressed by designing various strategies. Regulating the interfacial stability between the Li metal anode and the electrolyte is of paramount importance to achieve efficient and long cycle sustaining Li metal-based batteries [136]. The main strategies contain employing lithophilic hosts [137–140], designing novel electrolytes [141, 142], constructing modified solid electrolyte interphase (SEI) [131, 143], and so on. Among them, constructing robust SEI film on the anode with characteristic properties of electronic insulation and permissive to Li^+ ion is the most effective way to offer the battery with desired outcome [144]. The SEI layer constructed either by an *in situ* strategy in the presence of electrolyte additives or by *ex situ* artificially engineering endows the Li metal with improved coulombic efficiency, cycle stability, and safety. The SEI layer can be applied.

3.5.2.1 SEI Layer Construction by Electrolyte Additives

As proposed for the first time by Peled in 1979 [145], the SEI layer is generally formed during the initial immediate reaction between metal and electrolyte. Also, the properly formulated electrolyte with characteristic film-forming additives can induce an effective protective coating on the Li metal via preferential decomposition on the surface during the electrochemical processes. Some electrolyte additives, in addition to SEI forming, can play a flame retarding role which further improves the safety of the battery [146–148].

Adding a few amounts of additives can exhibit a significant effect on the formation of uniform SEI film during the initial stage of the battery cycling [26], which in turn guarantees safe and stable cycling of the Li metal, and protects from further side reactions with electrolytes in subsequent cycles [149, 150]. Owing to the high reactivity of Li metal, almost no species of the organic electrolytes and additives remain inert toward Li metal. Ideally, robust SEI film-forming additives should have higher highest occupied molecular orbital (HOMO) and lower lowest unoccupied molecular orbital (LUMO) energies against Li metal [131]. The electrolyte additives, for example, nitro compounds (organic/inorganic nitrates) containing N—O bond were reported for the first time by Mikhaylik in Li-S battery to enhance the electrochemical performance and coulombic efficiency [151]. Among them, the $LiNO_3$ additive attracted more attention due to efficient surface film-forming capabilities on Li metal [152, 153]. The protective SEI film formed in the presence of $LiNO_3$ is composed of both inorganic and organic compounds including LiN_xO_y ROLi, and $ROCO_2Li$ [154]. Besides, combined additives (lithium polysulfides and $LiNO_3$) were also investigated in Li-S batteries (Figure 3.4) [155, 156].

With optimal lithium polysulfides and $LiNO_3$ in electrolyte, a uniform and robust SEI film consisting of Li_2SO_3 and Li_2SO_4 was formed on the Li metal. Consequently, the Li dendrite and the polysulfide issues were effectively suppressed. The dendritic problem and morphology change of Li metal in Li-S battery were also mitigated with the "Solvent-in-Salt" electrolyte [157]. The ultrahigh salt concentration was proved to inhibit the dissolution of polysulfides and the formation of Li dendrites, resulting in enhanced electrochemical performance. Moreover, exploration has been studied with other electrolyte additive salts beyond $LiNO_3$ for Li-S battery. The lithium bis(oxalato) borate (LiBOB) in electrolyte can help form smooth and dense Li metal as well as improve cycle stability and discharge capacity [158]. Currently, the LiBOB additive is attracting more interest due to the advantages of high thermal stability up to 300 °C, ability to form robust SEI film on Li metal, noncorrosive to the current collector, and cheap as well as environmental safety [159, 160]. Furthermore, P_2S_5 can also serve as an electrolyte additive in Li-S battery by passivating the Li metal surface and assisting the dissolution of Li_2S, which effectively suppresses the Li dendrite and the polysulfide issues, resulting in improved cycle stability and high coulombic efficiency [161].

3.5.2.2 SEI Layer Construction by Artificial Engineering

An alternative strategy to construct the SEI film is artificial engineering that coating an artificial SEI layer on Li metal anode before cycling, which can avoid the functional additive-driven complications of the *in situ* SEI layer.

The approach eases the understanding of the characteristic nature of the SEI composition provided that the Li metal is allowed to interact with preferred additives. More interestingly, the artificial SEI layer provides operational safety, suppression of

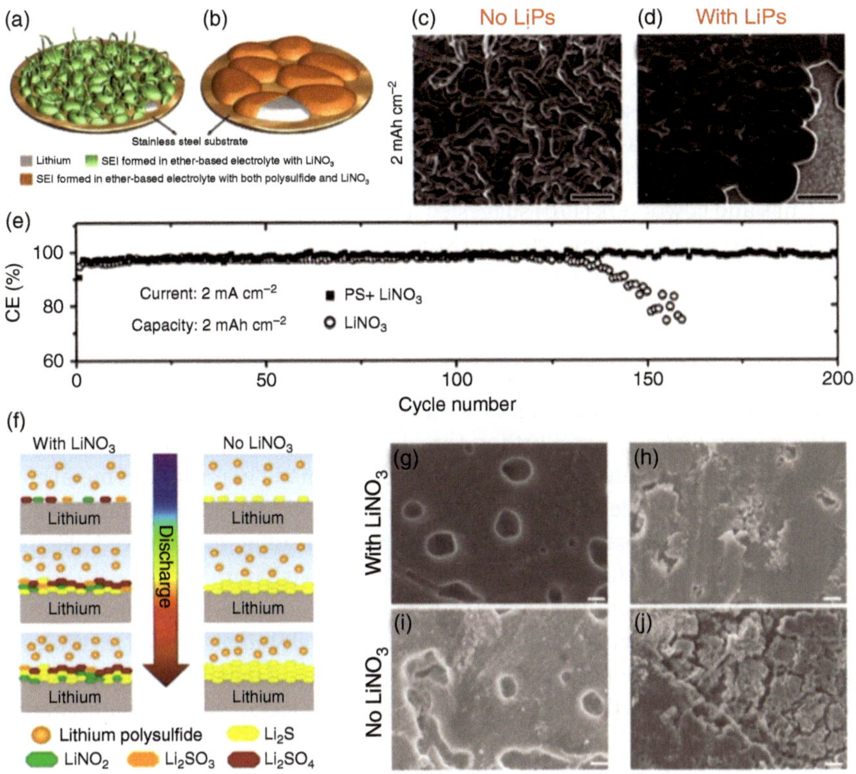

Figure 3.4 (a,b) Schematic representation of the Li-deposition morphology on stainless-steel substrate in the presence of only $LiNO_3$ and both $LiNO_3$ and polysulfide. SEM image of Li plated on bare stainless steel (c) with $LiNO_3$ alone and (d) with both $LiNO_3$ and lithium polysulfide at a current density of 2 mA cm^{-2}. (e) The corresponding electrolyte additives' effect on the cycling efficiency of the Li-plating/stripping. Source: Reproduced with permission of Li et al. [155]. © 2015/Springer Nature. (f) Scheme showing SEI formation on Li metal during discharge in the presence (left) and absence (right) of $LiNO_3$ electrolyte additive. SEM images of the Li metal morphology with $LiNO_3$ additive (g) after the first discharge and (h) after the first charge; (i) after the first discharge and (j) after the first charge for electrolyte without $LiNO_3$. Source: Reproduced with permission of Zhang et al. [156]. © 2017/Elsevier.

the corrosion reaction of Li metal, dendrite inhibition, efficient and homogeneous Li-plating/stripping, and robust Li metal anode [162–167].

Zhang and coworkers constructed an SEI layer on Li-anode using an electrochemical approach [168]. In a symmetric Li/Li cell with a functional film-forming fluoroethylene carbonate (FEC) additive, a protective robust SEI film was formed on the Li surface during the two cycling processes. The film composed of Li_2CO_3, LiF, polyene, and C—F bond-containing compounds was found to effectively protect the Li metal from corroding of organic solvent. Similarly, a Lewis acid AlI_3 was also used to construct SEI layer via *in situ* electrochemical reactions between the aprotic electrolytes and the Li metal in the symmetric cell. The aprotic dioxolane (DOL) underwent surface polymerization reaction on the Li metal surface initiated by the

I^- ions due to their strong surface affinity, forming coatings of oligomer thin film, LiI, and Li-Al alloy layer [169], which showed improved Li metal stability and dendrite suppression. However, difficulties in disassembly and subsequent assembly of the batteries drew researchers' attention away from using the electrochemical approach-based SEI film-construction technique [131]. The coating materials can also be applied directly to the surface of the Li metal to create SEI films. In this regard, Cui's and coworkers [170] suggested a rationally engineered SEI layer made of Cu_3N and styrene-butadiene rubber (SBR) that could be directly applied to the Li surface. The artificial SEI demonstrated high mechanical robustness, high ionic conductivity with uniform ion flux, and versatile enough to withstand interfacial fluctuation during cycling. Specifically, the Cu_3N nanoparticles can react with Li to form Li_3N, which was considered one of the fastest Li-ion conductors; while the high flexibility of SBR endowed the SEI film with good structural integrity throughout Li depositing/stripping processes. Furthermore, Jing et al. used a spin coating approach to plate a porous Al_2O_3 interlayer on the surface of Li metal to prevent Li metal corrosion and polysulfide side reactions [135]. The obtained uniform and crack-free Li metal can effectively suppress dendritic propagation, leading to improved electrochemical efficiency as well as uniform Li^+ plating and striping.

So far, the approaches outlined are only a few of the advancements made toward achieving chemically, electrochemically, and mechanically stable as well as efficient ionically conductive homogeneous SEI on the surface of Li metal anodes. However, more focus is required on this direction to realize the Li metal anode in high-energy Li-S batteries.

3.6 Aprotic Electrolytes for Li-S Batteries

Like the cathode and anode, electrolytes are also the prominent component of the electrochemical devices that play the media role for the ion conduction and effective wetting of the electrode surfaces for enhanced electrode/electrolyte interfacial processes. At room temperature, the high ionic conductivity of aprotic electrolytes can reach up to 10^{-3} to 10^{-2} S cm^{-1} [171].

Excellent electrolytes should satisfy the following certain criteria: (i) Wide potential window: an excellent electrolyte must have fast Li^+ transport and electrochemical robustness within the operating potential window. In this regard, the Li-S batteries operate in a relatively narrower potential range of 1.7–3.0 V vs. Li/Li^+, compared with Li-ion batteries. Thus, the electrolyte decomposition-induced safety issues and battery failure associated with high voltage are not a concern. Instead, a special focus is the stability of electrolytes to resist the highly reactive Li metal. (ii) Low flammability, low volatility, and low toxicity [171].

Electrolytes used in electrochemical cells, such as Li-S batteries, should have low viscosity and surface tension. As a result, these properties confer strong wettability on the electrolyte, resulting in more easy interaction between the electrolyte and the electrode surface, thus enabling low interfacial resistance [172]. The polysulfide and the Li metal anode in Li-S batteries have exceptionally high chemical reactivities, thus choosing an electrolyte that meets the desired requirements is challenging. The kinetics of sulfur and Li_2S should be considered during the selection of electrolytes as they affect the electronic and ionic transport pathways [172]. Given these problems, a distinct requirement in electrolyte design for Li-S batteries is still critical [173, 174].

3.6.1 Carbonate Electrolytes

Carbonate electrolytes, such as ethylene carbonate (EC), diethyl carbonate (DEC), propylene carbonate (PC), and dimethyl carbonate (DMC), are widely used in Li-ion batteries to dissolve lithium salts such as $LiPF_6$, $LiCF_3SO_3$, and $LiClO_4$. The resulting electrolytes have strong Li^+ ion diffusion, are excellent at passivating the anode, and have a large electrochemical potential window [171]. However, the chemical stability of carbonate electrolytes in Li-S batteries against the deleterious nucleophilic polysulfides is very weak, which would result in very poor electrochemical performance [175–179]. On discharging the Li-S battery, via disproportionation reaction of S_6^{2-} anion, the more stable sulfide radical ($S_3^{\cdot-}$) [180] is produced and sustained till the end of discharge. The carbonate electrolytes (for example, EC/DMC) could undergo nucleophilic reaction with polysulfide radical anions and make a further reduction to Li_2S_2/Li_2S impossible [181], leading to poor electrochemical efficiency, and even instant termination of battery. Thus, carbonate electrolytes are not suitable for Li-S batteries.

3.6.2 Ether Electrolytes

Unlike carbonate electrolytes, ether electrolytes are the prior choice for Li-S as they are stable at relatively lower operating voltage (1.7–3.0 V vs. Li/Li^+) [182, 183] and compatible with polysulfide anions. The high donor number enables ether electrolytes to coordinate with Lewis acid Li^+ ion [184]. Various electrolytes have been studied for Li-S batteries based on the ether solvents such as 1,2-dimethoxy ethane (DME) [185], 1,3-DOL [185, 186], tetrahydrofuran (THF) [41, 187], tetra ethylene glycol dimethyl ether (TEGDME) [188], poly(ethylene glycol) dimethyl ether (PEGDME) [182, 189], tri(ethylene glycol) dimethyl ether, diglyme (DGM) [185], and partly silanized ether [190]. Because of its high relative polarity, high dielectric constant, and low viscosity, DME has sparked much interest in Li-S batteries. It also has strong solubility for polysulfides. The DOL form an SEI layer on the Li metal surface via ring-opening reactions, which protects the anode surface from further reactions and maintains stable electrochemical results [191]. Among the ether solvents, the one with the highest oxidative potential stability, boiling, flashpoints, and resistance to combustion is preferred for use in Li-S batteries. Long-chain ether solvents are critical in this respect. Among various ether-based electrolytes with different structures, TEGDME was found to provide more exciting electrochemical results in Li-S batteries in a study performed by Barchasz et al. [182], which was attributed to the existence of strongly solvating oxygen atoms in the TEGDME structure.

Table 3.1 List of widely used solvents and corresponding properties for Li-S battery. (Where M_w, molecular weight; T_b, normal boiling point; T_m, normal melting point; μ, viscosity at 25 °C; ϵ, dielectric constant.)

Solvents		M_w	T_b (°C)	T_m (°C)	μ (cP)	ϵ
DME	1,2-Dimethoxyethane	90.12	85	−69	0.46	7.3
TEGDME	Tetra ethylene glycol dimethyl ether	222.28	275	−30	4.05	7.9
DOL	1,3-Dioxolane	74.08	78	−95	0.6	7.1
EEE	2-Ethoxyethyl ether	162.63	188	−44.3	1.4	—
PEGDME	Polyethylene glycol dimethyl ether	~250	>250	−23	7	—
Triglyme	Triethylene glycol dimethyl ether	178.23	216	−45	1.95	—
DEGDME	Diethylene glycol dimethyl ether	134.18	162	−68	0.99	7.3
THF	Tetrahydrofuran	72.11	66	−108.4	0.48	7.6

Source: Reproduced with permission from Refs. [182, 194–196].

3.6.3 Mixed Solvent Electrolytes

Referring to the aforementioned basic criteria of excellent electrolytes, a single solvent-based electrolyte cannot meet all the requirements. As a result, electrolytes with mixed solvent can be seen as an alternative to pursue electrolyte properties with such tradeoffs. In this case, the mixed solvent (binary or ternary mixtures)-based electrolytes gain properties from the constituent solvents and may contribute to certain optimized properties. For example, DOL was used as a co-solvent in conjunction with THF and toluene (TOL) to increase the Li^+-ion conduction [186]. In the approach, the electrolyte was prepared by dissolving 0.1 M Li_2S_6 and 2 M $LiClO_4$ in ternary mixture of THF/TOL/DOL (v/v/v = 1 : 1 : 8), in which TOL was used to reduce the dissolution of polysulfides [186, 192].

Also, a mixture of DME and DOL with other co-solvents was also claimed to be available in Li-S batteries [193]. However, there is an inherent disadvantage resulting from high vapor pressure and flammability of the constituent solvents (THF and DME). Despite this, DME is still widely used in various Li-S batteries today. Table 3.1 provides a list of ether solvents that are currently being used in various Li-S battery researches, either as a single or as a mixed (binary or ternary mixtures) solvent-based electrolyte in varying proportions. Because of its low viscosity, high Li^+ ion conductivity, high polysulfide dissolving capacity, and strong tendency toward forming protective SEI film, the binary mixture of DME/DOL (1 : 1 v/v) containing 1 M LiTFSI salt and $LiNO_3$ as an electrolyte additive has gained more frequent use as a standard in Li-S batteries [171, 194].

3.7 Separators and Functional Interlayers

The separator is an electrochemical cell component that isolates the direct electrical contact between the two electrodes (cathode and anode) while allowing ions to pass through the porous structures. Polymeric membranes, for example, microporous polypropylene and polyethylene, are used in most Li-S batteries. These separators offer advantages of low cost, versatility, high strength, and good chemical and electrochemical stability as well as regulated porosity. For Li-S batteries, the separators should be strong enough to prevent from being pierced by the possibly formed Li dendrites; meanwhile, they need effectively to hinder the percolation of polysulfides [4]. Thus, various modification strategic approaches have been designed to secure the advantages of the separators, ranging from coating one face of the traditional separators with conductive carbons [197], graphene [198], or other polysulfide adsorbents [102] to employing other types of polymeric [199–201] candidates as a potential modifier to allow the separators to play multifunctional roles in preventing polysulfide diffusion and dendrite penetration. Furthermore, the inevitable polysulfide shuttle problems in Li-S batteries are triggering enormous ingenious strategies from research communities all over the world to solve the problem.

Manthiram's research group [202] designed a new Li-S battery in 2012 by using microporous carbon paper as a bifunctional interlayer between the cathode and the separator. Since then, numerous studies have been carried out based on this novel configuration [67, 202–209]. The functional interlayer can be modified by altering the separator or by coating it on the cathode. It can also be sandwiched configured between the separator and the cathode/anode, as shown in Figure 3.5 [210]. Excellent interlayer should meet the following features such as (i) hierarchical porosity with high Li^+ ion and electronic conductivity, (ii) capability of trapping polysulfides by physical or chemical interactions, and (iii) chemical and mechanical stability that allows tolerating volume change of the cathode during battery operation. The interlayers with these ideal features can provide efficient sulfur utilization by minimizing shuttling problems and preventing corrosion of Li metal anode, which results in improved electrochemical performance. Furthermore, the high electrical conductivity of interlayer can make it serve as an "upper current collector," lowering the internal charge transfer resistance [210]. Promising results have been obtained using carbon-based materials with high porosity and conductivity [202–205] as part of the progress made in Li-S batteries using interlayer configurations. This can be attributed to the effective physical interaction between carbon and polysulfides. To further enhance the adsorption ability, doping carbonaceous materials with heteroatoms (nitrogen, oxygen, sulfur, boron, etc.) is an attractive

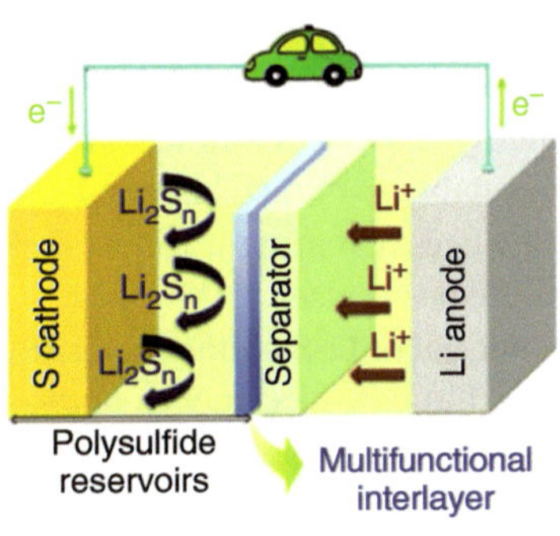

Figure 3.5 Schematic demonstration of multifunctional interlayer incorporated Li-S battery. Source: Reproduced with permission Fan et al. [210]. Copyright 2018, Royal Society of Chemistry.

strategy. With heteroatom-doped carbonaceous interlayers, the Li-S battery can exhibit improved polysulfide binding, shuttle inhibition, sulfur utilization, and consequently enhanced long cycle stability and high capacity [67, 206–208]. In addition, interlayers can be made from hybrid materials combining carbonaceous materials with nanostructured polar inorganic compounds (TMOs, sulfides, and carbides) [71, 94, 211–214]. Polymer-based conductive materials with carefully designed porosity and conductivity have the desired benefits in addressing polysulfide problems. They provide good mechanical strength for maintaining cathode stability and integrity during continuous charge/discharge cycles. The porous structure allows it to trap the polysulfide while selectively allowing Li^+ ions to pass through [215, 216].

3.8 Conclusion and Perspective

Li-S battery, with tempting properties such as low cost, nontoxicity, abundance, and, above all, high discharge capacity (1675 mAh g^{-1}) and energy density (2600 Wh kg^{-1}), is considered one of the ultimate power sources. However, its practicality is severely challenged by the cathodic dissolution, poor electron/ion conductivity of sulfur/Li_2S, and the highly reactive Li metal anode, collectively inducing poor battery performance as well as a safety concern. To address these issues, tremendous efforts have been made. Numerous innovative techniques ranging from material designs and implementations to adjustment of the cathode, anode, electrolyte, and separator have been successfully applied to improve the Li-S battery performance. The sulfur cathode hosted in/with various material groups with distinct physicochemical properties (electronic conductivity, porosity, surface area, and surface polarity) such as carbonaceous architectures, inorganic metal sulfides, oxides, nitrides, and polymer-based designs have been investigated. These material designs have greatly improved the battery's efficiency by suppressing the polysulfide diffusion, catalyzing the polysulfide conversion kinetics, allowing rapid charge transfer, and encouraging sulfur utilization. In addition, new configurations that introducing multifunctional interlayers between the separator and electrodes (either cathode or anode) are also developed. The electrolyte design and formulation indeed play an important role in the electrochemical performance and safety of the battery. The aprotic solvent-based liquid electrolytes can bring high ionic conductivity at room temperature, which are suitable for the fast charging/discharging processes at high current density. However, the stability of the electrolytes against highly reactive Li-anode and polysulfide anions is cumbersome. On the aspect of enabling robust Li metal anode, different approaches have shown promising advances. Creating a robust SEI layer on the Li metal via either *ex situ* or *in situ* strategies is useful. All in all, to realize the practical application and extract the potential of Li-S battery, revisiting the holistic approaches on the sulfur cathode, Li metal anode, electrolytes and electrolyte additives, separators, and other strategic designs with in-depth consideration of the intrinsic properties and understanding of polysulfides shuttle, Li metal dendrite formation is paramount important.

References

1 Yang, Z., Zhang, J., Kintner-Meyer, M.C.W. et al. (2011). Electrochemical energy storage for green grid. *Chemical Reviews* 111: 3577–3613.

2 Manthiram, A., Fu, Y., Chung, S.-H. et al. (2014). Rechargeable lithium–sulfur batteries. *Chemical Reviews* 114: 11751–11787.

3 Kumar, R., Liu, J., Hwang, J.-Y., and Sun, Y.-K. (2018). Recent research trends in Li–S batteries. *Journal of Materials Chemistry A* 6: 11582–11605.

4 Deng, N., Kang, W., Liu, Y. et al. (2016). A review on separators for lithium-sulfur battery: progress and prospects. *Journal of Power Sources* 331: 132–155.

5 Armand, M. and Tarascon, J.M. (2008). Building better batteries. *Nature* 451: 652–657.

6 Choi, J.W. and Aurbach, D. (2016). Promise and reality of post-lithium-ion batteries with high energy densities. *Nature Reviews Materials* 1 (4): 16013.

7 Guo, Y.-G., Hu, J.-S., and Wan, L.-J. (2008). Nanostructured materials for electrochemical energy conversion and storage devices. *Advanced Materials* 20: 2878–2887.

8 Yin, Y.-X., Xin, S., Guo, Y.-G., and Wan, L.-J. (2013). Lithium–sulfur batteries: electrochemistry, materials, and prospects. *Angewandte Chemie International Edition* 52: 13186–13200.

9 Yin, Y.-X., Yao, H.-R., and Guo, Y.-G. (2016). Scientific and technological challenges toward application of lithium–sulfur batteries. *Chinese Physics B* 25: 018801–018807.

10 Goodenough, J.B. and Park, K.-S. (2013). The li-ion rechargeable battery: a perspective. *Journal of the American Chemical Society* 135: 1167–1176.

11 Cao, A., Hu, J., and Wan, L. (2012). Morphology control and shape evolution in 3D hierarchical superstructures. *Science China Chemistry* 55: 2249–2256.

12 Su, D., Zhou, D., Wang, C., and Wang, G. (2018). Toward high performance lithium–sulfur batteries based on Li_2S cathodes and beyond: status, challenges, and perspectives. *Advanced Functional Materials* 28: 1800154–1800177.

13 Chung, S.-H., Chang, C.-H., and Manthiram, A. (2018). Progress on the critical parameters for lithium-sulfur batteries to be practically viable. *Advanced Functional Materials* 28: 1801188–1801208.

14 Aurbach, D. (2018). Introduction to the focus issue on lithium-sulfur batteries: materials, mechanisms, modeling, and applications. *Journal of the Electrochemical Society* 165: Y1.

15 Peng, H.-J., Huang, J.-Q., Cheng, X.-B., and Zhang, Q. (2017). Review on high-loading and high-energy lithium–sulfur batteries. *Advanced Energy Materials* 7: 1700260–1700314.

16 Li, F., Liu, Q., Hu, J. et al. (2019). Recent advances in cathode materials for rechargeable lithium-sulfur batteries. *Nanoscale* 11: 15418–15439.

17 Xu, R., Lu, J., and Amine, K. (2015). Progress in mechanistic understanding and characterization techniques of Li-S batteries. *Advanced Energy Materials* 5: 1500408–1500430.

18 Mikhaylik, Y.V. and Akridge, J.R. (2004). Polysulfide shuttle study in the Li/S battery system. *Journal of the Electrochemical Society* 151: A1969–A1976.

19 Ding, F., Xu, W., Graff, G.L. et al. (2013). Dendrite-free lithium deposition via self-healing electrostatic shield mechanism. *Journal of the American Chemical Society* 135: 4450–4456.

20 Jeong, G., Kim, Y.-U., Kim, H. et al. (2011). Prospective materials and applications for Li secondary batteries. *Energy & Environmental Science* 4: 1986–2002.

21 Takehara, Z.-I. (1997). Future prospects of the lithium metal anode. *Journal of Power Sources* 68: 82–86.

22 Aurbach, D., Zinigrad, E., Cohen, Y., and Teller, H. (2002). A short review of failure mechanisms of lithium metal and lithiated graphite anodes iin liquid electrolyte solutions. *Solid State Ionics* 148: 405–416.

23 Limthongkul, P., Jang, Y.-I., Dudney, N.J., and Chiang, Y.-M. (2003). Electrochemically-driven solid-state amorphization in lithium–metal anodes. *Journal of Power Sources* 119–121: 604–609.

24 Rosso, M., Brissot, C., Teyssot, A. et al. (2006). Dendrite short-circuit and fuse effect on Li/polymer/Li cells. *Electrochimica Acta* 51: 5334–5340.

25 Ryou, M.-H., Lee, D.J., Lee, J.-N. et al. (2012). Excellent cycle life of lithium-metal anodes in lithium-ion batteries with mussel-inspired polydopamine-coated separators. *Advanced Energy Materials* 2: 645–650.

26 Zhao, H., Deng, N., Yan, J. et al. (2018). A review on anode for lithium-sulfur batteries: progress and prospects. *Chemical Engineering Journal* 347: 343–365.

27 Pei, F., An, T., Zang, J. et al. (2016). From hollow carbon spheres to N-doped hollow porous carbon bowls: rational design of hollow carbon host for Li-S batteries. *Advanced Energy Materials* 6: 1502539–1502547.

28 Bruce, P.G., Freunberger, S.A., Hardwick, L.J., and Tarascon, J.-M. (2012). Li–O_2 and Li–S batteries with high energy storage. *Nature Materials* 11: 19–29.

29 Gröbmeyer, A., Becking, J., Bieker, P.M. et al. (2019). A facile preparation of S_8/C composite cathode for lithium-sulfur cells: influence of intrinsic and extrinsic cathode properties on the electrochemical performance. *Energy Technology* 7: 1800789–1800797.

30 Eftekhari, A. and Kim, D.-W. (2017). Cathode materials for lithium–sulfur batteries: a practical perspective. *Journal of Materials Chemistry A* 5: 17734–17776.

31 Demir-Cakan, R., Morcrette, M., Nouar, F. et al. (2011). Cathode composites for Li–S batteries via the use of oxygenated porous architectures. *Journal of the American Chemical Society* 133: 16154–16160.

32 Zhao, Y., Zhu, W., Chen, G.Z., and Cairns, E.J. (2016). Polypyrrole/TiO_2 nanotube arrays with coaxial heterogeneous structure as sulfur hosts for lithium sulfur batteries. *Journal of Power Sources* 327: 447–456.

33 Pang, Q., Kundu, D., Cuisinier, M., and Nazar, L.F. (2014). Surface-enhanced redox chemistry of polysulphides on a metallic and polar host for lithium-sulphur batteries. *Nature Communications* 5: 4759–4767.

34 Cui, Z., Zu, C., Zhou, W. et al. (2016). Mesoporous titanium nitride-enabled highly stable lithium-sulfur batteries. *Advanced Materials* 28: 6926–6932.

35 Tao, C., Lianbo, M., Baorui, C. et al. (2017). Metallic and polar Co_9S_8 inlaid carbon hollow nanopolyhedra as efficient polysulfide mediator for lithium–sulfur batteries. *Nano Energy* 38: 239–248.

36 Zheng, J., Tian, J., Wu, D. et al. (2014). Lewis acid–base interactions between polysulfides and metal organic framework in lithium sulfur batteries. *Nano Letters* 14: 2345–2352.

37 Kaiser, M.R., Han, Z., Liang, J. et al. (2019). Lithium sulfide-based cathode for lithium-ion/sulfur battery: recent progress and challenges. *Energy Storage Materials* 19: 1–15.

38 Yang, Y., Zheng, G., and Cui, Y. (2013). Nanostructured sulfur cathodes. *Chemical Society Reviews* 42 (7): 3018–3032.

39 Liang, J., Sun, Z.-H., Li, F., and Cheng, H.-M. (2016). Carbon materials for Li–S batteries: functional evolution and performance improvement. *Energy Storage Materials* 2: 76–106.

40 He, G., Mandlmeier, B., Schuster, J. et al. (2014). Bimodal mesoporous carbon nanofibers with high porosity: freestanding and embedded in membranes for lithium–sulfur batteries. *Chemistry of Materials* 26: 3879–3886.

41 Rauh, R.D., Abraham, K.M., Pearson, G.F. et al. (1979). A lithium/dissolved sulfur battery with an organic electrolyte. *Journal of the Electrochemical Society* 126: 523–527.

42 Kang, K., Meng, Y.S., Bréger, J. et al. (2006). Electrodes with high power and high capacity for rechargeable lithium batteries. *Science* 311 (5763): 977–980.

43 Chu, M.-Y. (1997). Rechargeable Positive Electrodes. US Patent 5686201A.

44 Peramunage, D. and Licht, S. (1993). A solid sulfur cathode for aqueous batteries. *Science* 261 (5124): 1029–1032.

45 Evers, S. and Nazar, L.F. (2013). New approaches for high energy density lithium–sulfur battery cathodes. *Accounts of Chemical Research* 46 (5): 1135–1143.

46 Ji, X., Lee, K.T., and Nazar, L.F. (2009). A highly ordered nanostructured carbon-sulphur cathode for lithium-sulphur batteries. *Nature Materials* 8: 500–506.

47 Liang, C., Dudney, N.J., and Howe, J.Y. (2009). Hierarchically structured sulfur/carbon nanocomposite material for high-energy lithium battery. *Chemistry of Materials* 21: 4724–4730.

48 Jayaprakash, N., Shen, J., Moganty, S.S. et al. (2011). Porous hollow carbon@sulfur composites for high-power lithium-sulfur batteries. *Angewandte Chemie International Edition* 50: 5904–5908.

49 Guo, J., Xu, Y., and Wang, C. (2011). Sulfur-impregnated disordered carbon nanotubes cathode for lithium-sulfur batteries. *Nano Letters* 11: 4288–4294.

50 Zheng, G., Yang, Y., Cha, J.J. et al. (2011). Hollow carbon nanofiber-encapsulated sulfur cathodes for high specific capacity rechargeable lithium batteries. *Nano Letters* 11: 4462–4467.

51 Huang, J.-Q., Liu, X.-F., Zhang, Q. et al. (2013). Entrapment of sulfur in hierarchical porous graphene for lithium–sulfur batteries with high rate performance from −40 to 60°C. *Nano Energy* 2 (2): 314–321.

52 Yan, L., Xiao, M., Wang, S. et al. (2017). Edge sulfurized graphene nanoplatelets via vacuum mechano-chemical reaction for lithium–sulfur batteries. *Journal of Energy Chemistry* 26: 522–529.

53 Yuan, Z., Peng, H.-J., Hou, T.-Z. et al. (2016). Powering lithium-sulfur battery performance by propelling polysulfide redox at sulfiphilic hosts. *Nano Letters* 16: 519–527.

54 Shao, Q., Wu, Z.-S., and Chen, J. (2019). Two-dimensional materials for advanced Li-S batteries. *Energy Storage Materials* 22: 284–310.

55 Xu, G., Yuan, J., Tao, X. et al. (2015). Absorption mechanism of carbon-nanotube paper-titanium dioxide as a multifunctional barrier material for lithium-sulfur batteries. *Nano Research* 8: 3066–3074.

56 Gu, X., Tong, C.-J., Lai, C. et al. (2015). A porous nitrogen and phosphorous dual doped graphene blocking layer for high performance Li–S batteries. *Journal of Materials Chemistry A* 3 (32): 16670–16678.

57 Gu, X., Zhang, S., and Hou, Y. (2016). Graphene-based sulfur composites for energy storage and conversion in Li-S batteries. *Chinese Journal of Chemistry* 34 (1): 13–31.

58 Kim, J.W., Ocon, J.D., Kim, H.-S., and Lee, J. (2015). Improvement of energy capacity with vitamin C treated dual-layered graphene–sulfur cathodes in lithium–sulfur batteries. *ChemSusChem* 8: 2883–2891.

59 Eftekhari, A. and Yazdani, B. (2010). Initiating electropolymerization on graphene sheets in graphite oxide structure. *Journal of Polymer Science Part A: Polymer Chemistry* 48: 2204–2213.

60 Banks, C.E., Moore, R.R., Davies, T.J., and Compton, R.G. (2004). Investigation of modified basal plane pyrolytic graphite electrodes: definitive evidence for the electrocatalytic properties of the ends of carbon nanotubes. *Chemical Communications* 1804–1805.

61 Davies, T.J., Hyde, M.E., and Compton, R.G. (2005). Nanotrench arrays reveal insight into graphite electrochemistry. *Angewandte Chemie International Edition* 44: 5121–5126.

62 Banks, C.E., Crossley, A., Salter, C. et al. (2006). Carbon nanotubes contain metal impurities which are responsible for the "electrocatalysis" seen at some nanotube-modified electrodes. *Angewandte Chemie International Edition* 45: 2533–2537.

63 Ji, L., Rao, M., Zheng, H. et al. (2011). Graphene oxide as a sulfur immobilizer in high performance lithium/sulfur cells. *Journal of the American Chemical Society* 133 (46): 18522–18525.

64 Huang, X., Sun, B., Li, K. et al. (2013). Mesoporous graphene paper immobilised sulfur as a flexible electrode for lithium–sulfur batteries. *Journal of Materials Chemistry A* 1 (43): 13484–13489.

65 Wang, C., Wang, X., Wang, Y. et al. (2015). Macroporous free-standing nano-sulfur/reduced graphene oxide paper as stable cathode for lithium-sulfur battery. *Nano Energy* 11: 678–686.

66 Han, K., Shen, J., Hao, S. et al. (2014). Free-standing nitrogen-doped graphene paper as electrodes for high-performance lithium/dissolved polysulfide batteries. *ChemSusChem* 7 (9): 2545–2553.

67 Qiu, Y., Li, W., Zhao, W. et al. (2014). High-rate, ultralong cycle-life lithium/sulfur batteries enabled by nitrogen-doped graphene. *Nano Letters* 14: 4821–4827.

68 Deng, W., Hu, A., Chen, X. et al. (2016). Sulfur-impregnated 3D hierarchical porous nitrogen-doped aligned carbon nanotubes as high-performance cathode for lithium-sulfur batteries. *Journal of Power Sources* 322: 138–146.

69 Wang, C., Zhang, F., Wang, X. et al. (2016). Preparation of a graphitic N-doped multi-walled carbon nanotube composite for lithium–sulfur batteries with long-life and high specific capacity. *RSC Advances* 6: 76568–76574.

70 Wang, S., Liao, J., Yang, X. et al. (2019). Designing a highly efficient polysulfide conversion catalyst with paramontroseite for high-performance and long-life lithium-sulfur batteries. *Nano Energy* 57: 230–240.

71 Liu, X., Huang, J.-Q., Zhang, Q., and Mai, L. (2017). Nanostructured metal oxides and sulfides for lithium–sulfur batteries. *Advanced Materials* 29: 1601759–1601784.

72 Lu, X., Wang, G., Zhai, T. et al. (2012). Stabilized TiN nanowire arrays for high-performance and flexible supercapacitors. *Nano Letters* 12 (10): 5376–5381.

73 Pang, Q., Kwok, C.Y., Kundu, D. et al. (2019). Lightweight metallic MgB_2 mediates polysulfide redox and promises high-energy-density lithium-sulfur batteries. *Joule* 3: 136–148.

74 Peng, H.-J., Zhang, G., Chen, X. et al. (2016). Enhanced electrochemical kinetics on conductive polar mediators for lithium-sulfur batteries. *Angewandte Chemie International Edition* 55: 12990–12995.

75 Al Salem, H., Chitturi, V.R., Babu, G. et al. (2016). Stabilizing polysulfide-shuttle in a Li–S battery using transition metal carbide nanostructures. *RSC Advances* 6 (111): 110301–110306.

76 Zhong, Y., Yin, L., He, P. et al. (2018). Surface chemistry in cobalt phosphide-stabilized lithium–sulfur batteries. *Journal of the American Chemical Society* 140: 1455–1459.

77 Huang, S., Lim, Y.V., Zhang, X. et al. (2018). Regulating the polysulfide redox conversion by iron phosphide nanocrystals for high-rate and ultrastable lithium-sulfur battery. *Nano Energy* 51: 340–348.

78 Luo, Y., Luo, N., Kong, W. et al. (2018). Multifunctional interlayer based on molybdenum diphosphide catalyst and carbon nanotube film for lithium–sulfur batteries. *Small* 14: 1702853–1702862.

79 Jin, H., Guo, C., Liu, X. et al. (2018). Emerging two-dimensional nanomaterials for electrocatalysis. *Chemical Reviews* 118: 6337–6408.

80 He, J. and Manthiram, A. (2019). A review on the status and challenges of electrocatalysts in lithium-sulfur batteries. *Energy Storage Materials* 20: 55–70.

81 Pu, J., Shen, Z., Zheng, J. et al. (2017). Multifunctional Co_3S_4@sulfur nanotubes for enhanced lithium-sulfur battery performance. *Nano Energy* 37: 7–14.

82 Lin, H., Yang, L., Jiang, X. et al. (2017). Electrocatalysis of polysulfide conversion by sulfur-deficient MoS_2 nanoflakes for lithium–sulfur batteries. *Energy & Environmental Science* 10: 1476–1486.

83 Ye, C., Zhang, L., Guo, C. et al. (2017). A 3D hybrid of chemically coupled nickel sulfide and hollow carbon spheres for high performance lithium–sulfur batteries. *Advanced Functional Materials* 27: 1702524–1702533.

84 Babu, G., Masurkar, N., Al Salem, H., and Arava, L.M.R. (2017). Transition metal dichalcogenide atomic layers for lithium polysulfides electrocatalysis. *Journal of the American Chemical Society* 139: 171–178.

85 Shin, W., Lu, J., and Ji, X. (2019). ZnS coating of cathode facilitates lean-electrolyte Li-S batteries. *Carbon Energy* 1: 165–172.

86 Gao, X., Yang, X., Li, M. et al. (2019). Cobalt-doped SnS_2 with dual active centers of synergistic absorption-catalysis effect for high-S loading Li-S batteries. *Advanced Functional Materials* 29: 1806724–1806732.

87 Lu, X., Zhang, Q., Wang, J. et al. (2019). High performance bimetal sulfides for lithium-sulfur batteries. *Chemical Engineering Journal* 358: 955–961.

88 Guo, B., Bandaru, S., Dai, C. et al. (2018). Self-supported $FeCo_2S_4$ nanotube arrays as binder-free cathodes for lithium–sulfur batteries. *ACS Applied Materials & Interfaces* 10: 43707–43715.

89 Dong, C., Gao, W., Jin, B., and Jiang, Q. (2018). Advances in cathode materials for high-performance lithium-sulfur batteries. *iScience* 6: 151–198.

90 Balach, J., Linnemann, J., Jaumann, T., and Giebeler, L. (2018). Metal-based nanostructured materials for advanced lithium–sulfur batteries. *Journal of Materials Chemistry A* 6: 23127–23168.

91 Song, M.-S., Han, S.-C., Kim, H.-S. et al. (2004). Effects of nanosized adsorbing material on electrochemical properties of sulfur cathodes for Li/S secondary batteries. *Journal of the Electrochemical Society* 151: A791–A795.

92 Ji, X., Evers, S., Black, R., and Nazar, L.F. (2011). Stabilizing lithium-sulphur cathodes using polysulphide reservoirs. *Nature Communications* 2: 325–332.

93 Yu, M., Ma, J., Song, H. et al. (2016). Atomic layer deposited TiO_2 on a nitrogen-doped graphene/sulfur electrode for high performance lithium–sulfur batteries. *Energy & Environmental Science* 9 (4): 1495–1503.

94 Liang, X., Hart, C., Pang, Q. et al. (2015). A highly efficient polysulfide mediator for lithium-sulfur batteries. *Nature Communications* 6: 5682–5690.

95 Zheng, C., Niu, S., Lv, W. et al. (2017). Propelling polysulfides transformation for high-rate and long-life lithium–sulfur batteries. *Nano Energy* 33: 306–312.

96 Song, Y., Zhao, W., Zhu, X. et al. (2018). Vanadium dioxide-graphene composite with ultrafast anchoring behavior of polysulfides for lithium–sulfur batteries. *ACS Applied Materials & Interfaces* 10: 15733–15741.

97 Li, W., Hicks-Garner, J., Wang, J. et al. (2014). V_2O_5 polysulfide anion barrier for long-lived Li–S batteries. *Chemistry of Materials* 26: 3403–3410.

98 Imtiaz, S., Ali Zafar, Z., Razaq, R. et al. (2018). Electrocatalysis on separator modified by molybdenum trioxide nanobelts for lithium–sulfur batteries. *Advanced Materials Interfaces* 5: 1800243–1800252.

99 Lin, H., Zhang, S., Zhang, T. et al. (2018). Elucidating the catalytic activity of oxygen deficiency in the polysulfide conversion reactions of lithium-sulfur batteries. *Advanced Energy Materials* 8: 1801868–1801879.

100 Ma, L., Chen, R., Zhu, G. et al. (2017). Cerium oxide nanocrystal embedded bimodal micromesoporous nitrogen-rich carbon nanospheres as effective sulfur host for lithium–sulfur batteries. *ACS Nano* 11: 7274–7283.

101 Ren, W., Ma, W., Umair, M.M. et al. (2018). CoO/Co-activated porous carbon cloth cathode for high performance Li–S batteries. *ChemSusChem* 11: 2695–2702.

102 Zhang, Z., Lai, Y., Zhang, Z. et al. (2014). Al_2O_3-coated porous separator for enhanced electrochemical performance of lithium sulfur batteries. *Electrochimica Acta* 129: 55–61.

103 Xiang, M., Wu, H., Liu, H. et al. (2017). A flexible 3D multifunctional MgO-decorated carbon foam@CNTs hybrid as self-supported cathode for high-performance lithium-sulfur batteries. *Advanced Functional Materials* 27: 1702573–1702586.

104 Zhou, Y., Zhou, C., Li, Q. et al. (2015). Enabling prominent high-rate and cycle performances in one lithium–sulfur battery: designing permselective gateways for Li+ transportation in Holey-CNT/S cathodes. *Advanced Materials* 27: 3774–3781.

105 Li, X., Zhang, L., Ding, Z., and He, Y. (2017). Ultrafine Nd_2O_3 nanoparticles doped carbon aerogel to immobilize sulfur for high performance lithium–sulfur batteries. *Journal of Electroanalytical Chemistry* 799: 617–624.

106 Zhong, Y., Chao, D., Deng, S. et al. (2018). Confining sulfur in integrated composite scaffold with highly porous carbon fibers/vanadium nitride arrays for high-performance lithium-sulfur batteries. *Advanced Functional Materials* 28: 1706391–1706400.

107 Suo, L., Zhu, Y., Han, F. et al. (2015). Carbon cage encapsulating nano-cluster Li_2S by ionic liquid polymerization and pyrolysis for high performance Li–S batteries. *Nano Energy* 13: 467–473.

108 Sun, Q. and Fu, Z.-W. (2008). Vanadium nitride as a novel thin film anode material for rechargeable lithium batteries. *Electrochimica Acta* 54 (2): 403–409.

109 Nandi, D.K., Sen, U.K., Choudhury, D. et al. (2014). Atomic layer deposited molybdenum nitride thin film: a promising anode material for Li ion batteries. *ACS Applied Materials & Interfaces* 6 (9): 6606–6615.

110 Park, H.-C., Lee, K.-H., Lee, Y.-W. et al. (2014). Mesoporous molybdenum nitride nanobelts as an anode with improved electrochemical properties in lithium ion batteries. *Journal of Power Sources* 269: 534–541.

111 Gillot, F., Oró-Solé, J., and Palacín, M.R. (2011). Nickel nitride as negative electrode material for lithium ion batteries. *Journal of Materials Chemistry* 21 (27): 9997–10002.

112 Choi, D. and Kumta, P.N. (2011). Synthesis and characterization of nanostructured niobium and molybdenum nitrides by a two-step transition metal halide approach. *Journal of the American Ceramic Society* 94 (8): 2371–2378.

113 Zheng, H., Fang, S., Tong, Z. et al. (2015). Stabilized titanium nitride nanowire supported silicon core–shell nanorods as high capacity lithium-ion anodes. *Journal of Materials Chemistry A* 3 (23): 12476–12481.

114 Das, B., Reddy, M.V., Rao, G.V.S., and Chowdari, B.V.R. (2012). Synthesis of porous-CoN nanoparticles and their application as a high capacity anode for lithium-ion batteries. *Journal of Materials Chemistry* 22 (34): 17505–17510.

115 Deng, C., Wang, Z., Wang, S., and Yu, J. (2019). Inhibition of polysulfide diffusion in lithium–sulfur batteries: mechanism and improvement strategies. *Journal of Materials Chemistry A* 7: 12381–12413.

116 Mosavati, N., Salley, S.O., and Ng, K.Y.S. (2017). Characterization and electrochemical activities of nanostructured transition metal nitrides as cathode materials for lithium sulfur batteries. *Journal of Power Sources* 340: 210–216.

117 Zhang, S., Liu, M., Ma, F. et al. (2015). A high energy density Li_2S@C nanocomposite cathode with a nitrogen-doped carbon nanotube top current collector. *Journal of Materials Chemistry A* 3 (37): 18913–18919.

118 Chung, S.-H. and Manthiram, A. (2019). A Li_2S-TiS_2-electrolyte composite for stable Li_2S-based lithium–sulfur batteries. *Advanced Energy Materials* 9: 1901397–1901406.

119 Wu, F., Lee, J.T., Magasinski, A. et al. (2014). Solution-based processing of graphene–Li_2S composite cathodes for lithium-ion and lithium–sulfur batteries. *Particle & Particle Systems Characterization* 31 (6): 639–644.

120 Guo, J., Yang, Z., Yu, Y. et al. (2013). Lithium–sulfur battery cathode enabled by lithium–nitrile interaction. *Journal of the American Chemical Society* 135 (2): 763–767.

121 Nan, C., Lin, Z., Liao, H. et al. (2014). Durable carbon-coated Li_2S core–shell spheres for high performance lithium/sulfur cells. *Journal of the American Chemical Society* 136 (12): 4659–4663.

122 Ye, F., Hou, Y., Liu, M. et al. (2015). Fabrication of mesoporous Li_2S–C nanofibers for high performance Li/Li_2S cell cathodes. *Nanoscale* 7 (21): 9472–9476.

123 Cai, K., Song, M.-K., Cairns, E.J., and Zhang, Y. (2012). Nanostructured Li_2S–C composites as cathode material for high-energy lithium/sulfur batteries. *Nano Letters* 12 (12): 6474–6479.

124 Zhang, K., Wang, L., Hu, Z. et al. (2014). Ultrasmall Li_2S nanoparticles anchored in graphene nanosheets for high-energy lithium-ion batteries. *Scientific Reports* 4: 6467–6474.

125 Li, S., Leng, D., Li, W. et al. (2020). Recent progress in developing Li_2S cathodes for Li–S batteries. *Energy Storage Materials* 27: 279–296.

126 Jeong, S., Bresser, D., Buchholz, D. et al. (2013). Carbon coated lithium sulfide particles for lithium battery cathodes. *Journal of Power Sources* 235: 220–225.

127 Seh, Z.W., Wang, H., Hsu, P.-C. et al. (2014). Facile synthesis of Li_2S–polypyrrole composite structures for high-performance Li_2S cathodes. *Energy & Environmental Science* 7: 672–676.

128 Whittingham, M.S. (2004). Lithium batteries and cathode materials. *Chemical Reviews* 104: 4271–4302.

129 Wang, T., Li, Y., Zhang, J. et al. (2020). Immunizing lithium metal anodes against dendrite growth using protein molecules to achieve high energy batteries. *Nature Communications* 11 (1): 5429–5438.

130 Chen, Y., Luo, Y., Zhang, H. et al. (2019). The challenge of lithium metal anodes for practical applications. *Small Methods* 1800551–1800574.

131 Cheng, X.-B., Zhang, Q., Zhang, R. et al. (2017). Toward safe lithium metal anode in rechargeable batteries: a review. *Chemical Reviews* 117: 10403–10473.

132 Xu, W., Wang, J., Ding, F. et al. (2014). Lithium metal anodes for rechargeable batteries. *Energy & Environmental Science* 7: 513–537.

133 Huang, C., Xiao, J., Shao, Y. et al. (2014). Manipulating surface reactions in lithium–sulphur batteries using hybrid anode structures. *Nature Communications* 5 (1): 3015–3023.

134 Ma, G., Wen, Z., Wu, M. et al. (2014). A lithium anode protection guided highly-stable lithium–sulfur battery. *Chemical Communications* 50 (91): 14209–14212.

135 Jing, H.-K., Kong, L.-L., Liu, S. et al. (2015). Protected lithium anode with porous Al_2O_3 layer for lithium–sulfur battery. *Journal of Materials Chemistry A* 3: 12213–12219.

136 Xu, K. (2010). Electrolytes and interphasial chemistry in Li ion devices. *Energies* 3 (1): 135–154.

137 Liu, B., Zhang, Y., Pan, G. et al. (2019). Ordered lithiophilic sites to regulate Li plating/stripping behavior for superior lithium metal anodes. *Journal of Materials Chemistry A* 7 (38): 21794–21801.

138 Yue, X.-Y., Li, X.-L., Wang, W.-W. et al. (2019). Wettable carbon felt framework for high loading Li-metal composite anode. *Nano Energy* 60: 257–266.

139 Yue, X.-Y., Wang, W.-W., Wang, Q.-C. et al. (2019). Cuprite-coated Cu foam skeleton host enabling lateral growth of lithium dendrites for advanced Li metal batteries. *Energy Storage Materials* 21: 180–189.

140 Zhang, D., Dai, A., Wu, M. et al. (2020). Lithiophilic 3D porous CuZn current collector for stable lithium metal batteries. *ACS Energy Letters* 5 (1): 180–186.

141 Zhou, D., Liu, R., He, Y.-B. et al. (2016). SiO_2 hollow nanosphere-based composite solid electrolyte for lithium metal batteries to suppress lithium dendrite growth and enhance cycle life. *Advanced Energy Materials* 6 (7): 1502214–1502224.

142 Lu, Q., He, Y.-B., Yu, Q. et al. (2017). Dendrite-free, high-rate, long-life lithium metal batteries with a 3D cross-linked network polymer electrolyte. *Advanced Materials* 29 (13): 1604460–1604468.

143 Song, R., Ge, Y., Wang, B. et al. (2019). A new reflowing strategy based on lithiophilic substrates towards smooth and stable lithium metal anodes. *Journal of Materials Chemistry A* 7 (30): 18126–18134.

144 Jiang, Y., Wang, B., Liu, P. et al. (2020). Modified solid-electrolyte interphase toward stable Li metal anode. *Nano Energy* 77: 105308–105336.

145 Peled, E. (1979). The electrochemical behavior of alkali and alkaline earth metals in nonaqueous battery systems—the solid electrolyte interphase model. *Journal of the Electrochemical Society* 126 (12): 2047–2051.

146 Wang, J., Lin, F., Jia, H. et al. (2014). Towards a safe lithium–sulfur battery with a flame-inhibiting electrolyte and a sulfur-based composite cathode. *Angewandte Chemie International Edition* 53: 10099–10104.

147 Hao, J., Wang, J., Lin, F. et al. (2014). TPPi as a flame retardant for rechargeable lithium batteries with sulfur composite cathodes. *Chemical Communications* 50: 7011–7013.

148 Shiga, T., Kato, Y., Kondo, H., and Okuda, C.-A. (2017). Self-extinguishing electrolytes using fluorinated alkyl phosphates for lithium batteries. *Journal of Materials Chemistry A* 5: 5156–5162.

149 Chen, X., Hou, T.-Z., Li, B. et al. (2017). Towards stable lithium-sulfur batteries: mechanistic insights into electrolyte decomposition on lithium metal anode. *Energy Storage Materials* 8: 194–201.

150 Ota, H., Akai, T., Namita, H. et al. (2003). XAFS and TOF–SIMS analysis of SEI layers on electrodes. *Journal of Power Sources* 119–121: 567–571.

151 Mikhaylik, Y. (2008). Electrolytes for Lithium Sulfur Cells. U.S. Patent 2008, US007354680B2, pp. 1–24.

152 Liang, X., Wen, Z., Liu, Y. et al. (2011). Improved cycling performances of lithium sulfur batteries with $LiNO_3$-modified electrolyte. *Journal of Power Sources* 196: 9839–9843.

153 Zhang, S.S. (2012). Role of $LiNO_3$ in rechargeable lithium/sulfur battery. *Electrochimica Acta* 70: 344–348.

154 Xiong, S., Xie, K., Diao, Y., and Hong, X. (2012). Properties of surface film on lithium anode with $LiNO_3$ as lithium salt in electrolyte solution for lithium–sulfur batteries. *Electrochimica Acta* 83: 78–86.

155 Li, W., Yao, H., Yan, K. et al. (2015). The synergetic effect of lithium polysulfide and lithium nitrate to prevent lithium dendrite growth. *Nature Communications* 6: 7436–7444.

156 Zhang, L., Ling, M., Feng, J. et al. (2018). The synergetic interaction between $Lino_3$ and lithium polysulfides for suppressing shuttle effect of lithium-sulfur batteries. *Energy Storage Materials* 11: 24–29.

157 Suo, L., Hu, Y.-S., Li, H. et al. (2013). A new class of solvent-in-salt electrolyte for high-energy rechargeable metallic lithium batteries. *Nature Communications* 4: 1481–1490.

158 Xiong, S., Kai, X., Hong, X., and Diao, Y. (2012). Effect of LiBOB as additive on electrochemical properties of lithium–sulfur batteries. *Ionics* 18: 249–254.

159 Täubert, C., Fleischhammer, M., Wohlfahrt-Mehrens, M. et al. (2010). LiBOB as electrolyte salt or additive for lithium-ion batteries based on $LiNi_{0.8}Co_{0.15}Al_{0.05}O_2$/graphite. *Journal of the Electrochemical Society* 157: A721–A728.

160 Kaneko, H., Sekine, K., and Takamura, T. (2005). Power capability improvement of LiBOB/PC electrolyte for Li-ion batteries. *Journal of Power Sources* 146: 142–145.

161 Lin, Z., Liu, Z., Fu, W. et al. (2013). Phosphorous pentasulfide as a novel additive for high-performance lithium-sulfur batteries. *Advanced Functional Materials* 23: 1064–1069.

162 Gauthier, M., Carney, T.J., Grimaud, A. et al. (2015). Electrode–electrolyte interface in Li-ion batteries: current understanding and new insights. *The Journal of Physical Chemistry Letters* 6: 4653–4672.

163 Jozwiuk, A., Berkes, B.B., Weiß, T. et al. (2016). The critical role of lithium nitrate in the gas evolution of lithium–sulfur batteries. *Energy & Environmental Science* 9: 2603–2608.

164 Han, Y., Duan, X., Li, Y. et al. (2016). Improved cycling performances with high sulfur loading enabled by pre-treating lithium anode. *Ionics* 22: 151–157.

165 Ma, G., Wen, Z., Wang, Q. et al. (2014). Enhanced cycle performance of a Li–S battery based on a protected lithium anode. *Journal of Materials Chemistry A* 2: 19355–19359.

166 Cheng, X.-B., Yan, C., Chen, X. et al. (2017). Implantable solid electrolyte interphase in lithium-metal batteries. *Chem* 2: 258–270.

167 Cheng, X.-B., Yan, C., Huang, J.-Q. et al. (2017). The gap between long lifespan Li-S coin and pouch cells: the importance of lithium metal anode protection. *Energy Storage Materials* 6: 18–25.

168 Liu, Q.-C., Xu, J.-J., Yuan, S. et al. (2015). Artificial protection film on lithium metal anode toward long-cycle-life lithium–oxygen batteries. *Advanced Materials* 27: 5241–5247.

169 Ma, L., Kim, M.S., and Archer, L.A. (2017). Stable artificial solid electrolyte interphases for lithium batteries. *Chemistry of Materials* 29: 4181–4189.

170 Liu, Y., Lin, D., Yuen, P.Y. et al. (2017). An artificial solid electrolyte interphase with high Li-ion conductivity, mechanical strength, and flexibility for stable lithium metal anodes. *Advanced Materials* 29: 1605531–1605539.

171 Zhang, S., Ueno, K., Dokko, K., and Watanabe, M. (2015). Recent advances in electrolytes for lithium-sulfur batteries. *Advanced Energy Materials* 5: 1500117–1500145.

172 Scheers, J., Fantini, S., and Johansson, P. (2014). A review of electrolytes for lithium–sulphur batteries. *Journal of Power Sources* 255: 204–218.

173 Gao, J., Lowe, M.A., Kiya, Y., and Abruña, H.D. (2011). Effects of liquid electrolytes on the charge–discharge performance of rechargeable lithium/sulfur batteries: electrochemical and in-situ X-ray absorption spectroscopic studies. *Journal of Physical Chemistry C* 115: 25132–25137.

174 Zhang, S.S. (2013). Liquid electrolyte lithium/sulfur battery: fundamental chemistry, problems, and solutions. *Journal of Power Sources* 231: 153–162.

175 Zhang, S.S. (2014). Understanding of sulfurized polyacrylonitrile for superior performance lithium/sulfur battery. *Energies* 7: 4588–4600.

176 Zhang, S.S. (2013). Sulfurized carbon: a class of cathode materials for high performance lithium/sulfur batteries. *Frontiers in Energy Research* 1: 1–9.

177 Wang, L., He, X., Li, J. et al. (2012). Charge/discharge characteristics of sulfurized polyacrylonitrile composite with different sulfur content in carbonate based electrolyte for lithium batteries. *Electrochimica Acta* 72: 114–119.

178 Xin, S., Gu, L., Zhao, N.-H. et al. (2012). Smaller sulfur molecules promise better lithium–sulfur batteries. *Journal of the American Chemical Society* 134: 18510–18513.

179 Zhang, Z., Li, Z., Hao, F. et al. (2014). 3D interconnected porous carbon aerogels as sulfur immobilizers for sulfur impregnation for lithium-sulfur batteries with high rate capability and cycling stability. *Advanced Functional Materials* 24: 2500–2509.

180 Xiao, J., Hu, J.Z., Chen, H. et al. (2015). Following the transient reactions in lithium-sulfur batteries using an in situ nuclear magnetic resonance technique. *Nano Letters* 15: 3309–3316.

181 Lochala, J., Liu, D., Wu, B. et al. (2017). Research progress toward the practical applications of lithium-sulfur batteries. *ACS Applied Materials & Interfaces* 9: 24407–24421.

182 Barchasz, C., Leprêtre, J.-C., Patoux, S., and Alloin, F. (2013). Electrochemical properties of ether-based electrolytes for lithium/sulfur rechargeable batteries. *Electrochimica Acta* 89: 737–743.

183 Yamin, H. and Peled, E. (1983). Electrochemistry of a nonaqueous lithium/sulfur cell. *Journal of Power Sources* 9: 281–287.

184 Angulakshmi, N. and Stephan, A.M. (2015). Efficient electrolytes for lithium-sulfur batteries. *Frontiers in Energy Research* 3: 1–8.

185 Kim, S., Jung, Y., and Lim, H.S. (2004). The effect of solvent component on the discharge performance of lithium–sulfur cell containing various organic electrolytes. *Electrochimica Acta* 50: 889–892.

186 Peled, E., Sternberg, Y., Gorenshtein, A., and Lavi, Y. (1989). Lithium-sulfur battery: evaluation of dioxolane-based electrolytes. *Journal of the Electrochemical Society* 136: 1621–1625.

187 Rauh, R.D., Shuker, F.S., Marston, J.M., and Brummer, S.B. (1977). Formation of lithium polysulfides in aprotic media. *Journal of Inorganic and Nuclear Chemistry* 39: 1761–1766.

188 Ryu, H.S., Ahn, H.J., Kim, K.W. et al. (2006). Self-discharge characteristics of lithium/sulfur batteries using tegdme liquid electrolyte. *Electrochimica Acta* 52: 1563–1566.

189 Wang, Y.-X., Chou, S.-L., Liu, H.-K., and Dou, S.-X. (2013). The electrochemical properties of high-capacity sulfur/reduced graphene oxide with different electrolyte systems. *Journal of Power Sources* 244: 240–245.

190 Cheon, S.-E., Ko, K.-S., Cho, J.-H. et al. (2003). Rechargeable lithium sulfur battery: II. Rate capability and cycle characteristics. *Journal of The Electrochemical Society* 150: A800–A805.

191 Aurbach, D., Pollak, E., Elazari, R. et al. (2009). On the surface chemical aspects of very high energy density, rechargeable Li–sulfur batteries. *Journal of the Electrochemical Society* 156: A694–A702.

192 Yamin, H., Gorenshtein, A., Penciner, J. et al. (1988). Lithium sulfur battery: oxidation/reduction mechanisms of polysulfides in THF solutions. *Journal of the Electrochemical Society* 135: 1045–1048.

193 Chu, M.-Y., De Jonghe, L.C., Visco, S.J., and Katz, B.D. (2000). Liquid Electrolyte Lithium-Sulfur Batteries. US Patent 2000, US 6030720A (08/948,969), pp. 1–28.

194 Choi, J.-W., Kim, J.-K., Cheruvally, G. et al. (2007). Rechargeable lithium/sulfur battery with suitable mixed liquid electrolytes. *Electrochimica Acta* 52: 2075–2082.

195 Kodama, D., Kanakubo, M., Kokubo, M. et al. (2011). Density, viscosity, and solubility of carbon dioxide in glymes. *Fluid Phase Equilibria* 302: 103–108.

196 Barchasz, C., Leprêtre, J.-C., Patoux, S., and Alloin, F. (2013). Revisiting TEGDME/DIOX binary electrolytes for lithium/sulfur batteries: importance of solvation ability and additives. *Journal of the Electrochemical Society* 160: A430–A436.

197 Chung, S.-H., Han, P., Singhal, R. et al. (2015). Electrochemically stable rechargeable lithium–sulfur batteries with a microporous carbon nanofiber filter for polysulfide. *Advanced Energy Materials* 5: 1500738–1500750.

198 Zhou, G., Pei, S., Li, L. et al. (2014). A graphene–pure-sulfur sandwich structure for ultrafast, long-life lithium–sulfur batteries. *Advanced Materials* 26: 625–631.

199 Tang, H. and Pan, M. (2015). Nafion. In: *EPST* (ed. H.F. Mark), 1–14. John Wiley & Sons, Inc.

200 Wu, F., Chen, J., Chen, R. et al. (2011). Sulfur/polythiophene with a core/shell structure: synthesis and electrochemical properties of the cathode for rechargeable lithium batteries. *Journal of Physical Chemistry C* 115: 6057–6063.

201 Yu, L.-Y., Xu, Z.-L., Shen, H.-M., and Yang, H. (2009). Preparation and characterization of PVDF–SiO_2 composite hollow fiber UF membrane by sol–gel method. *Journal of Membrane Science* 337: 257–265.

202 Su, Y.-S. and Manthiram, A. (2012). Lithium–sulphur batteries with a microporous carbon paper as a bifunctional interlayer. *Nature Communications* 3: 1166–1172.

203 Su, Y.-S. and Manthiram, A. (2012). A new approach to improve cycle performance of rechargeable lithium–sulfur batteries by inserting a free-standing MWCNT interlayer. *Chemical Communications* 48: 8817–8819.

204 Lee, C.-L. and Kim, I.-D. (2015). A hierarchical carbon nanotube-loaded glass-filter composite paper interlayer with outstanding electrolyte uptake properties for high-performance lithium–sulphur batteries. *Nanoscale* 7: 10362–10367.

205 Zeng, Q., Leng, X., Wu, K.-H. et al. (2015). Electroactive cellulose-supported graphene oxide interlayers for Li–S batteries. *Carbon* 93: 611–619.

206 Song, J., Xu, T., Gordin, M.L. et al. (2014). Nitrogen-doped mesoporous carbon promoted chemical adsorption of sulfur and fabrication of high-areal-capacity sulfur cathode with exceptional cycling stability for lithium-sulfur batteries. *Advanced Functional Materials* 24: 1243–1250.

207 Pang, Q., Tang, J., Huang, H. et al. (2015). A nitrogen and sulfur dual-doped carbon derived from polyrhodanine@cellulose for advanced lithium–sulfur batteries. *Advanced Materials* 27: 6021–6028.

208 Yang, J., Chen, F., Li, C. et al. (2016). A free-standing sulfur-doped microporous carbon interlayer derived from luffa sponge for high performance lithium–sulfur batteries. *Journal of Materials Chemistry A* 4: 14324–14333.

209 Bizuneh, G.G., Fan, J., Sun, C. et al. (2019). $LaLiO_2$-based multi-functional interlayer for enhanced performance of Li-S batteries. *Journal of the Electrochemical Society* 166: A68–A73.

210 Fan, W., Zhang, L., and Liu, T. (2018). Multifunctional second barrier layers for lithium–sulfur batteries. *Materials Chemistry Frontiers* 2: 235–252.

211 Liang, X., Kwok, C.Y., Lodi-Marzano, F. et al. (2016). Tuning transition metal oxide–sulfur interactions for long life lithium sulfur batteries: the "goldilocks" principle. *Advanced Energy Materials* 6: 1501636–1501645.

212 Seh, Z.W., Li, W., Cha, J.J. et al. (2013). Sulphur–TiO_2 yolk–shell nanoarchitecture with internal void space for long-cycle lithium–sulphur batteries. *Nature Communications* 4: 1331–1337.

213 Tao, X., Wang, J., Ying, Z. et al. (2014). Strong sulfur binding with conducting magneli-phase Ti_nO_{2n-1} nanomaterials for improving lithium-sulfur batteries. *Nano Letters* 14: 5288–5294.

214 Wang, X., Li, G., Li, J. et al. (2016). Structural and chemical synergistic encapsulation of polysulfides enables ultralong-life lithium–sulfur batteries. *Energy & Environmental Science* 9: 2533–2538.

215 Cui, Y., Wen, Z., Liang, X. et al. (2012). A tubular polypyrrole based air electrode with improved O_2 diffusivity for Li–O_2 batteries. *Energy & Environmental Science* 5: 7893–7897.

216 Ma, G., Wen, Z., Jin, J. et al. (2014). Enhanced cycle performance of Li–S battery with a polypyrrole functional interlayer. *Journal of Power Sources* 267: 542–546.

4

Metal-CO_2 Batteries: Mechanisms and Advanced Materials

Chang Guo, Keyu Xie, and Xiao Han

Northwestern Polytechnical University, School of Materials Science and Engineering, 127 Youyi West Road, Xi'an, 710072, PR China

4.1 Introduction

With the rapid development of modern social economy and technology, the demand for energy is increasing year by year. In terms of energy consumption, traditional energy consumption still exceeds 80% [1]. Due to the nonrenewable nature of traditional energy, countries all over the world are looking for alternatives. The use of solar energy, hydro energy, wind energy, and nuclear energy can generate more electricity, but these renewable energy sources have a series of problems such as uneven distribution, limited time, and safety. Therefore, it is particularly important to find excellent energy conversion equipment. Rechargeable lithium-ion batteries have become one of the better choices [2]. However, the theoretical specific capacity of traditional lithium-ion batteries is low due to the limited electrochemical capacity of anode materials [3], so it is imperative to research and develop more novel secondary batteries with larger capacity.

Environment-friendly Li-air batteries with high theoretical capacity have attracted the attention of scholars [4–9]. It is worth noting that the gas used is pure oxygen in most Li-air batteries [10, 11]. Subsequently, some researchers have studied the possibility of using carbon dioxide. In 2011, Asaoka and coworkers first found that the capacity of the Li-O_2 battery is greatly improved when CO_2 is introduced into the battery system [12]. More importantly, metal-CO_2 batteries can not only absorb the greenhouse gas CO_2 to delay climate warming but also efficiently convert CO_2 into valuable electricity. In some CO_2-rich environments, such as Mars and the deep ocean, metal-CO_2 batteries can capture CO_2 directly without artificial additions [13]. The application of these special occasions greatly broadens the application prospects of metal-CO_2 batteries.

Metal-CO_2 batteries include Li-CO_2 batteries, Na-CO_2 batteries, K-CO_2 batteries, Mg-CO_2 batteries, Zn-CO_2 batteries, and Al-CO_2 batteries. Currently, Li-CO_2 batteries are the most studied due to their wide electrochemical window and highest specific discharge capacity [14–23]. Compared with metallic Li, the reserves of

Electrochemical Energy Storage Devices: Non-Conventional Technologies and Materials, First Edition. Edited by Yongbing Tang and Luojiang Zhang.

metallic Na and K in the earth's crust are higher [24]. Due to the relative cheapness of metals Na and K, Na-CO_2 and K-CO_2 batteries have also received extensive attention [25–28]. In comparison, Zn, Mg, and Al have low chemical reactivity, so they are usually used in aqueous electrolytes [29–31]. These three types of cells also exhibit lower voltage windows but high product selectivity. Therefore, the charge–discharge mechanisms will show great differences with different anode materials.

Metal-CO_2 batteries have been greatly developed in the past decade, especially Li-CO_2 batteries with high specific capacity. However, some current issues of metal-CO_2 batteries hinder their commercial application. The problem mainly exhibits in (i) lack of cheap catalysts with high catalytic activity to achieve 100% reversible decomposition of products, (ii) unclear charge–discharge mechanism of batteries, and (iii) lack of electrolytes with high ionic conductivity and low volatility. To this end, we must understand in detail the electrochemical mechanisms of various metal-CO_2 batteries and find more advanced cathode materials and electrolytes. In this chapter, we first analyze the electrochemical reaction mechanisms of various metal-CO_2 batteries in detail. Furthermore, the advanced cathode materials and typical electrolytes in various metal-CO_2 batteries are summarized. Finally, the future development direction of metal-CO_2 batteries is prospected.

4.2 The Electrochemistry Mechanism of Metal-CO_2 Battery

Different metal anodes produce different electrochemical reaction mechanisms. According to the valence of the anode, common metal-CO_2 battery metal anodes can be divided into monovalent metals, namely alkali metal elements (Li, Na, K), divalent metals (Mg, Zn), and trivalent metals (Al). Due to the high activity of alkali metals, the corresponding electrolytes are generally organic electrolytes [32]. In the organic electrolyte, the discharge products of the battery are usually Me_2CO_3 and C. The battery corresponding to the alkali metal anode has the advantages of high discharge voltage and large theoretical specific capacity. Compared with alkali metals, divalent metals and trivalent metals have lower chemical activity, so they generally correspond to aqueous electrolytes or ionic liquids [29–31]. The battery corresponding to the divalent metal and trivalent metal anode has the advantages of product diversity. However, disproportionation products are also generated in metal-CO_2 batteries used in some special cathode materials [33–35]. For example, CO or oxalate appears in Li-CO_2 and Na-CO_2 batteries. Therefore, different types of electrolytes and the composition of cathode materials and anode materials will cause different charge–discharge reactions. Given that, understanding the electrochemical mechanisms corresponding to various anodes is very necessary for the study of metal-CO_2 batteries. In this section, various metal-CO_2 battery electrochemical mechanisms are described.

4.2.1 Discharge/Charge Mechanisms of Li-CO_2 Battery

4.2.1.1 Discharge Mechanisms of Li-CO_2 Battery

In 2011, Asaoka and coworkers first found that the Li-O_2/CO_2 battery with the introduction of CO_2 gas exhibited a larger discharge capacity than the Li-O_2 battery [12]. Two years later, Archer and coworkers designed a Li-CO_2 battery by using ionic liquid electrolyte and commercially activated carbon cathode [20]. Of noteworthy is that, this is the first primary model structure of Li-CO_2 battery. Since then, more and more Li-CO_2 batteries are constructed in the past decade [36–44]. As shown in Figure 4.1, a typical Li-CO_2 battery includes a metal Li anode, an organic electrolyte, and a porous catalyst cathode, i.e. a gas electrode. During the discharge process, the metallic Li anode loses electrons to form Li^+ ions, which then dissolve into the electrolyte [17]. Furthermore, Li^+ ions in the electrolyte diffuse to the surface of the gas cathode under the driving force of the potential difference. At the cathode/electrolyte interface, dissolved CO_2 molecules capture electrons and combine with Li^+ to form the discharge product of a mixture of Li_2CO_3 and carbon.

In Li-CO_2 batteries, the presence or absence of O_2 makes a big difference in the reaction mechanism. When oxygen is present, O_2 molecules first get electrons to form superoxide radicals ($O_2^{\cdot-}$) (Eq. 4.1) during the discharge process. Subsequently, $O_2^{\cdot-}$ captures two CO_2 molecules in the electrolyte to generate peroxyoxalate ($C_2O_6^{2-}$) (Eqs. 4.2 and 4.3). Finally, $C_2O_6^{2-}$ further reacts with $O_2^{\cdot-}$ and Li^+ to produce Li_2CO_3 and release O_2 (Eq. 4.4) [45, 46]. It can be seen that in the Li-O_2/CO_2 electrochemical discharge reaction, O_2 is firstly reduced instead of CO_2 [47]. The specific reaction process is shown in Eqs. (4.1)–(4.4). Notably, Li-O_2/CO_2 batteries with 30 to 70% CO_2 exhibit higher full discharge capacity than both Li-O_2 and Li-CO_2 batteries. The high discharge capacity of Li-O_2/CO_2 batteries is due to the controlled reaction rate.

$$O_2 + e^- \rightarrow O_2^{\cdot-} \tag{4.1}$$

$$O_2^{\cdot-} + CO_2 \rightarrow CO_4^{\cdot-} \tag{4.2}$$

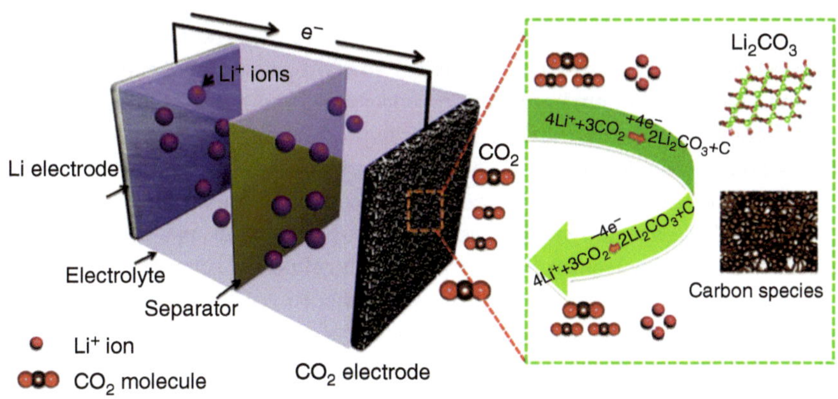

Figure 4.1 Schematic diagram of a typical Li-CO_2 battery. Source: Reproduced with permission Liu et al. [17]. Copyright 2019, The Royal Society of Chemistry.

$$CO_4^{\cdot -} + CO_2 + O_2^{\cdot -} \rightarrow C_2O_6^{2-} + O_2 \tag{4.3}$$

$$C_2O_6^{2-} + 2O_2^{\cdot -} + 4Li^+ \rightarrow 2Li_2CO_3 + 2O_2 \tag{4.4}$$

In typical Li-CO_2 batteries, O_2 is not involved in the overall reaction process. Nemeth and Srajer proposed that dissolved CO_2 molecules can capture electrons in the cathode and reduce to $C_2O_4^{2-}$ through a one-electron reaction (Eq. 4.5) [48]. Referring to the disproportionation reaction of Li_2O_2 in Li-O_2 batteries, it can be speculated that a similar disproportionation reaction can occur during CO_2 reduction in Li-CO_2 batteries [34, 49]. When the battery is discharged, CO_2 gains electrons at the electrolyte/cathode interface to form oxalate ($C_2O_4^{2-}$). Subsequently, the unstable $C_2O_4^{2-}$ undergoes a two-step disproportionation reaction to form carbonate (CO_3^{2-}) and carbon, as shown in Eqs. (4.6) and (4.7). Finally, CO_3^{2-} produced by the reaction couples with Li^+ to form Li_2CO_3 crystals (Eq. 4.8). Therefore, the discharge reaction equations for nonaqueous Li-CO_2 batteries can be summarized in Eq. (4.9).

$$2CO_2 + 2e^- \rightarrow C_2O_4^{2-} \tag{4.5}$$

$$C_2O_4^{2-} \rightarrow CO_2^{2-} + CO_2 \tag{4.6}$$

$$C_2O_4^{2-} + CO_2^{2-} \rightarrow 2CO_3^{2-} + C \tag{4.7}$$

$$CO_3^{2-} + 2Li^+ \rightarrow Li_2CO_3 \tag{4.8}$$

$$4Li^+ + 4e^- + 3CO_2 \rightarrow 2Li_2CO_3 + C \qquad E^0_{rev} = 2.82\,V \tag{4.9}$$

Zhou and coworkers found that when the Li-CO_2 battery is deeply discharged below 2.0 V, a new discharge plateau different from normal discharge plateau may appear around 1.90 V, indicating different electrochemical reactions at this potential [50]. By using *in situ* Raman analysis and TEM, the discharge products are identified as lithium oxide (Li_2O) and amorphous carbon (Eq. 4.10), which is different from Li_2CO_3 produced in normal discharge plateau. At the same time, the formation of Li_2O can be regulated by controlling the CO_2 reaction rate (current density) and supply rate (CO_2 partial pressure, CO_2 diffusivity). In addition, some special catalyst can also change the electrochemical reaction process and the final discharge product. Wang and coworkers found that the discharge products on the three-dimensional porous metal Zn cathode are Li_2CO_3 and CO gas, rather than amorphous solid carbon. The corresponding CO_2 electroreduction mechanism is exhibited in Eq. (4.11) [33]. Many studies have shown that molybdenum-based catalysts and their derivatives underwent a two-electron transfer process during the discharge reaction, and the discharge product is detected as lithium oxalate. In 2017, Chen and coworkers first discovered that when Mo_2C/CNTs are used as a cathode catalyst, the discharge product is lithium oxalate [34]. Combined with Raman spectroscopic analysis, it can be seen that molybdenum acted as an electron donator for charge transfer. The relevant reaction process is shown in Eq. (4.12). Recently, Wang and coworkers demonstrated theoretically and experimentally that MoN nanofibers possessed outstanding catalytic properties to CO_2RR and CO_2ER for Li-CO_2 batteries [51]. Similar to the molybdenum carbide catalyst, the corresponding product of the MoN nanofibers is also lithium oxalate. Further research showed that MoN can stabilize intermediate lithium oxalate ($Li_2C_2O_4$), which is an unstable substance in discharge process. The two-electron reaction process that occurred in the MoN cathode is shown in Eq. (4.13).

$$4Li^+ + CO_2 + 4e^- \rightarrow 2Li_2O + C\ E^0_{rev} = 1.89\ V \tag{4.10}$$

$$2Li^+ + 2CO_2 \leftrightarrow Li_2CO_3 + CO\ E^0_{rev} = 2.50\ V \tag{4.11}$$

$$2CO_2 + 2e^- \rightarrow C_2O_4{}^{2-}$$

$$C_2O_4{}^{2-} + 2Li^+ + Mo_2C \rightarrow Li_2C_2O_4 + Mo_2C \tag{4.12}$$

$$CO_2RR\ (\text{discharge}):\ 2CO_2 + 2e^- \rightarrow C_2O_4{}^{2-} \tag{4.13}$$

Besides the aprotic electrolytes mentioned above, Li-CO_2 batteries exhibit different reaction mechanisms in aqueous electrolytes. Ma and coworkers used nanoporous Pd film as cathode and $Li_{1.5}Al_{0.5}Ge_{1.5}(PO_4)_3$ (LAGP) as solid-state electrolyte for aqueous Li-CO_2 batteries [52]. In the batteries, CO_2 can be electrochemically reduced to soluble liquid products during discharge. During the subsequent charging process, the soluble liquid product can be oxidized and reversibly generate CO_2. Nuclear magnetic resonance (NMR) spectral analysis showed that the discharge products are formate products rather than lithium carbonate. According to the Nernst equation, it can be inferred that the possible reactions of the cathode and anode are shown in Eqs. (4.14)–(4.16).

$$\text{Cathode:}\ CO_2 + 2H^+ + 2e^- \leftrightarrow HCOOH\ e_c = -0.24\ V \tag{4.14}$$

$$\text{Anode:}\ Li - e^- \rightarrow Li^+ \tag{4.15}$$

$$\text{Overall reaction:}\ CO_2 + 2H^+ + 2Li \leftrightarrow HCOOH + 2Li^+\ E^0 = 2.80\ V \tag{4.16}$$

4.2.1.2 Charge Mechanisms of Li-CO_2 Battery

According to the decomposition degree of the discharge products, the charging process of Li-CO_2 batteries is divided into reversible reactions and irreversible reactions. Li-CO_2 batteries can be considered reversible if both the discharge products Li_2CO_3 and carbon can be completely decomposed. Otherwise, they are rechargeable but not reversible. For the former, Li-CO_2 batteries have promising applications in renewable energy conversion and storage [26, 50, 53]. For the latter, Li-CO_2 batteries provide an efficient strategy for CO_2 fixation and CO_2 reduction.

In the irreversible reactions, the battery charging process only involves the decomposition of Li_2CO_3 without the degradation of amorphous carbon. Previous research speculated that there may be three decomposition processes for Li_2CO_3, with the following equations:

$$2Li_2CO_3 \rightarrow 2CO_2 + O_2 + 4Li^+ + 4e^- \tag{4.17}$$

$$2Li_2CO_3 \rightarrow 2CO_2 + O_2^{\cdot -} + 4Li^+ + 3e^- \tag{4.18}$$

$$2Li_2CO_3 \rightarrow 2CO_2 + {}^1O_2 + 4Li^+ + 4e^- \tag{4.19}$$

As shown in Eq. (4.17), the first path can be understood as the self-decomposition of Li_2CO_3. During this process, the decomposition of Li_2CO_3 can release CO_2 and O_2 [54]. Zhou and coworkers analyzed the charging process in Li_2CO_3-^{12}C and Li_2CO_3-^{13}C electrodes using isotope labeling and differential electrochemical mass spectrometry (DEMS). However, the results showed that CO_2 is the main charging product without O_2 release. Furthermore, O_2 release was also not detected in Li-CO_2 battery based on Mn-supported metal-organic framework (MOF) catalysts when the Li-CO_2 battery was charged [55]. Thus, the existence of the decomposition path (Eq. 4.17) can be basically denied.

Peng and coworkers proposed a second route for the decomposition of Li_2CO_3, as in Eq. (4.18) [54]. During battery charging, the decomposition of Li_2CO_3 produced CO_2 and superoxide radicals $O_2^{\cdot -}$, which can be dissolved in the electrolyte. Among them, superoxide radicals may undergo two reactions. When diffusing to the electrolyte/cathode interface, the superoxide radical will further lose electrons to release O_2 (Eq. 4.20). In addition, superoxide radicals can also directly corrode the electrolyte solvent, induce side reactions, and generate a series of indeterminate by-products.

$$O_2^{\cdot -} - e^- \rightarrow O_2 \tag{4.20}$$

In addition, Freunberger and coworkers used 9,10-dimethylanthracene (9,10-DMA) as a chemical probe to detect the possible existence of singlet oxygen ($^1O_2$) in the product of Li_2CO_3 decomposition reaction, as shown in Eq. (4.19) [56]. The charge products were analyzed by high-performance liquid chromatography (HPLC) and ^1H NMR, indicating that DMA-$^1O_2$ is formed in the electrolyte when the charging voltage rose above 3.8 V. It can be seen that this conclusion can explain why no O_2 is released during the charging process. However, many studies have shown that $^1O_2$ can also corrode battery components and electrodes during charging and cause severe side reactions [54].

The decomposition path of Li_2CO_3 is not random, but can be regulated by controlling kinetic factors such as current density [50]. Zhou et al. used DEMS to analyze the charging reaction process. Figure 4.2a shows initial discharge–charge curve of porous carbon cathode at the current density of 100 mA g^{-1}. When the charging current density is relatively low (<500 mA g^{-1}), the Li-CO_2 battery only released CO_2 at the charge-to-mass ratio of CO_2 (Figure 4.2b). Through calculation and analysis, it can be obtained that the charge-to-mass ratio is $3e^-/2CO_2$, indicating that the Li_2CO_3 decomposition reaction obeyed Eq. (4.18). When the charging current density is relatively high (>2000 mA g^{-1}), the battery released CO_2 and O_2, whose charge-to-mass ratios were $2e^-/CO_2$ and $4e^-/O_2$ (Figure 4.2c), respectively, indicating that the decomposition reaction of Li_2CO_3 obeyed Eqs. (4.17) and (4.19).

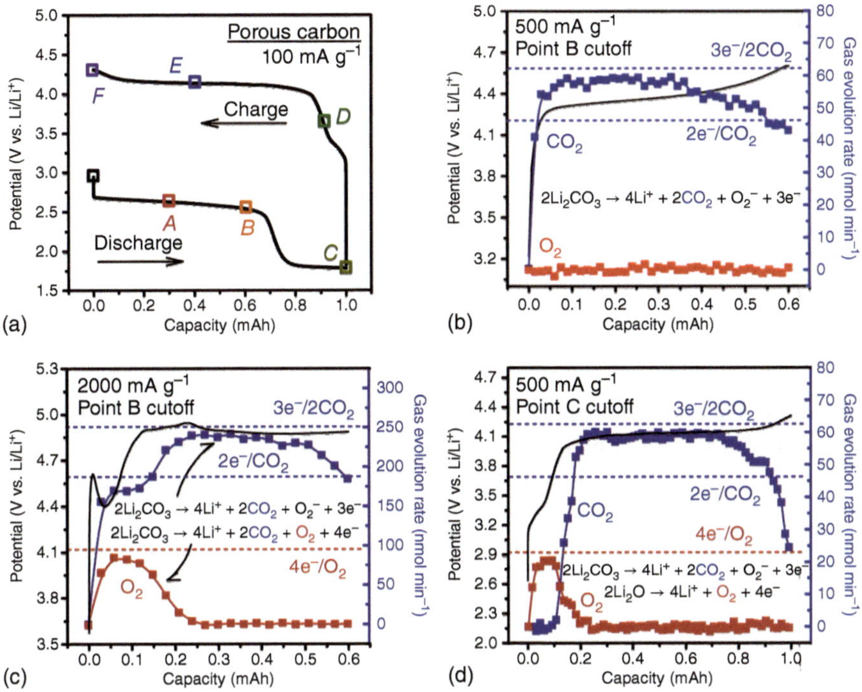

Figure 4.2 (a) Galvanostatic discharge profiles of porous carbon cathode for Li-CO_2 battery with a current density of 100 mA g^{-1} and a cutoff capacity of 1 mAh. (b,c) *In situ* DEMS results during charging process at 500 mA g^{-1} (b) and 2000 mA g^{-1} (c) current density after discharged to point B. (d) *In situ* DEMS results during charging process at 500 mA g^{-1} current density after discharged to point C. Source: Reproduced with permission Qiao et al. [50]. Copyright 2019, Elsevier.

Additionally, when started charge from point C, lithium oxide also produced oxygen through a four-electron reaction (Figure 4.2d).

For the reversible reaction, both amorphous carbon and Li_2CO_3 can participate in the decomposition reaction during the charging of Li-CO_2 batteries. The reaction equation is shown in Eq. (4.21). Thermodynamic calculations based on Gibbs free energy showed that the reaction has a low reversible potential of 2.82 V [50]. Therefore, designing along this pathway is crucial for reversible and rechargeable Li-CO_2 batteries. With the assistance of some special catalysts (such as Ru), the Li_2CO_3 battery successfully achieved a reversible reaction with high charge–discharge efficiency. Zhou and coworkers used magnetron sputtering to prepare Au electrodes and Au-Ru electrodes, respectively. Then, *in situ* surface-enhanced Raman spectroscopy (SERS) is used to detect the generation and decomposition of discharge products [50, 53].

Ru catalyst can significantly promote the co-decomposition of Li_2CO_3 and carbon. For the Au electrode, no matter whether the electrolyte is triethylene glycol dimethyl ether (TEGDME) or dimethyl sulfoxide (DMSO), carbon did not participate in the decomposition reaction. The reason is that the high electrocatalytic activity of the specific crystal planes of Ru thermodynamically lowered the reaction barrier, thereby facilitating the reversible reaction during charging. In addition, Zhou et al. also constructed Ni nanoparticle composites that were highly dispersed on N-doped graphene (Ni-NG) and were also shown to contribute to the decomposition of Li_2CO_3 and carbon [57].

$$2Li_2CO_3 + C \rightarrow 3CO_2 + 4Li^+ + 4e^- \; E^0_{rev} = 2.82\,V \tag{4.21}$$

There were two types of charging reactions that were different from the aforementioned typical charging reactions, namely, the decomposition of the lithium oxalate or formic acid. When the Mo_2C and MoN catalysts served as the cathode of Li-CO_2 battery, the discharge product obtained is lithium oxalate [34, 51]. The resulting lithium oxalate will be reversibly decomposed into Li^+ and carbon dioxide during the charging reaction. The specific reaction is shown in Eq. (4.22). It is worth noting that due to the instability of lithium oxalate and its low decomposition energy barrier, the Li-CO_2 battery exhibited low overpotential. Additionally, in the aqueous electrolyte, due to the presence of protons, the corresponding discharge product is formic acid. In the subsequent charging process, the oxidation reaction of formic acid is carried out (Eq. 4.23).

$$Li_2C_2O_4 \rightarrow 2Li^+ + 2CO_2 + 2e^- \tag{4.22}$$

$$HCOOH + 2Li^+ \rightarrow 2Li + 2H^+ + CO_2 \tag{4.23}$$

4.2.2 Discharge/Charge Mechanisms of Na-CO_2 Battery

In 2013, Archer et al. first introduced the application of Na-CO_2/O_2 batteries in different electrolytes [35]. In the pure CO_2 atmosphere, the full discharge capacities of Na-CO_2 batteries in 1 M $NaClO_4$/TEGDME electrolyte and 0.75 M $NaCF_3SO_3$/1-ethyl-3-methylimidazolium trifluoromethanesulfonate electrolyte were 183 and 173 mAh g^{-1}, respectively. Surprisingly, with the addition of an appropriate amount of oxygen, the discharge capacity will be increased by more than 10 times. Transmission electron microscope (TEM) together with X-ray diffraction (XRD) analysis results showed that the discharge products in TEGDME electrolyte are mixtures of Na_2CO_3 and NaC_2O_4, while the main discharge product in ionic liquid is sodium oxalate [35]. The authors believed that the reaction path that may occur in TEGDME is shown in Eqs. (4.24)–(4.28), while that in ionic liquids is shown in Eqs. (4.29) and (4.30).

$$4O_2 + 4e^- \rightarrow 4O_2^{\cdot-} \tag{4.24}$$

$$O_2^{\cdot-} + CO_2 \rightarrow CO_4^{\cdot-} \tag{4.25}$$

$$CO_4^{\cdot-} + CO_2 \rightarrow C_2O_6^{\cdot-} \tag{4.26}$$

$$C_2O_6^{\cdot-} + O_2^{\cdot-} \rightarrow C_2O_6^{2-} + O_2 \tag{4.27}$$

$$C_2O_6^{2-} + 2O_2^{\cdot -} + 4Na^+ \rightarrow 2Na_2CO_3 + 2O_2 \quad (4.28)$$

$$CO_2 + O_2^{2-} \rightarrow CO_4^{2-} \quad (4.29)$$

$$CO_4^{2-} + CO_2 + 2Na^+ \rightarrow Na_2C_2O_4 + O_2 \quad (4.30)$$

Chen and coworkers constructed Na-CO_2 battery by using multiwalled carbon nanotube (MWCNT) treated with TEGDME (tetraglyme) as the gas electrode and $NaClO_4$/TEGDME (tetraglyme) as the electrolyte [58]. The treated gas electrode not only possessed extremely high electronic conductivity and sufficient discharge product storage space but also had excellent electrolyte wettability. Due to the unique three-dimensional structure of the gas electrode, the battery showed effectively reduced polarization and long cycle stability. By analyzing the TEM results of MWCNT cathodes after different discharge depths, it can be observed that more and more nanoparticles (discharge products) are deposited on the carbon tubes. Furthermore, the interplanar spacings of products are consistent with the (201) and (401) interplanar spacings of Na_2CO_3, thus indicating the generation of Na_2CO_3. The energy-dispersive spectrometer (EDS) results also showed that the atomic composition ratio of Na and C is 2.5 : 1, which indirectly explained the C element came from not only Na_2CO_3 but also the carbon element of the product [58]. Reversible charge–discharge reaction (Eq. 4.31) is demonstrated for the first time by various *in-situ* and *ex-situ* characterizations.

$$4Na + 3CO_2 \rightarrow 2Na_2CO_3 + C \quad (4.31)$$

Zhang and coworkers used *in situ* environmental transmission electron microscopy (ETEM) to further reveal the electrochemical reactions for Na-CO_2 nanobatteries [59]. To facilitate the discharge/charge reaction process, single-atom Pt-doped carbon nanotubes with nitrogen doped (Pt@NCNT) are used as cathodes. *In situ* TEM results showed that nanosphere products are generated in Pt@NCNT cathode during the discharge process, while the nanosphere products completely disappeared during the charging process. Selected area electron diffraction analysis showed that the composition of the nanospheres is Na_2CO_3. The reaction mechanism during the discharge process may be as follows: when negative voltage is applied to the Pt@NCNT side, the discharge is induced by the intercalation of Na^+ ions into the nitrogen-doped carbon nanotubes (NCNT) and the transport of Na^+ on the surface. After a period of discharge, the intercalation rate slowed down, and then Na^+ reacted with CO_2 to form Na_2CO_3 on the surface of NCNT. Na_2CO_3 reactive spheres were formed on the surface of Pt@NCNT. During the charging process, the product began to shrink, and the Na_2CO_3 spheres decomposed at the surface into Na and the deintercalation of Na ions, which corresponded to the reaction reported in Na-CO_2 batteries: $2Na_2CO_3 + C \rightarrow 4Na + 3CO_2$.

4.2.3 Discharge/Charge Mechanisms of K-CO_2 Battery

Among the three common monovalent metal-CO_2 battery systems, the high cost and scarcity of lithium materials (only 0.0017 wt% storage in the earth's crust) severely limit the sustainable large-scale application of Li-CO_2 batteries. Compared with Na-CO_2 batteries, the discharge potential of K-CO_2 batteries (2.48 V) is higher than the discharge potential of Na-CO_2 batteries (2.35 V). Therefore, K-CO_2 batteries have attracted the attention of scholars due to their higher theoretical discharge voltage and lower product decomposition energy. Zhang et al. first revealed the discharge–charge principle of K/CO_2 nanobatteries using *in situ* aberration-corrected (AC) ETEM [27]. During the discharge process, K_2CO_3 hollow spheres and a large number of nanobubbles are formed in the carbon nanotube cathode. During the charging process, K_2CO_3 is decomposed into K and CO_2, resulting in the consumption of CNTs. The corresponding charge–discharge reaction is shown in Eqs. (4.32) and (4.33). Since the carbon monoxide (CO) that generated in the discharge process did not participate in the charging reaction, this reaction is not completely reversible in principle. The results of this study provide a preliminary understanding of the working mechanism of K-CO_2 batteries.

$$2K^+ + 2e^- + 2CO_2 \rightarrow K_2CO_3 + CO \tag{4.32}$$

$$2K_2CO_3 + C \rightarrow 4K + 3CO_2 \tag{4.33}$$

To overcome the side reactions and dendrite deposition of potassium anode during the cycle reaction process, Chen and coworkers constructed a novel K-CO_2 battery with KSn alloy as the anode and carboxyl-containing multiwalled carbon nanotubes (MWCNTs-COOH) as the cathode catalyst [60]. Different from the conclusion of Zhang et al. that CO gas is generated in the discharge reaction, Chen and coworkers used *in situ* GC (gas chromatograph) to detect the gas of the discharge reaction and found that there was no CO gas in the discharge reaction at different discharge depths. Subsequently, the Raman patterns in carbon-free (silver nanowires) cathode material further confirmed that the discharge products contained carbon and potassium carbonate.

The entire battery reaction is $4KSn + 3CO_2 = 2K_2CO_3 + C + 4Sn$. Wang and coworkers greatly improved the stability of K-CO_2 batteries and extended cycle life by constructing dense solid electrolyte interface film and using efficient catalysts in 2021 [28]. The battery had high discharge capacity (9436 mAh g^{-1}), good rate performance, small voltage gap (0.81 V at 50 mA g^{-1}), and long cycle life (450 cycles with cut capacity of 500 mA g^{-1}). The DEMS results showed that only CO_2 gas is generated during the discharge–charge process of the K-CO_2 battery. More importantly, the corresponding e^-/CO_2 ratio is $4e^-/3CO_2$, following the reaction path shown in Eq. (4.34).

$$4K^+ + 3CO_2 + 4e^- \leftrightarrow 2K_2CO_3 + C \tag{4.34}$$

4.2.4 Discharge/Charge Mechanisms of Mg-CO_2 Battery

Metal-CO_2 batteries have been widely developed as a new energy storage method that combines high energy density and CO_2 utilization. Although the monovalent metal-CO_2 battery mentioned in the previous section is developing rapidly, the active lithium, sodium, and potassium anode materials bring security risks to the battery that cannot be ignored. Magnesium, as a representative multivalent metal, has the advantages of low redox potential similar to monovalent metals, not easy to grow dendrites, and abundant reserves in the crust. It has broad development prospects in the field of multivalent metal-carbon dioxide batteries.

In 2013, Archer and coworkers first proposed Mg-O_2/CO_2 battery based on the Na-O_2/CO_2 battery system [35]. In pure carbon dioxide atmosphere, the discharge capacity of Mg-CO_2 batteries is very poor, but when oxygen is introduced, the discharge capacity would be greatly improved. However, it was a pity that Archer et al. did not provide explanation on the mechanism of Mg-O_2/CO_2 cells. Kim et al. successfully converted the nonrechargeable Mg-CO_2 battery into a secondary battery by using membrane-free and aqueous electrolyte (containing 1 M KOH and 1 M NaCl) (Figure 4.3a). Interestingly, although the membrane-free Mg-CO_2 battery is a secondary battery, different reactions took place on the anode and cathode, respectively. In discharge process, hydrogen evolution reaction occurred at cathode (Eq. 4.35),

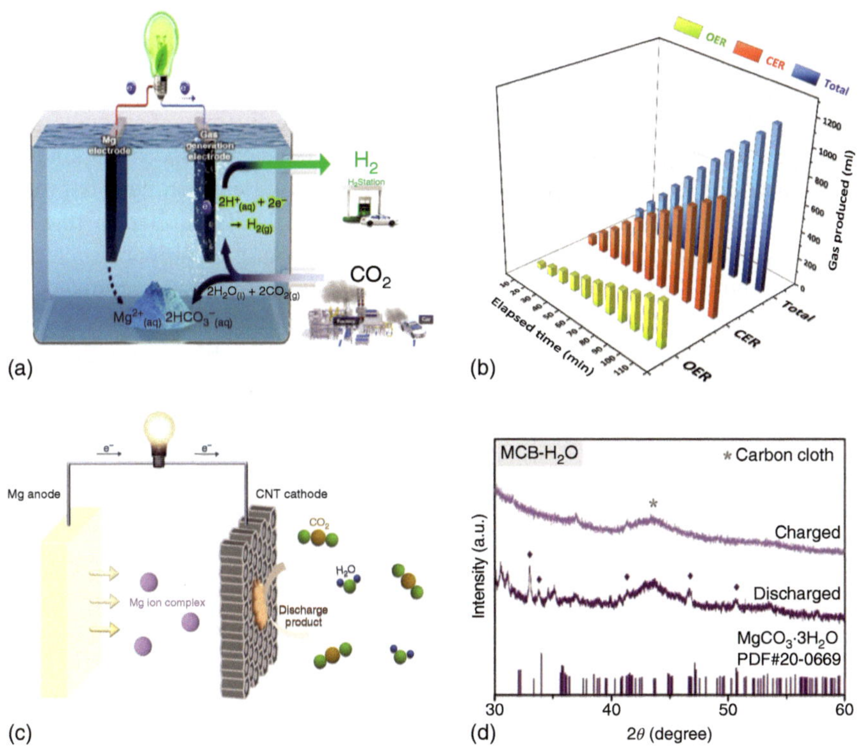

Figure 4.3 (a) Schematic diagram of discharge–charge electrochemical mechanism of Mg-CO_2 battery using aqueous electrolyte. (b) The amount of gas production during the charge process. Source: Reproduced with permission Kim et al. [61]. Copyright 2021, Elsevier. (c) Schematic diagram of discharge mechanism of moisture-assisted Mg-CO_2 battery. (d) The XRD pattern of discharged product and charge product. Source: Reproduced with permission Zhang et al. [29]. Copyright 2022, Wiley-VCH.

while the evolution of oxygen and chlorine gas mainly occurred (Eq. 4.36) during charging (Figure 4.3b) [61]. The results of the study showed that the membrane-free CO_2 battery can be stably cycled over 800 minutes with an almost constant voltage gap. Recently, Luo and coworkers greatly improved the cyclability and discharge capacity of the battery by introducing a humid atmosphere into the Mg-CO_2 battery system (Figure 4.3c). Nonaqueous Mg-CO_2 battery can be stably cycled for more than 250 hours and the discharge voltage can be maintained above 1V continuously [29]. According to the results obtained from discharge products analysis (Figure 4.3d) and theoretical calculation, the reversible reaction is shown in Eq. (4.37).

$$2H^+{}_{(aq)} + 2e^- \rightarrow H_{2(g)} \tag{4.35}$$

$$2H_2O \rightarrow O_{2(g)} + 4H^+ + 4e^-$$

$$2Cl^- \rightarrow Cl_{2(g)} + 2e^- \tag{4.36}$$

$$2Mg + 3CO_2 + 6H_2O \leftrightarrow 2MgCO_3 \cdot 3H_2O + C \tag{4.37}$$

4.2.5 Discharge/Charge Mechanisms of Zn-CO_2 Battery

Zn-CO_2 batteries have attracted much attention due to the diversity of products and abundant resources of metal Zn. Different types of cathode catalysts correspond to different discharge–charge mechanisms. Since its working environment is protic electrolyte, the discharge mechanism of Zn-CO_2 batteries is similar to that of electrocatalytic reduction of CO_2. Typically, in 2018, Wang et al. constructed a reversible aqueous Zn-CO_2 battery for the first time using mutually cross-linked 3D porous Pd nanosheets as cathode materials [62]. The 3D porous Pd nanosheets with highly active surface acted as bifunctional electrocatalysts, which can efficiently reduce CO_2 and oxidize HCOOH, realizing the efficient and reversible conversion between CO_2 and HCOOH. The electrode reaction and overall reaction are shown in Eqs. (4.38)–(4.40).

$$\text{Cathode:}\ CO_2 + 2H^+ + 2e^- \leftrightarrow HCOOH$$

$$(1\ \text{M NaCl}, 0.1\ \text{M HCOONa, sat:}\ CO_2),\ E_c = -0.294\ \text{V} \tag{4.38}$$

$$\text{Anode:}\ Zn - 2e^- \leftrightarrow Zn^{2+}\ (1\ \text{M KOH with}\ 0.02\ \text{M}\ Zn(CH_3COO)_2)$$

$$Zn^{2+} + 4OH^- \leftrightarrow Zn(OH)_4{}^{2-}$$

$$Zn(OH)_4{}^{2-} \leftrightarrow ZnO + 2OH^- + H_2O\ E_a = -1.249\ \text{V} \tag{4.39}$$

$$\text{Overall reaction:}\ Zn + CO_2 + 2OH^- + 2H^+ \leftrightarrow ZnO + HCOOH + H_2O$$

$$E_{theo} = E_c - E_a = 0.955\ \text{V} \tag{4.40}$$

In 2019, Wang et al. proposed and realized a reversible aqueous Zn-CO_2 battery using bifunctional Ir@Au catalyst as cathode [63]. 3D porous Ir@Au bifunctional catalyst had high catalytic activity in CO_2 reduction and oxygen evolution reactions. As shown in Eqs. (4.41) and (4.42), during the discharge reaction of the battery, CO_2 is reduced to carbon monoxide, while the oxygen evolution reaction occurred during the charge process. The authors believed that CO_2RR and oxygen evolution reaction (OER) are the main electrochemical mechanisms of this rechargeable battery. Due to excellent cathode materials, the aqueous Zn-CO_2 battery had an open circuit potential of 0.7 V and a discharge capacity of 583 Wh kg^{-1}. Aqueous Zn-CO_2 batteries offered a more potential and greener alternative for CO_2 utilization and energy conversion and storage.

$$\text{Discharge:}\ \ CO_2 + 2H^+ + 2e^- \rightarrow CO + H_2O \tag{4.41}$$

$$\text{Charge:}\ \ H_2O \rightarrow \frac{1}{2}O_2 + 2H^+ + 2e^- \tag{4.42}$$

Cao and coworkers successfully assembled Zn-CO_2 battery using 3D porous hollow carbon nanotube fibers (CHFs) as the cathode, 1-ethyl-3-methylimidazolium tetrafluoroborate ([EMIM][BF_4]) as the electrolyte, and zinc wire as the anode [30]. Different from the corresponding CO_2 reduction mechanisms in the different catalysts and aqueous electrolytes mentioned above, CO_2 is reduced to methane in ionic liquid electrolytes. Specifically, residual water in ionic liquid shuttled between the Zn anode and the ionic liquid ensured the stable generation of H^+ and e^- required for CO_2 reduction, and the anion radical $CO_2{}^-$ disproportionated with CO_2 molecules to generate CO_{ads} and $CO_3{}^{2-}$, Finally, CO_{ads} generated methane under the action of protons and electrons. The discharge reaction mechanism is shown in Eq. (4.43).

$$CO_2 + 8H^+ + 8e^- \rightarrow CH_4 + 2H_2O \tag{4.43}$$

4.2.6 Discharge/Charge Mechanisms of Al-CO_2 Battery

As the most abundant metal on earth, aluminum is an ideal anode material for metal-CO_2 batteries. As early as 2013, Archer et al. verified the possibility of using Al-CO_2 batteries as primary batteries [20]. This battery used Al foils as the anode, conductive carbon as the cathode, and ionic liquid containing $AlCl_3$ and 1-ethyl-3-methylimidazolium chloride as the electrolyte. The results showed that the discharge capacity of the battery is only 162 mAh g^{-1} with a cutoff voltage of 0 V. Al Sadat and Archer considered that with the increase in temperature and the discovery of new cathode materials with higher specific surface area, the performance of Al-CO_2 batteries will be greatly improved. In 2016, Al Sadat and Archer et al. found the addition of O_2 can greatly improve the discharge capacity of Al-CO_2 batteries [64]. By using various characterization methods (direct analysis in real-time mass spectrometry [DARTMS], scanning electron microscope- energy dispersive spectrometry [SEM-EDS], X-ray photoelectron spectroscopy [XPS], thermogravimetric analysis [TGA]–Fourier transform infrared [TGA-FTIR]), the discharge product is identified as aluminum oxalate. In the discharge process, oxygen is first reduced at the cathode and further formed a superoxide intermediate. Subsequently, the superoxide reacted with carbon dioxide to produce aluminum oxalate. The specific reaction and the overall reaction are shown in Eqs. (4.44)–(4.50). Ma et al. in 2018 proposed and realized a rechargeable Al-CO_2 battery by using aluminum foil as the anode, ionic liquid as the electrolyte, and palladium-coated porous nano-gold (NPG@Pd) as the cathode [31]. The overpotential of the battery is only 0.091 V and the energy efficiency is as high as 87.7%. By using Raman spectroscopy, XPS, FTIR, and electron energy loss spectroscopy (EELS), the discharge products were confirmed to be aluminum carbonate and carbon. The reaction mechanism is shown in Eq. (4.51).

$$2Al \leftrightarrow 2Al^{3+} + 6e^- \tag{4.44}$$

$$6O_2 + 6e^- \leftrightarrow 6O_2^{\cdot -} \tag{4.45}$$

$$3CO_2 + 3O_2^{\cdot -} \leftrightarrow 3CO_4^{\cdot -} \tag{4.46}$$

$$3CO_4^{\cdot -} + 3O_2^{\cdot -} \leftrightarrow 3CO_4^{2-} + 3O_2 \tag{4.47}$$

$$3CO_4^{2-} + 3CO_2 \leftrightarrow 3C_2O_4^{2-} + 3O_2 \tag{4.48}$$

$$2Al^{3+} + 3C_2O_4^{2-} \leftrightarrow Al_2(C_2O_4)_3 \tag{4.49}$$

Overall reaction:

$$2Al + 6CO_2 \leftrightarrow Al_2(C_2O_4)_3 \tag{4.50}$$

$$4Al + 9CO_2 \leftrightarrow 2Al_2(CO_3)_3 + 3C \tag{4.51}$$

Li and coworkers first proposed an aqueous Al-CO_2 primary battery in 2020. They used Bi_2S_3-loaded gas diffusion layer as cathode and 1 M KOH as electrolyte in the Al-CO_2 battery. The results showed that the battery had a high working voltage (0.6 V) and good discharge stability. More importantly, a peak power density of 11 mW cm^{-2} and a formate yield of 0.5 mmol cm^{-2} h^{-1} were recorded at 0.3 V [65]. The cathode, anode, and overall reaction equation are shown in Eqs. (4.52)–(4.54). It is worth noting that the aqueous Al-CO_2 battery was only a primary battery and cannot be charged. The authors did not further analyze the discharge–charge mechanism of the aqueous Al-CO_2 battery in detail.

Cathode:

$$CO_2 + 2e^- + H_2O \rightarrow HCOO^- + OH^- \tag{4.52}$$

Anode:

$$Al + 4OH^- \rightarrow AlO_2^- + 2H_2O + 3e^- \tag{4.53}$$

Overall reaction:

$$2Al + 3CO_2 + 5OH^- \leftrightarrow 2AlO_2^- + 3HCOO^- + H_2O \tag{4.54}$$

4.3 The Cathode Materials of Metal-CO_2 Battery

In metal-CO_2 batteries, the cathode is an important part, which usually consists of a support (current collector or self-supporting electrode matrix) and a cathode active material, that is, a catalyst. Among them, carbon paper is usually used as the current collector, because carbon paper has high electrical conductivity to ensure small internal resistance of the battery and high air permeability to ensure the gas mass transfer required for the discharge–charge reaction. The failure of rechargeable metal-CO_2 batteries is mainly due to the incomplete separation of discharge products (Me_2CO_3 and C for metal (Li, Na, K, Al, Mg)-CO_2 batteries, HCOOH or CO for Zn-CO_2 battery). Therefore, the demand for superior catalysts has increased dramatically. In general, an ideal solid catalyst should have the following properties: (i) large specific surface area, which can provide effective storage space for the discharge product to ensure high discharge capacity; (ii) moderate porosity, which is conducive to the diffusion of Me^+ and CO_2 to ensure that the battery has high rate performance; (iii) good electrical conductivity to ensure fast electron transfer; (iv) high electrocatalytic activity for discharge product decomposition, which is beneficial to reduce charge–discharge polarization to improve battery energy efficiency, (v) electrochemically and thermally stable, and not corrode or decompose during battery operation; and (vi) cost-effective and environmental friendly.

In recent years, the research on catalysts has made great progress, and the overall performance of the metal-CO_2 battery has also been greatly improved. At present, the main catalysts for metal-CO_2 batteries mainly include the following: (i) carbon-based catalysts, mainly including commercial activated carbon (Super P, Ketjen Black, etc.) and nano-carbon materials (graphene, carbon nanotubes, etc.); (ii) noble metals (Ru, Ir, Au, etc.) and their derivatives; (iii) transition metals and their derivatives (NiO, MnO, Mo_2C, etc.); and (iv) porous framework-based catalysts (MOF and covalent organic framework [COF]).

4.3.1 Carbon-Based Catalysis and Additive Catalysis

Carbon materials have the advantages of abundant resources, low price, lightweight, high electrical conductivity, large specific surface area, and controllable pore structure. Therefore, carbon-based catalysis has been widely used in metal-CO_2 batteries [20, 35, 66–68]. Traditional carbon-based catalysts, such as carbon nanotubes, graphene, and conductive carbon black, were first used in metal-CO_2 batteries. In 2013, Archer and coworkers found for the first time that Super P can exhibit a high discharge capacity in Li-CO_2 batteries (6062 $mAh\,g^{-1}$ at 100 $mA\,g^{-1}$) [20]. Later, with the advancement of carbon material preparation technology, some new carbon-based catalysts (Ketjen Black, Graphdiyne) have also been used in metal-air batteries. Wang and coworkers used graphdiyne as a cathode catalyst for Li-CO_2 batteries and showed high specific capacity (18,416 $mAh\,g^{-1}$ at 100 $mA\,g^{-1}$) [67]. As a proof-of-concept, Wang et al. fabricated a flexible belt battery using the graphdiyne cathode that also exhibited extremely high energy density (165.5 $Wh\,kg^{-1}$). However, untreated commercial carbon materials have lower catalytic activity for battery reactions, mainly manifested in high voltage gap.

To solve this problem, various modification methods have been applied to carbon-based catalysts, such as heteroatom doping and defect fabrication [22, 69–72]. Heteroatom doping includes inorganic nonmetallic atomic doping and metal single-atom doping, among which inorganic nonmetallic atomic doping is dominated by N, B, S, and P, while metal single-atom doping is dominated by Fe and Ni.

Dai et al. designed two bifunctional graphene-based materials with sufficient defect structures, namely porous graphene (hG) and B, N-doped porous graphene BN-hG (Figure 4.4a) and used them in Li-CO_2 batteries [70]. Compared with the hG cathode, the BN-hG cathode exhibited extremely high reversible capacity and low polarization (Figure 4.4b,c). Dai and coworkers also manufactured carbon quantum dots (CQDs) with edge defects supported on porous graphene (CQDs/hG)

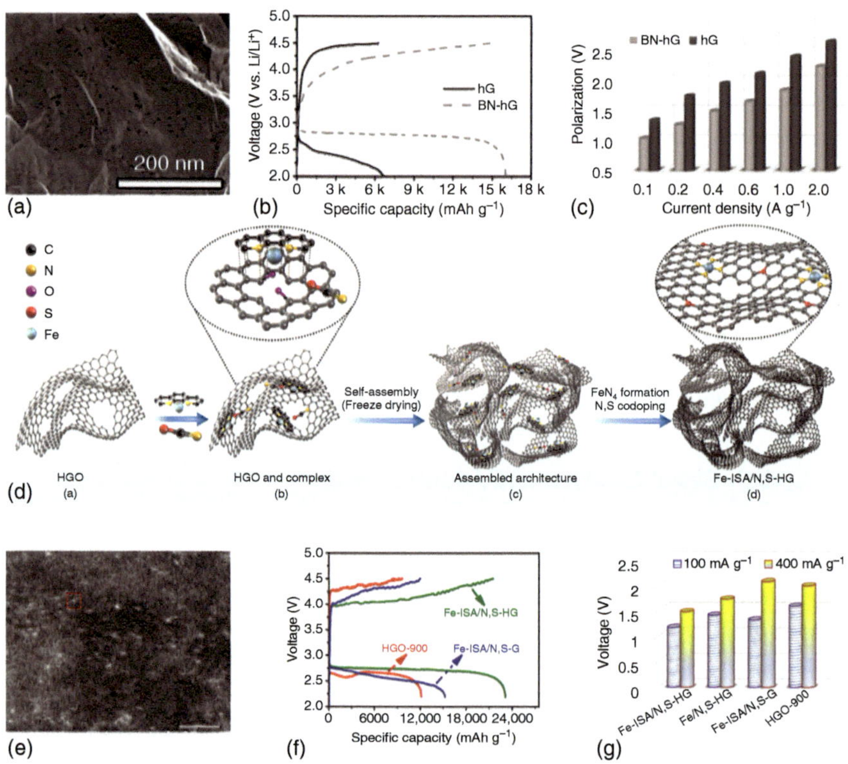

Figure 4.4 (a) SEM image of BN-hG. (b) Full discharge capacity of hG and BN-hG. (c) The polarization of hG and BN-hG at different current density. Source: Reproduced with permission of Qie et al. [70]. © 2017/John Wiley & Sons. (d) The synthesis process of Fe-ISA/N,S-hG material. (e) TEM image of Fe-ISA/N,S-hG material (scale bar: 50 nm). (f) Full discharge capacity of HGO-900, Fe-ISA/N,S-G, and Fe-ISA/N,S-HG. (g) The voltage gap of four cathodes at the different densities of 100 and 400 mA g^{-1}. Source: Reproduced with permission of Hu et al. [71]. © 2020/John Wiley & Sons.

as cathode for enhancing the electrochemical performance of Li-CO_2 batteries [69]. Due to the excellent catalytic activity of CQDs and the high electron/CO_2 transport ability of porous graphene, the CQDs/hG-0.3 cathode for Li-CO_2 batteries also displayed good cycling stability and rate performance. Wang and coworkers used chemical vapor deposition to prepare bifunctional bamboo-like N-doped carbon nanotubes catalysts with high catalytic activity and applied them to K-CO_2 batteries. It also showed good cycling stability (1500 hours at 50 mA g^{-1}) and low voltage gap (about 0.8 V at 50 mA g^{-1}) [28].

Xie and coworkers successfully prepared graphene loaded with Co, Ni, and Fe single atoms by coordination precipitation and thermal reduction. Due to the unique bonding structure and electron distribution, single atoms can significantly improve the catalytic activity of the material, reduce the overpotential of the battery, and improve the cycle performance of battery. Among them, Fe-SA/N-rGO cathode exhibited high full discharge capacity (16,835 mAh g^{-1} at 100 mA g^{-1}), low battery charge–discharge overpotential (1.45 V at 200 mA g^{-1}), and long cycle life (170 cycles at 400 mA g^{-1}) in Li-CO_2 batteries [72]. Dai and coworkers combined heteroatoms with carbon defects to design a 3D hollow N-S doped graphene-supported single-atom Fe cathode efficient catalyst (Fe-ISA/N, S-HG) for Li-CO_2 batteries (Figure 4.4d,e). Due to the N-S co-doping and the unique electron spin effect of Fe-N_4, it exhibited high discharge capacity and low polarization (Figure 4.4f,g).

4.3.2 Noble Metal-Based Catalysis

Compared with carbon-based catalysts, noble metal catalysts have the advantages of high catalytic activity and good stability and have been used in the field of new energy. Zhou et al. first used Ru nanoparticles (approximately 5–10 nm in diameter) deposited on Super P as the cathode for Li-CO_2 batteries [53]. The results showed that in the first discharge–charge process, it can provide a discharge capacity of 8229 mAh g^{-1} and an energy conversion efficiency of 86.2%.

Furthermore, the discharge products decomposition process of carbon-free Au electrodes and Au-Ru electrodes was investigated by using *in situ* surface-enhanced Raman technique, demonstrating that metallic Ru can effectively promote the reversible reaction of Li_2CO_3 and carbon, rather than the simple electrochemical decomposition of Li_2CO_3. Wang et al. synthesized a three-dimensional nickel foam supported Ru nanoparticles catalyst (Ru/Ni) using the simple displacement reaction method and used it as a binder-free cathode for Li-CO_2 batteries [73]. The highly dispersed Ru nanosheets in the Ru/Ni cathode effectively promoted the decomposition of the discharge product Li_2CO_3, thereby reducing the charge overpotential. It exhibited a discharge capacity of 9502 mAh g^{-1} and a Coulombic efficiency of 95.4% at a current density of 100 mA g^{-1}, which is much better than those of the KB/Ni electrode (5840 mAh g^{-1} and a Coulombic efficiency of 13.8%). Meanwhile, the Ru/Ni catalytic cathode also showed good rate capability (3177 mAh g^{-1} at a current density of 500 mA g^{-1}).

Ru derivatives (such as RuO_2 and RuP_2) also exhibited high catalytic activity and chemical stability in metal-CO_2 batteries. Wang and coworkers efficiently reduced charge voltage to 3.25 V by using RuO_2/layered double oxides material. Due to confined RuO_2 catalyst dispersed in the interlayer of bimetal layered double oxide (RuO_2/LDO), RuO_2 with good catalytic activity is monodispersed on the supporter. This structure involving monodisperse catalyst and layered supporter is beneficial to the decomposition of discharge products, which can significantly reduce the polarization of the battery and improve the energy efficiency [74]. Due to the excellent catalytic activity of Ru and its derivatives, Wang et al. reported for the first time that Ru/KB cathode is used in Na-CO_2 batteries [68]. Compared with KB cathode, Ru/KB exhibited higher discharge capacity (11,537 mAh g^{-1} at 100 mA g^{-1}) and higher rate capability (over 1000 hours at 200 mA g^{-1}). In addition to Ru-based catalysts, Pd, Ir, and Pt materials have also been used as cathodes for metal-CO_2 batteries. Ma et al. used palladium-coated nanoporous gold (NPG@Pd) as the cathode for Al-CO_2 batteries, which showed a lower voltage gap (0.091 V) and more superior cycling performance compared with pure nanoporous gold (NPG) [31]. Wang et al. prepared ultrathin Ir nanosheets anchored on the surface of carbon nanofibers (Ir-NSs-CNFs) as cathodes for Li-CO_2 batteries [75], which were cycled for more than 400 times at a current density of 500 mA g^{-1}. During the discharge process, the Ir-NSs-CNFs can stabilize the amorphous granular intermediate and delay intermediate transformation into Li_2CO_3, while during the charging process, it is favorable for the complete decomposition of Li_2CO_3. Kim et al. constructed a Mg-CO_2 aqueous battery with IrO_2/PtC as the cathode catalyst. In the reaction, the IrO_2/PtC catalyst can not only accelerate the kinetics of the reaction but also inhibit the solidification of carbonate products during the charging reaction, thereby greatly improving the utilization rate of CO_2 [61]. Additionally, in aqueous Zn-CO_2 batteries, different noble metal catalysts displayed different product selectivity. For example, when Ir/Au catalyst is used as the Zn-CO_2 battery cathode, the discharge product is CO, but when the catalyst is Pd, it is HCOOH [45].

Noble metal alloy catalysts with single-phase structure are also of great interest in metal-CO_2 batteries. Guo and coworkers designed ultrathin RuRh nanosheets as cathode for Li-CO_2 batteries [76]. It exhibited a charge voltage of 3.75 V and a voltage gap of 1.0 V at a current density of 200 mA g^{-1}. To reduce the usage of noble metals and maintain the high catalytic activity of the catalysts, various alloy catalysts that consist of noble metal and transition metal are used in metal-air batteries. Fan and coworkers prepared three alloy catalysts (RuCo, RuNi, and RuCu) by solvothermal method [77]. Among them, RuCo nanosheets show the lowest charging voltage (3.74 V at 100 mA g^{-1}) and overpotential (0.94 V at 100 mA g^{-1}), which is better than the performance of Ru nanoparticles mentioned above. Similarly, the RuCu nanoalloy particles reported by Zhou et al. also exhibited extremely low overpotential (0.9 V at 100 mA g^{-1}) and charging voltage (3.7 V at 100 mA g^{-1}) [37]. Low-cost NiFe alloy catalysts have also been reported for Li-CO_2 batteries. Tsiakaras and coworkers successfully anchored the NiFe core–shell structure on a 3D N-doped carbon nanonetwork structure, thereby obtaining a self-supporting battery cathode [23]. The initial discharge–charge test results showed that the Li-CO_2 battery with NiFe alloy exhibited low overpotential with a charge voltage of

3.78 V and a discharge voltage of 2.76 V. However, in subsequent cycle process, the battery exhibited instability and large polarization. This may be related to the low intrinsic activity of nonprecious metals.

4.3.3 Transition Metal-Based Catalysts

Although noble metal-based materials exhibit excellent catalytic activity for the formation/decomposition of discharge products, they are costly and scarce in resources. Therefore, the development of ideal and efficient nonprecious metal catalysts is a feasible strategy to solve these problems. At present, transition metal catalyst elements commonly used in battery cathodes mainly include Cu, Fe, Co, Ni, Mo, and Mn [34, 38–42, 51].

Mn-based catalysts have many applications in various monovalent metal-CO_2 batteries. Most studies showed that MnO and its derivatives had a better decomposition effect on discharge products. Liu and coworkers applied porous Mn_2O_3 materials to Li-CO_2 batteries and obtained excellent performance with an overpotential of only 1.40 V at a current density of 50 mA g^{-1} [40]. Subsequently, Ge et al. prepared Co-doped MnO_2 cathode catalysts. The Li-CO_2 battery based on this catalyst exhibited lower overpotential (0.73 V at 100 mA g^{-1}) and ultrahigh cycling performance (more than 500 cycles with a current density of 100 mA g^{-1}) [41]. In addition, Mn-based catalysts are used in Na-CO_2 batteries. Sui et al. used a two-step method (hydrothermal followed by annealing) to prepare MnO/CF and $CoMnO_2$/CF materials as high-efficiency cathodes for Na-CO_2 batteries [68]. Compared with MnO/CF cathode, $CoMnO_2$/CF cathode exhibited lower overpotential and higher cycling stability.

Among the transition metal-based catalysis, the brightest "star catalysts" are two molybdenum-based catalysts, molybdenum carbide and molybdenum nitride. Molybdenum carbide (Mo_2C) has an electronic structure similar to that of noble metals and exhibits excellent electrical conductivity and electrocatalytic properties. Chen et al. synthetized Mo_2C nanoparticles and further dispersed nanoparticles onto a three-dimensional network of carbon nanotubes (Mo_2C/CNTs) by ball milling and carbothermic reduction [34]. Mo_2C/CNT as a cathode catalyst exhibited extremely low charging voltage. What is more, they found that Mo_2C material can change the battery's charge–discharge reaction path. During the chemical reaction, Mo_2C can interact with the metastable intermediate $Li_2C_2O_4$ to form $Li_2C_2O_4$-Mo_2C, which prevents the further disproportionation of $Li_2C_2O_4$ to thermodynamically stable Li_2CO_3. In general, metastable $Li_2C_2O_4$ has much higher electrical conductivity and electrochemical activity than Li_2CO_3 crystals and can be completely decomposed at relatively low voltages. In 2022, Wang and coworkers fabricated a self-supporting CC@MoN flexible cathode for Li-CO_2 battery using solvothermal method and NH_3 annealing treatment [51]. The constructed battery had a charging voltage platform as low as 3.19 V and an initial energy efficiency of over 88%. Through experiments and simulation calculations, it can be seen that the MoN catalyst had better surface adsorption performance for $Li_2C_2O_4$, which created favorable conditions for its stable discharge intermediates. Besides, MoFeNi alloy and MoS_2 catalysts have also been adopted as cathode materials for metal-CO_2 batteries.

In addition to the application of transition metal catalysts in monovalent metal-CO_2 batteries, there are also a small amount of Fe-based and Ni-based derivative catalysts used in Zn-CO_2 battery. Liu et al. used nitrogen-doped polyhedral-supported Fe-P nanocrystal catalysts for the cathode in Zn-CO_2 battery. The catalysts exhibited high catalytic performance for CO_2RR with CO faradaic efficiency over 95% (vs. Reversible Hydrogen Electrode) at −0.55 V [78]. The peak power density of Zn-CO_2 battery is 0.85 mW cm^{-2}, which is comparable to other Zn-CO_2 battery performance based on noble metal cathode. The high selectivity and high efficiency of the catalysts were attributed to the presence of highly catalytic Fe-P nanocrystals in the N-doped carbon matrix, which can effectively increase the number of catalytically active sites and interfacial charge transfer. Zhang and coworkers synthesized a nickel binuclear bridged catalyst (Ni_2-N_4-C_2) by a two-step method [79]. When the catalyst is used as the cathode material of the Zn-CO_2 battery, the battery can be stably cycled for more than 130 times (almost 40 hours) with a high peak power density of 2.43 mW cm^{-2} at a constant current density of 0.5 mA cm^{-2}. The experimental results and density functional theory calculations showed that the porous carbon fiber with unique atom bridging active site is beneficial in improving the activity of carbon dioxide reduction and adsorption of intermediates by adjusting the electronic structure.

4.3.4 Porous Framework-Based Catalysts

MOFs and COFs with large surface areas, structural tunability, and well-defined porosity have received extensive attention in the field of catalysis. Wang and coworkers prepared eight MOF material catalysts with different catalytic points (Ni_2(dobdc), Co_2(dobdc), Mn_2(dobdc), Mn(bdc), Fe(bdc), Cu(bdc), $Mn(C_2H_2N_3)_2$, and $Mn(HCOO)_2$) using traditional methods [55]. Compared with other metal MOF-based catalysts, the Mn-based catalyst MOF exhibited lower charging potential in Li-CO_2 battery, especially the Mn_2(bdc) catalyst whose corresponding charging voltage is 3.96 V. They believed that the catalytic activity of Mn(II) and the highly porous MOF structure jointly promoted the electrochemical performance of Li-CO_2 battery.

Subsequently, Wang and coworkers designed uniformly dispersed untralthin MnO nanoparticles on N-doped 3D graphene [80]. It is worth noting that nitrogen-doped carbon framework is prepared by thermal decomposition of Mn-MOF linked in graphene oxide. Different from the Mn(bdc) material with poor conductivity in previous work, the nitrogen-doped carbon framework exhibited better conductivity, which further facilitated the transfer of electrons. The Li-CO_2 batteries with MnO@NC-G as cathode exhibited extremely low overpotential (0.88 V at 50 mA g^{-1}) and long cycle life (more than 200 cycles at 1 A g^{-1}).

COFs are periodic network structures, which can be formed by small molecular building units linked by covalent bonds. Strong covalent bonds endow them with high stability and ease of regulation, making COFs potentially useful as emitters, catalysis, and energy storage materials. For metal-CO_2 batteries, ordered porosity in COFs provided specific one-dimensional (1-D) channels for gas storage/separation and ion conduction. In 2020, Xie and coworkers prepared COF-Ru@CNT catalysts with high catalytic activity by solvothermal method, realizing high discharge capacity, and long cycle performance of Li-CO_2 batteries [81]. Among them, COF materials acted as gas reservoirs and ion diffusion channels, while Ru nanoparticles supported on CNTs acted as efficient catalysts. The amide functional group on the edge of the hole enabled the COF to coordinate with the Ru atoms on the CNT. In addition, its one-dimensional channel facilitated the transport of CO_2 gas and Li^+, further resulting in the rapid formation and decomposition of Li_2CO_3/C at the interface of COF and Ru@CNT.

4.4 The Electrolyte of Metal-CO_2 Battery

As a crucial part of metal-CO_2 batteries, the electrolyte has an important influence on the electrochemical performance of the battery, involving the battery's discharge capacity, energy density, and cycle stability. Therefore, it is of vital importance that develop an electrolyte with high dielectric constant and high ionic conductivity. Hitherto, there are two typical electrolytes (liquid electrolyte and solid-state electrolyte) commonly used in metal-CO_2 batteries. Further, liquid electrolyte can be divided into aprotic electrolyte and aqueous electrolyte.

4.4.1 The Nonaqueous Liquid Electrolyte

The main aprotic liquid electrolyte solvents reported in metal-CO_2 batteries include esters, ethers, sulfones, and ionic liquids. Asaoka and coworkers first tried to apply ester solvents in metal-air batteries [12]. However, the results showed that the esters solution was susceptible to nucleophilic attack by superoxide ions, resulting in a large number of by-products appearing on the surface of the cathode. These by-products can seriously affect the cycle life and energy efficiency of the battery. Surprisingly, sulfones (DMSO) and ether electrolytes (TEGDME) remained inert to superoxide ions and did not appear in many side reactions in the charge–discharge reaction (Figure 4.5a), so they are widely used as solvents for monovalent metal-CO_2 batteries [82, 84]. In addition, ionic liquids are also used in various metal-CO_2 batteries due to high vapor pressure, high thermal stability, and wide electrochemical performance window [20].

The performance of the liquid electrolyte is related to not only the solvent but also the concentration of electrolytes, additives, and other factors. For example, some redox mediator additives (LiBr and LiI) will improve the overall rate capability and discharge capacity of the battery [83]. Unlike cathode catalysts that do not bind products to participate in charge–discharge reactions, halogen material (LiBr) will react with the reactants and participates in CO_2RR and CO_2ER reactions (Figure 4.5b). The reactions processed are shown in Eqs. (4.55)–(4.57). Compared with the electrolyte without LiBr additive, the discharge capacity of Li-CO_2 battery

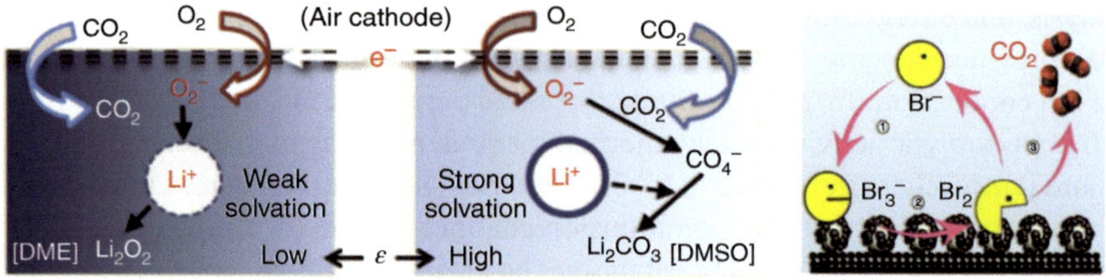

Figure 4.5 (a) Effect of electrolyte solvent dielectric constant on reaction pathways in Li-O_2/CO_2 batteries. Source: Reproduced with permission Lim et al. [82]. Copyright 2013, American Chemical Society. (b) Proposed mechanism for the charging process with LiBr as the redox mediator in Li-CO_2 batteries. Source: Reproduced with permission Wang et al. [83]. Copyright 2017, Wiley-VCH.

with LiBr-added electrolyte would be increased by five times [83]. In addition to halogen material additives, organic material additives are also added to the electrolyte. Grimaud and coworkers added 2,5-di-tert-butyl-1,4-benzoquinone (DBBQ) to the electrolyte to promote the reduction of CO_2 in Li-CO_2 batteries [85]. However, abundant by-products are appeared in the electrolyte that consisted of quinone-based solvents, further deteriorating the electrochemical performance of the battery. Therefore, whether quinone electrolytes can be used as electrolytes for metal-CO_2 batteries is worthy of further study.

$$3Br^- \rightarrow Br_3^- + e^- \tag{4.55}$$

$$2Br_3^- \rightarrow 3Br_2 + 2e^- \tag{4.56}$$

$$2Li_2CO_3 + C + 2Br_2 \leftrightarrow 3CO_2 + 4LiBr \tag{4.57}$$

4.4.2 The Aqueous Electrolyte

Aqueous electrolytes are commonly used in Zn-CO_2 battery because of their low reactivity with water. In aqueous electrolytes, due to the presence of protons, various reaction paths may occur when different kinds of catalysts serve as the cathode of Zn-CO_2 battery. In Zn-CO_2 battery, the product corresponding to the two-electron transfer reaction is carbon monoxide, and the product corresponding to the eight-electron transfer reaction is methane. Additionally, a small amount of aqueous electrolytes are also used in Li-CO_2 batteries and made great progress [52, 86]. The most typical Li-CO_2 battery aqueous electrolyte is water-in-salt electrolyte, which consists of a high concentration of salt and an aqueous solution. Such electrolytes are characterized by stable performance over a wide electrochemical window and good ion transport properties.

Xia and coworkers assembled a Li-CO_2 battery using water-in-salt (LiTFSI/H_2O) as the electrolyte and Mo_2C as the cathode. Further studies have shown that the reaction path of the battery is disproportionated in the water-in-salt electrolyte, and the discharge product is changed from refractory lithium carbonate to lithium oxalate [86]. More importantly, lithium oxalate can stably exist in the aqueous electrolyte and on the surface of cathode. Impressively, Wang and coworkers recently used a mixed aqueous electrolyte of LiCl and HCOOLi as the electrolyte of Li-CO_2 battery. The results showed the electrolyte not only can achieve reversible cycling of high-performance Li-CO_2 batteries but also can generate a new energy product, formic acid [52]. Kang and coworkers first used water-in-salt electrolyte ($NaClO_4$ aqueous solution) in Na-CO_2 batteries. In Na-CO_2 batteries, high concentration of aqueous electrolyte can effectively delay the occurrence of hydrogen evolution reaction (HER) and enlarge the electrochemical stability window of the battery [87].

4.4.3 The Solid-State Electrolyte

To solve the problem that the solvent in the liquid electrolyte is easy to volatilize, some scholars have studied the possibility of solid electrolyte as the electrolyte of metal-CO_2 secondary battery in recent years.

Compared with liquid electrolytes, solid electrolytes have the advantages of high safety, low volatility, and stability. Different types of solid-state electrolytes have different ionic conductivity and change with temperature. According to the chemical composition as the classification basis, the solid electrolytes commonly used in metal-CO_2 batteries can be divided into inorganic solid electrolytes and polymer solid electrolytes. The NASICON-type LAGP and LLZTO inorganic all-solid-state electrolytes have been used in metal-CO_2 batteries. Zhou et al. used LAGP as electrolyte in a Li-CO_2 battery for the first time and found that the battery could cycle stably for 30 cycles at a current density of 500 mA g^{-1} and exhibited a discharge capacity of 2499 mAh g^{-1} at the temperature of 60 °C [88]. Additionally, in 2021, Liu et al. used another NASICON-type electrolyte $Na_3Zr_2Si_2PO_{12}$ (NZSP) in Na-CO_2 batteries. The battery exhibited high full discharge capacity (28,830 mAh g^{-1} with a current density of 100 mA g^{-1}) and low voltage gap (1.4 V with a current density of 50 mA g^{-1}) [89, 90].

Different from inorganic solid electrolytes, polymer solid electrolytes have good flexibility and elasticity, which make them well adapted to the volume change of Li metal anodes during the long-time discharge–charge process [90–92]. Generally speaking, polymer solid electrolytes are composed of polymer matrices and lithium salts. The polymers used must contain Lewis base sites where lithium salts can be dissolved. In 2017, Chen and coworkers reported a liquid-free solid-state Li-CO_2 battery. By doping 3% SiO_2 particles in poly(methacrylate) /poly(ethylene glycol) (PEG)-$LiClO_4$ polymer, they prepared a composite solid electrolyte with an ion conductivity of 7.14×10^{-5} S cm^{-1} (at 55 °C) [91]. Later, by doping $Li_7La_3Zr_{1.4}Ta_{0.6}O_{12}$ particles in polyethylene oxide , they obtained a composite solid-state electrolyte. The Li-CO_2 battery can be stably run for 70 cycles at a constant capacity of 1000 mAh g^{-1} at 70 °C [90]. Polymer electrolytes are also used in Na-CO_2 batteries. Sun et al. used a magnesium-doped $Na_3Zr_2Si_2PO_{12}$ (NZSP) solid-state electrolyte. This study has shown that Mg ion doping can greatly enhance the ionic conductivity of the electrolyte and thus improve the overall electrochemical performance of the Na-CO_2

battery [92]. In addition, polyvinylidene fluoride-co-hexafluoropropylene composite polymer electrolyte (CPE) is also applied in Na-CO_2 batteries [93].

4.5 Summary and Outlook

This chapter briefly and systematically summarizes the charge/discharge mechanisms of various types of metal-CO_2 batteries in recent years and focuses on the effects of cathode materials and electrolytes on reaction kinetics and electrochemical performance. Although metal-CO_2 batteries have achieved certain development, the following issues and some development opportunities should be paid more attention to in future exploration research.

It is crucial to in-depth research on the reaction mechanism and reaction kinetics of divalent metal/trivalent metal-CO_2 batteries during charge and discharge process. Currently, some advanced *in situ* characterization methods have been used in the elucidation of the reaction mechanism of monovalent metal-CO_2 batteries, such as *in situ* DEMS, *in situ* TEM, and *in situ* Raman. However, in divalent metal/trivalent metal-CO_2 batteries, advanced *in-situ* approaches are rarely applied. More importantly, the reaction mechanism of divalent metal/trivalent metal-CO_2 batteries is currently unclear. Therefore, it is necessary to combine advanced *in situ* characterization to understand the nature of the catalytic reaction and to clarify the relationship between the internal structure of the catalyst and the electrochemical performance.

At present, there are many kinds of cathode materials for metal-CO_2 batteries. However, how to select the material with the best battery performance among thousands of catalysts has become an arduous task. Machine learning and density functional theory, which are effective screening and computational methods in the field of materials, provide a new idea. Through density functional theory, the adsorption energy of different catalysts for intermediate products and the decomposition energy barriers of products can be accurately calculated, and then a database of excellent catalysts can be obtained theoretically. Then, various algorithm models of machine learning are used for training and testing, and finally the optimal cathode material is screened out.

The development of new liquid electrolytes with low volatility and good electrochemical stability is urgent for metal-CO_2 battery. Moreover, it is necessary to find suitable additives for improving battery reaction kinetics and rate performance. At the same time, for solid electrolytes, overall battery failure due to interfacial chemical reaction cannot be ignored. Hence, the ionic conductivity of the solid electrolyte needs to be improved, and the interface contact between the solid electrolyte and the cathode needs to be further optimized, thereby reducing the interface impedance. In addition, some flexible soft-pack solid-state batteries have also been used in monovalent metal-CO_2 batteries, but they show the characteristics of low specific capacity. Thus, fabrication of lightweight solid electrolytes has to be done so as to achieve high specific capacity metal-CO_2 battery.

Metal-CO_2 battery with excellent performance is inseparable from a cathode material with high catalytic activity and an electrolyte with high ionic conductivity. It is necessary to comprehensively consider the interaction between each part to achieve breakthrough progress. We believe that with the discovery of more advanced materials, metal-CO_2 batteries will surely become the high-performance green energy in the field of next-generation energy storage.

References

1 Holechek, J.L., Geli, H.M.E., Sawalhah, M.N., and Valdez, R. (2022). A global assessment: can renewable energy replace fossil fuels by 2050? *Sustainability* 14 (8): 4792.

2 Armand, M. and Tarascon, J.M. (2008). Building better batteries. *Nature* 451: 652–657.

3 Nitta, N., Wu, F., Lee, J.T., and Yushin, G. (2015). Li-ion battery materials: present and future. *Materials Today* 18 (5): 252–264.

4 Tan, P., Wei, Z.H., Shyy, W. et al. (2016). A nano-structured RuO_2/NiO cathode enables the operation of non-aqueous lithium–air batteries in ambient air. *Energy & Environmental Science* 9 (5): 1783–1793.

5 Aurbach, D., McCloskey, B.D., Nazar, L.F., and Bruce, P.G. (2016). Advances in understanding mechanisms underpinning lithium-air batteries. *Nature Energy* 1: 16128.

6 Geng, D., Ding, N., Hor, T.S.A. et al. (2016). From lithium-oxygen to lithium-air batteries: challenges and opportunities. *Advanced Energy Materials* 6 (9): 1502164.

7 Wang, J., Li, Y., and Sun, X. (2013). Challenges and opportunities of nanostructured materials for aprotic rechargeable lithium–air batteries. *Nano Energy* 2 (4): 443–467.

8 Luntz, A.C. and McCloskey, B.D. (2014). Nonaqueous Li-air batteries: a status report. *Chemical Reviews* 114 (23): 11721–11750.

9 Dutta, A., Ito, K., and Kubo, Y. (2019). Establishing the criteria and strategies to achieve high power during discharge of a Li–air battery. *Journal of Materials Chemistry A* 7 (40): 23199–23207.

10 Park, J.-B., Lee, S.H., Jung, H.-G. et al. (2018). Redox mediators for Li–O_2 batteries: status and perspectives. *Advanced Materials* 30 (1): 1704162.

11 Li, Y., Wang, J., Li, X. et al. (2011). Superior energy capacity of graphene nanosheets for a nonaqueous lithium-oxygen battery. *Chemical Communications* 47 (33): 9438–9440.

12 Takechi, K., Shiga, T., and Asaoka, T. (2011). A Li–O_2/CO_2 battery. *Chemical Communications* 47 (12): 3463–3465.

13 Yu, X. and Manthiram, A. (2020). Recent advances in lithium–carbon dioxide batteries. *Small Structures* 1 (2): 2000027.

14 Huang, B. and Frapper, G. (2018). Pressure-induced polymerization of CO_2 in lithium–carbon dioxide phases. *Journal of the American Chemical Society* 140 (1): 413–422.

15 Goodarzi, M., Nazari, F., and Illas, F. (2018). Assessing the performance of cobalt phthalocyanine nanoflakes as molecular catalysts for Li-promoted oxalate formation in Li–CO_2–oxalate batteries. *The Journal of Physical Chemistry C* 122 (45): 25776–25784.

16 Xiao, X., Shang, W., Yu, W. et al. (2020). Toward the rational design of cathode and electrolyte materials for aprotic Li-CO_2 batteries: a numerical investigation. *International Journal of Energy Research* 44 (1): 496–507.

17 Liu, B., Sun, Y., Liu, L. et al. (2019). Recent advances in understanding Li–CO_2 electrochemistry. *Energy & Environmental Science* 12 (3): 887–922.

18 Mao, Y., Tang, C., Tang, Z. et al. (2019). Long-life Li–CO_2 cells with ultrafine IrO_2-decorated few-layered δ-MnO_2 enabling amorphous Li_2CO_3 growth. *Energy Storage Materials* 18: 405–413.

19 Song, L., Wang, T., Wu, C. et al. (2019). A long-life Li–CO_2 battery employing a cathode catalyst of cobalt-embedded nitrogen-doped carbon nanotubes derived from a Prussian blue analogue. *Chemical Communications* 55 (85): 12781–12784.

20 Xu, S., Das, S.K., and Archer, L.A. (2013). The Li–CO_2 battery: a novel method for CO_2 capture and utilization. *RSC Advances* 3 (18): 6656–6660.

21 Hu, A., Shu, C., Xu, C. et al. (2019). Design strategies toward catalytic materials and cathode structures for emerging Li–CO_2 batteries. *Journal of Materials Chemistry A* 7 (38): 21605–21633.

22 Li, X., Zhou, J., Zhang, J. et al. (2019). Bamboo-like nitrogen-doped carbon nanotube forests as durable metal-free catalysts for self-powered flexible Li–CO_2 batteries. *Advanced Materials* 31 (39): 1903852.

23 Liang, H., Zhang, Y., Chen, F. et al. (2019). A novel NiFe@NC-functionalized N-doped carbon microtubule network derived from biomass as a highly efficient 3D free-standing cathode for Li-CO_2 batteries. *Applied Catalysis B: Environmental* 244: 559–567.

24 Li, Y.G. and Lu, J. (2017). Metal air batteries: will they be the future electrochemical energy storage device of choice? *ACS Energy Letters* 2 (6): 1370–1377.

25 Xu, C., Zhang, K., Zhang, D. et al. (2020). Reversible hybrid sodium-CO_2 batteries with low charging voltage and long-life. *Nano Energy* 68: 104318.

26 Mu, X., Pan, H., He, P., and Zhou, H. (2020). Li–CO_2 and Na–CO_2 batteries: toward greener and sustainable electrical energy storage. *Advanced Materials* 32 (27): 1903790.

27 Zhang, L., Tang, Y., Liu, Q. et al. (2018). Probing the charging and discharging behavior of K-CO_2 nanobatteries in an aberration corrected environmental transmission electron microscope. *Nano Energy* 53: 544–549.

28 Li, X., Qi, G., Zhang, J. et al. (2022). Artificial solid-electrolyte interphase and bamboo-like N-doped carbon nanotube enabled highly rechargeable K-CO_2 batteries. *Advanced Functional Materials* 32 (2): 2105029.

29 Zhang, C., Wang, A., Guo, L. et al. (2022). A moisture-assisted rechargeable Mg–CO_2 battery. *Angewandte Chemie International Edition* 61 (17): e202200181.

30 Wang, K., Wu, Y., Cao, X. et al. (2020). A Zn–CO_2 flow battery generating electricity and methane. *Advanced Functional Materials* 30 (9): 1908965.

31 Ma, W., Liu, X., Li, C. et al. (2018). Rechargeable Al–CO_2 batteries for reversible utilization of CO_2. *Advanced Materials* 30 (28): 1801152.

32 Wang, F., Li, Y., Xia, X. et al. (2021). Metal–CO_2 electrochemistry: from CO_2 recycling to energy storage. *Advanced Energy Materials* 11 (25): 2100667.

33 Xie, J., Liu, Q., Huang, Y. et al. (2018). A porous Zn cathode for Li–CO_2 batteries generating fuel-gas CO. *Journal of Materials Chemistry A* 6 (28): 13952–13958.

34 Hou, Y., Wang, J., Liu, L. et al. (2017). Mo_2C/CNT: an efficient catalyst for rechargeable Li–CO_2 batteries. *Advanced Functional Materials* 27 (27): 1700564.

35 Das, S.K., Xu, S., and Archer, L.A. (2013). Carbon dioxide assist for non-aqueous sodium–oxygen batteries. *Electrochemistry Communications* 27: 59–62.

36 Qiao, Y., Liu, Y., Chen, C. et al. (2018). 3D-printed graphene oxide framework with thermal shock synthesized nanoparticles for Li-CO_2 batteries. *Advanced Functional Materials* 28 (51): 1805899.

37 Jin, Y., Chen, F., Wang, J., and Johnston, R.L. (2019). Tuning electronic and composition effects in ruthenium-copper alloy nanoparticles anchored on carbon nanofibers for rechargeable Li-CO_2 batteries. *Chemical Engineering Journal* 375: 121978.

38 Zhang, Z., Zhang, Z., Liu, P. et al. (2018). Identification of cathode stability in Li–CO_2 batteries with Cu nanoparticles highly dispersed on N-doped graphene. *Journal of Materials Chemistry A* 6 (7): 3218–3223.

39 Lu, S., Shang, Y., Ma, S. et al. (2019). Porous NiO nanofibers as an efficient electrocatalyst towards long cycling life rechargeable Li–CO_2 batteries. *Electrochimica Acta* 319: 958–965.

40 Ma, W., Lu, S., Lei, X. et al. (2018). Porous Mn_2O_3 cathode for highly durable Li–CO_2 batteries. *Journal of Materials Chemistry A* 6 (42): 20829–20835.

41 Ge, B., Sun, Y., Guo, J. et al. (2019). A Co-doped MnO_2 catalyst for Li-CO_2 batteries with low overpotential and ultrahigh cyclability. *Small* 15 (34): 1902220.

42 Pipes, R., Bhargav, A., and Manthiram, A. (2018). Nanostructured anatase titania as a cathode catalyst for Li–CO_2 batteries. *ACS Applied Materials & Interfaces* 10 (43): 37119–37124.

43 Pipes, R., He, J., Bhargav, A., and Manthiram, A. (2019). Efficient Li–CO_2 batteries with molybdenum disulfide nanosheets on carbon nanotubes as a catalyst. *ACS Applied Energy Materials* 2 (12): 8685–8694.

44 Wang, H., Xie, K., You, Y. et al. (2019). Realizing interfacial electronic interaction within ZnS quantum dots/N-rGO heterostructures for efficient Li–CO_2 batteries. *Advanced Energy Materials* 9 (34): 1901806.

45 Xie, J. and Wang, Y. (2019). Recent development of CO_2 electrochemistry from Li–CO_2 batteries to Zn–CO_2 batteries. *Accounts of Chemical Research* 52 (6): 1721–1729.

46 Zhou, J., Li, X., Yang, C. et al. (2019). A quasi-solid-state flexible fiber-shaped Li–CO_2 battery with low overpotential and high energy efficiency. *Advanced Materials* 31 (3): 1804439.

47 Freunberger, S.A., Chen, Y., Peng, Z. et al. (2011). Reactions in the rechargeable lithium–O_2 battery with alkyl carbonate electrolytes. *Journal of the American Chemical Society* 133 (20): 8040–8047.

48 Németh, K. and Srajer, G. (2014). CO_2/oxalate cathodes as safe and efficient alternatives in high energy density metal–air type rechargeable batteries. *RSC Advances* 4 (4): 1879–1885.

49 Xing, Y., Yang, Y., Li, D. et al. (2018). Crumpled Ir nanosheets fully covered on porous carbon nanofibers for long-life rechargeable lithium–CO_2 batteries. *Advanced Materials* 30 (51): 1803124.

50 Qiao, Y., Yi, J., Wu, S. et al. (2017). Li-CO_2 electrochemistry: a new strategy for CO_2 fixation and energy storage. *Joule* 1 (2): 359–370.

51 Qi, G., Zhang, J., Chen, L. et al. (2022). Binder-free MoN nanofibers catalysts for flexible 2-electron oxalate-based Li-CO_2 batteries with high energy efficiency. *Advanced Functional Materials* 32 (22): 2112501.

52 Xue, H., Gong, H., Lu, X. et al. (2021). Aqueous Formate-based Li-CO_2 battery with low charge overpotential and high working voltage. *Advanced Energy Materials* 11 (35): 2101630.

53 Yang, S., Qiao, Y., He, P. et al. (2017). A reversible lithium–CO_2 battery with Ru nanoparticles as a cathode catalyst. *Energy & Environmental Science* 10 (4): 972–978.

54 Zhao, Z., Huang, J., and Peng, Z. (2018). Achilles' heel of lithium–air batteries: lithium carbonate. *Angewandte Chemie International Edition* 57 (15): 3874–3886.

55 Li, S., Dong, Y., Zhou, J. et al. (2018). Carbon dioxide in the cage: manganese metal–organic frameworks for high performance CO_2 electrodes in Li–CO_2 batteries. *Energy & Environmental Science* 11 (5): 1318–1325.

56 Mahne, N., Renfrew, S.E., McCloskey, B.D., and Freunberger, S.A. (2018). Electrochemical oxidation of lithium carbonate generates singlet oxygen. *Angewandte Chemie International Edition* 57 (19): 5529–5533.

57 Zhang, Z., Wang, X.-G., Zhang, X. et al. (2018). Verifying the rechargeability of Li-CO_2 batteries on working cathodes of Ni nanoparticles highly dispersed on N-doped graphene. *Advanced Science* 5 (2): 1700567.

58 Hu, X., Sun, J., Li, Z. et al. (2016). Rechargeable room-temperature Na–CO_2 batteries. *Angewandte Chemie International Edition* 55 (22): 6482–6486.

59 Zhu, Y., Feng, S., Zhang, P. et al. (2020). Probing the electrochemical evolutions of Na–CO_2 nanobatteries on Pt@NCNT cathodes using in-situ environmental TEM. *Energy Storage Materials* 33: 88–94.

60 Lu, Y., Cai, Y., Zhang, Q. et al. (2021). Rechargeable K-CO_2 batteries with a KSn anode and a carboxyl-containing carbon nanotube cathode catalyst. *Angewandte Chemie International Edition* 60 (17): 9540–9545.

61 Kim, J., Seong, A., Yang, Y. et al. (2021). Indirect surpassing CO_2 utilization in membrane-free CO_2 battery. *Nano Energy* 82: 105741.

62 Xie, J., Wang, X., Lv, J. et al. (2018). Reversible aqueous zinc–CO_2 batteries based on CO_2–HCOOH interconversion. *Angewandte Chemie International Edition* 57 (52): 16996–17001.

63 Wang, X., Xie, J., Ghausi, M.A. et al. (2019). Rechargeable Zn–CO_2 electrochemical cells mimicking two-step photosynthesis. *Advanced Materials* 31 (17): 1807807.

64 Al Sadat, W.I. and Archer, L.A. (2016). The O_2-assisted Al/CO_2 electrochemical cell: a system for CO_2 capture/conversion and electric power generation. *Science Advances* 2 (7): e1600968.

65 Ding, P., Zhang, J., Han, N. et al. (2020). Simultaneous power generation and CO_2 valorization by aqueous Al–CO_2 batteries using nanostructured Bi_2S_3 as the cathode electrocatalyst. *Journal of Materials Chemistry A* 8 (25): 12385–12390.

66 Zhang, X., Zhang, Q., Zhang, Z. et al. (2015). Rechargeable Li–CO_2 batteries with carbon nanotubes as air cathodes. *Chemical Communications* 51 (78): 14636–14639.

67 Zhang, J., Wang, F., Qi, G. et al. (2021). Rechargeable Li-CO_2 batteries with graphdiyne as efficient metal-free cathode catalysts. *Advanced Functional Materials* 31 (26): 2101423.

68 Sui, D., Chang, M., Wang, H. et al. (2021). A brief review of catalytic cathode materials for Na-CO_2 batteries. *Catalysts* 11 (5): 603.

69 Jin, Y., Hu, C., Dai, Q. et al. (2018). High-performance Li-CO_2 batteries based on metal-free carbon quantum dot/holey graphene composite catalysts. *Advanced Functional Materials* 28 (47): 1804630.

70 Qie, L., Lin, Y., Connell, J.W. et al. (2017). Highly rechargeable lithium-CO_2 batteries with a boron- and nitrogen-codoped holey-graphene cathode. *Angewandte Chemie International Edition* 56 (24): 6970–6974.

71 Hu, C., Gong, L., Xiao, Y. et al. (2020). High-performance, long-life, rechargeable Li–CO_2 batteries based on a 3D holey graphene cathode implanted with single iron atoms. *Advanced Materials* 32 (16): 1907436.

72 Zhou, L., Wang, H., Zhang, K. et al. (2021). Fast decomposition of Li_2CO_3/C actuated by single-atom catalysts for Li-CO_2 batteries. *Science China Materials* 64 (9): 2139–2147.

73 Zhao, H., Li, D., Li, H. et al. (2019). Ru nanosheet catalyst supported by three-dimensional nickel foam as a binder-free cathode for Li–CO_2 batteries. *Electrochimica Acta* 299: 592–599.

74 Xu, S.-M., Ren, Z.-C., Liu, X. et al. (2018). Carbonate decomposition: low-overpotential Li-CO_2 battery based on interlayer-confined monodisperse catalyst. *Energy Storage Materials* 15: 291–298.

75 Wang, C., Zhang, Q., Zhang, X. et al. (2018). Fabricating Ir/C nanofiber networks as free-standing air cathodes for rechargeable Li-CO_2 batteries. *Small* 14 (28): 1800641.

76 Xing, Y., Wang, K., Li, N. et al. (2020). Ultrathin RuRh alloy nanosheets enable high-performance lithium-CO_2 battery. *Matter* 2 (6): 1494–1508.

77 Wang, Y., Zhou, J., Lin, C. et al. (2022). Decreasing the overpotential of aprotic Li-CO_2 batteries with the in-plane alloy structure in ultrathin 2D Ru-based nanosheets. *Advanced Functional Materials* 32 (30): 2202737.

78 Liu, S., Wang, L., Yang, H. et al. (2022). Nitrogen-doped carbon polyhedrons confined Fe–P nanocrystals as high-efficiency bifunctional catalysts for aqueous Zn–CO_2 batteries. *Small* 18 (10): 2104965.

79 Cao, X., Zhao, L., Wulan, B. et al. (2022). Atomic bridging structure of nickel–nitrogen–carbon for highly efficient electrocatalytic reduction of CO_2. *Angewandte Chemie International Edition* 61 (6): e202113918.

80 Li, S., Liu, Y., Zhou, J. et al. (2019). Monodispersed MnO nanoparticles in graphene-an interconnected N-doped 3D carbon framework as a highly efficient gas cathode in Li–CO_2 batteries. *Energy & Environmental Science* 12 (3): 1046–1054.

81 Li, X., Wang, H., Chen, Z. et al. (2019). Covalent-organic-framework-based Li–CO_2 batteries. *Advanced Materials* 31 (48): 1905879.

82 Lim, H.K., Lim, H.D., Park, K.Y. et al. (2013). Toward a lithium-"Air" battery: the effect of CO_2 on the chemistry of a lithium-oxygen cell. *Journal of the American Chemical Society* 135 (26): 9733–9742.

83 Wang, X.-G., Wang, C., Xie, Z. et al. (2017). Improving electrochemical performances of rechargeable Li–CO_2 batteries with an electrolyte redox mediator. *ChemElectroChem* 4 (9): 2145–2149.

84 Guo, Z., Li, J., Qi, H. et al. (2019). A highly reversible long-life Li–CO_2 battery with a RuP_2-based catalytic cathode. *Small* 15 (29): 1803246.

85 Yin, W., Grimaud, A., Azcarate, I. et al. (2018). Electrochemical reduction of CO_2 mediated by quinone derivatives: implication for Li–CO_2 battery. *The Journal of Physical Chemistry C* 122 (12): 6546–6554.

86 Feng, N., Wang, B., Yu, Z. et al. (2021). Mechanism-of-action elucidation of reversible Li–CO_2 batteries using the water-in-salt electrolyte. *ACS Applied Materials & Interfaces* 13 (6): 7396–7404.

87 Im, E., Ryu, J.H., Baek, K. et al. (2021). "Water-in-salt" and NASICON electrolyte-based Na–CO_2 battery. *Energy Storage Materials* 37: 424–432.

88 Du, Y., Liu, Y., Yang, S. et al. (2021). A rechargeable all-solid-state Li–CO_2 battery using a $Li_{1.5}Al_{0.5}Ge_{1.5}(PO_4)_3$ ceramic electrolyte and nanoscale RuO_2 catalyst. *Journal of Materials Chemistry A* 9 (15): 9581–9585.

89 Tong, Z., Wang, S.-B., Fang, M.-H. et al. (2021). Na–CO_2 battery with NASICON-structured solid-state electrolyte. *Nano Energy* 85: 105972.

90 Wang, R., Zhang, X., Cai, Y. et al. (2019). Safety-reinforced rechargeable Li-CO_2 battery based on a composite solid state electrolyte. *Nano Research* 12 (10): 2543–2548.

91 Hu, X., Li, Z., and Chen, J. (2017). Flexible Li-CO_2 batteries with liquid-free electrolyte. *Angewandte Chemie International Edition* 56 (21): 5785–5789.

92 Lu, L., Sun, C., Hao, J. et al. (2023). A high-performance solid-state Na-CO_2 battery with poly(vinylidene-fluoride-co-hexafluoropropylene)-$Na_{3.2}Zr_{1.9}Mg_{0.1}Si_2PO_{12}$ electrolyte. *Energy & Environmental Materials* 6 (3): e12364.

93 Hu, X., Li, Z., Zhao, Y. et al. (2017). Quasi-solid state rechargeable Na-CO_2 batteries with reduced graphene oxide Na anodes. *Science Advances* 3 (2): e1602396.

5

Multivalent-Ion Batteries: Magnesium and Beyond

Qirong Liu[1,2] and Yongbing Tang[1,2]

[1] *Advanced Energy Storage Technology Research Center, Shenzhen Institutes of Advanced Technology, Chinese Academy of Sciences, 1068 Xueyuan Avenue, Shenzhen University Town, Shenzhen 518055, China*
[2] *University of Chinese Academy of Sciences, No.19 Yuquan Road, Shijingshan District, Beijing 100049, China*

With ever-increasing demand for lithium-ion batteries (LIBs) in different application fields such as electric vehicles, mobile electronic devices, and large-scale energy storage, the limited Li/Ni/Co resources and their uneven geographical distribution in the Earth's crust lead to a sharply increasing cost [1, 2]. Developing non-Li rechargeable batteries that rely on sustainable and earth-abundant metal elements, such as monovalent ions (Na^+ and K^+) and multivalent ions (Mg^{2+}, Ca^{2+}, Zn^{2+}, and Al^{3+}) [3–5], has attracted considerable attention (Figure 5.1). However, due to the larger ionic radii of these monovalent ions than Li^+ ions, non-Li monovalent-ion batteries generally suffer from relatively low specific capacities and energy densities [6]. In contrast, multivalent ions have multiple charges and their electrochemical reactions involve the transfer of multiple electrons in one charge–discharge process [7]. Meanwhile, multivalent ions allow for more uniform plating/stripping on the corresponding metal anode, limiting dendritic growth compared to Na and K metal anodes. Therefore, multivalent-ion batteries are expected to present the superiority of high energy densities over non-Li monovalent-ion batteries. For example, the high reserve of magnesium/calcium resources as the eighth/fifth most abundant element (2%/5%) in the Earth's crust endows the Mg^{2+}/Ca^{2+}-ion batteries with low cost and environmental benignity. Compared with LIBs, multivalent-ion batteries present several merits: (i) High theoretical volumetric energy density. Due to its bivalent nature, Mg anode can deliver a theoretical volumetric capacity of $3832\,mAh\,cm^{-3}$, which is much higher than that of Li anode ($2046\,mAh\,cm^{-3}$). In addition, the Mg anode has a low electrochemical redox potential of $-2.37\,V$ (vs. standard hydrogen electrode, SHE) [8, 9]. (ii) High safety originating from low sensitivity to the dendrite growth and propagation [3]. Compared to lithium, multivalent metals (e.g. Mg and Ca) may be deposited uniformly from appropriate electrolyte solutions, with little or no dendritic growth. Because the Mg anode has a relatively low reactive activity and is prone to a dendrite-free plating/stripping process, it is beneficial in reducing the possibility of dendrite penetration through

Electrochemical Energy Storage Devices: Non-Conventional Technologies and Materials, First Edition. Edited by Yongbing Tang and Luojiang Zhang.

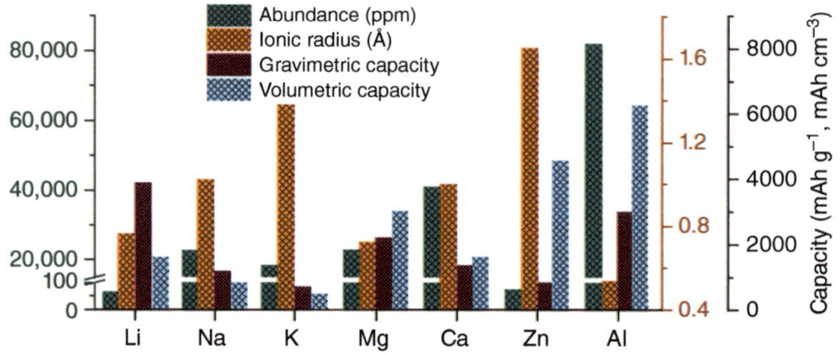

Figure 5.1 Comparison between elemental abundance, ionic radius, and gravimetric and volumetric capacities of different metal anodes. Source: Reproduced with permission [4].

the cell. In addition, the Mg metal anode has higher chemical stability than alkali metals. Similarly, the fifth most earth-abundant Ca resource features extensive distribution and nontoxicity. The Ca metal anode can deliver high volumetric and gravimetric capacities of 2073 mAh cm^{-3} and 1337 mAh g^{-1}, respectively, and the standard redox potential of Ca/Ca^{2+} (−2.87 V vs. SHE) is close to that of Li/Li^+ [10–12]. In this sense, research on multivalent-ion batteries provides an alternative pathway for developing next-generation rechargeable battery systems.

However, there are some hurdles that remain to be surmounted for these multivalent-ion batteries (Figure 5.2) such as the formation of a nonconductive passivation layer, the dendrite growth on the anode surface, the sluggish intercalation kinetics, and the degradation of cathode structure induced by the intercalation of multivalent ions with high charge densities [16–18]. The multiple charges of multivalent ions result in relatively higher charge densities, which generally lead to strong electrostatic interactions with the host lattices of cathodes, thus deteriorating their structural stability [19]. These multivalent-ion intercalation behaviors involve a local charge distribution in the lattice structure, which has a significant influence on the local electronic structure, to some extent, limits the diffusion kinetics of multivalent ions within cathodes. As a result, designing proper cathode materials is a common challenge in developing multivalent-ion batteries. In addition, there are some special issues that exist in different multivalent-ion battery systems.

For example, there is generally a lack of well-compatible electrolytes for Mg/Ca metal anode in Mg/Ca-ion batteries, because the formation of the nonconductive passivation layer leads to poor reversibility of Mg/Ca plating/stripping on the anode surface [13, 20, 21]. It has been demonstrated that the solid-electrolyte interphase (SEI) layer formed on the Ca metal anode mainly comprises CaF_2, $CaCO_3$, $Ca(OH)_2$, or calcium alkoxides, which is nonconductive to Ca^{2+} diffusion [22, 23]. Although there are some research showing the Ca^{2+} conductivity of the SEI consisting of CaH_2 or defect-containing CaF_2 [23, 24], the influence of the SEI on the Ca^{2+} plating/stripping remains undefined. Besides, insufficient cycling stability limits the practical application of aqueous Zn-ion batteries. The dendrite growth, the formation of irreversible by-products, and the corrosion issue of Zn anode raise

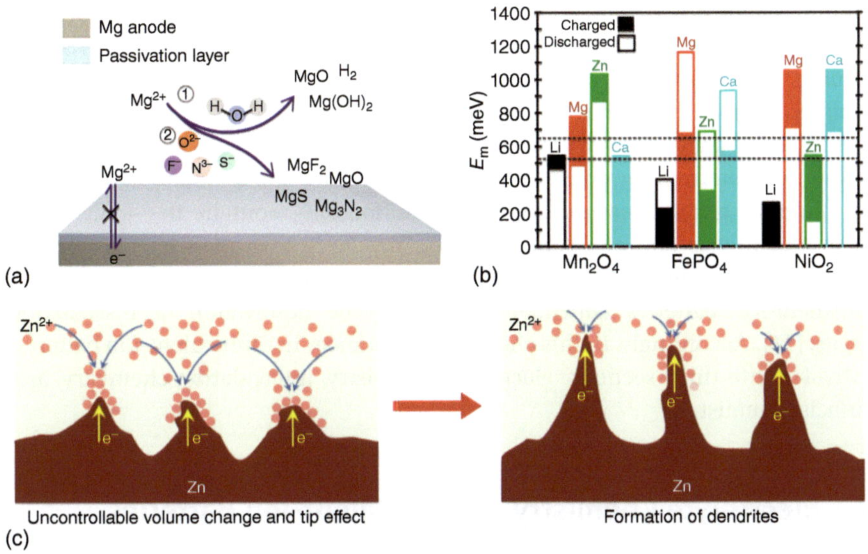

Figure 5.2 (a) Schematic illustration of the formation mechanism of passivation layer formation mechanism and the corresponding solutions. Source: Reproduced with permission [13]. Copyright 2023, Elsevier. (b) A summary of migration barriers (E_m) compared to the prescribed ~525–650 meV threshold (dashed). Source: Reproduced with permission [14]. Copyright 2015, American Chemical Society. (c) Schematic process of dendrite formation. Source: Reproduced with permission [15]. Copyright 2020, Springer Nature.

concerns about the severe irreversibility of Zn plating/stripping [25]. Similarly, Al-ion batteries also face some formidable challenges, such as a poor cycling lifespan, energetic parasitic reactions, unstable voltage platforms, and insufficient structural stability of cathodes [26, 27].

To surmount these hurdles of multivalent-ion batteries, various strategies have been developed to improve the reversibility of metal plating/stripping on metal anodes, to enhance the diffusion capability of multivalent ions in the bulk of materials, and to promote the interfacial kinetics at the electrode/electrolyte interfaces. (i) Electrolyte chemistry is regulated to improve diffusivity of multivalent ions and the electrochemical compatibility between electrolytes and electrodes [13, 28]. For example, the dissociation of electrolyte salts is promoted via introducing strong chelating solvent molecules and/or highly electronegative auxiliary anions, which is beneficial for improving the diffusivity of multivalent ions in electrolytes and further optimize their electro-/chemical compatibility with electrode materials [19]. (ii) Intercalation chemistry is finely designed to enhance intercalation kinetics of multivalent carriers. In contrast to a liquid electrolyte, the diffusion of multivalent ions in the bulk solid electrode environment remains more challenging. On the one hand, the high charge density of multivalent ions and relatively narrow spacing for ion hopping in rigid crystal structures result in strong electronegative interactions [29]. On the other hand, the intercalation of multivalent ions is accompanied by the change in local electronic structure and the distortion of lattice structure [30]. Therefore, the fine design of intercalation chemistry has been implemented to improve the diffusion kinetics of multivalent carriers in cathodes via either crystal engineering of cathode materials or screening for active ions with multiple charges. (iii) Manipulating interfacial chemistry enables reversible plating/stripping of multivalent ions and reduces the interfacial impedance within multivalent-ion batteries. It is demonstrated that the rate-determining processes of multivalent-ion storage may be associated with multiple electrochemical steps, which not only contain the ionic diffusion in electrolytes and electrode materials involving migration energy barriers of multivalent carriers but also relate to their interfacial mass transfer within multivalent-ion batteries dependent on either the desolvation or dissociation energies [31]. Accordingly, in this chapter, the advances in multivalent-ion batteries are divided into three sections: electrolyte chemistry, intercalation chemistry, and interfacial chemistry.

5.1 Electrolyte Chemistry of Multivalent-Ion Batteries

Multivalent-ion batteries share a similar working mechanism with LIBs, where the energy storage process involves the reversible transport of active ions and electrons in electrochemical redox reactions. Therefore, the highly efficient operation of multivalent-ion batteries is dependent on the reversible and smooth shuttling of multivalent-ion between the anode and the cathode through the electrolyte during charging/discharging processes [32]. However, the electrochemical redox kinetics of multivalent-ion batteries is limited by the insufficient mobility of multivalent ions in electrode materials, electrolytes, and interfaces. The intrinsic features of multivalent ions result in strong interplays between polar species or moieties, thus causing a distinct electrochemical polarization effect, which hinders the transport dynamics of multivalent ions [32]. As is well known, the electrolyte plays a vital role in delivering the electrochemical properties of batteries, because it is not only the medium for ion transport but also participates in the formation of interfaces. The interface chemistry is dictated by the electrolyte via dominating the parasitic reactions at the interfaces between electrolyte/electrodes and determining the electrochemical voltage stability window. The mobility of multivalent ions in electrolyte environment can be promoted via enhancing the dissociation ability and reducing desolvation barrier of electrolyte. For example, magnesium-ion batteries were first proposed as a primary battery at the beginning of the twentieth century, marking a milestone in the development of rechargeable Mg-ion batteries [20], which also inspired the study of reversible Mg electrochemistry based on different electrolytes. These different solvents and Mg salts were designed to improve the chemical and electrochemical compatibility of electrolytes, boosting the transport dynamics of Mg^{2+} in electrolytes.

One formidable problem of Mg metal anode is the intrinsic formation of an insulating and passivating layer upon exposure to moisture, which creates an impermeable physical barrier for the transport of Mg^{2+} ions [13]. Consequently, the passivation layer causes the irreversibility of electrochemical reactions on the Mg metal anode (Figure 5.3a). On the one hand, when the Mg metal anode comes into contact with H_2O or moisture in the air, products such as MgO and $Mg(OH)_2$ are formed on the surface of the Mg anode. On the other hand, anions in the electrolyte can be electrochemically reduced to generate passive interphases such as MgO and MgF_2. The transport paths of Mg^{2+} ion at the interface will be isolated by the passivation layer, which limits the reversibility of the Mg stripping/deposition process and reduces the Coulombic efficiency and cycling stability. Therefore, the electrolyte design should be elaborately implemented to limit the presence of passivation layer. The issue can be partially alleviated by finely optimizing conventional electrolyte recipes, for example, by developing new solvents, electrolyte salts, or functional additives [24, 37]. In the past decades, many efforts have been made to explore thermodynamically compatible electrolytes for Mg-ion batteries with Mg metal anodes, such as Grignard-based electrolytes and boron-based electrolytes [20]. Despite their high reduction stability at the Mg metal anode, these electrolytes generally suffer from intrinsically weak oxidation stability and insufficient compatibility with high-voltage cathode materials. The rational design of suitable electrolytes plays a vital role in overcoming these challenges. Generally speaking, the electrolytes should exhibit good compatibility with metal anodes for multivalent-ion batteries, a high ionic conductivity for multivalent-ion transport, and a relatively wide electrochemical voltage stability window.

Rationally selecting solvents is a prerequisite to designing compatible electrolyte components of multivalent-ion batteries. The selection of electrolyte solvents should take into consideration the following characteristics: (i) the ability to enable high salt solubility and good dissociation of electrolyte salts to enhance the mobility of multivalent ions; (ii) high electro-/chemical compatibility with electrode materials; and (iii) low desolvation energy of solvated cations to promote the interfacial kinetics of multivalent ions. For example, the commonly used solvents for electrolytes in Mg-ion batteries are ethers such as ethylene glycol dimethyl ether (DME), tetrahydrofuran (THF), dimethyl sulfoxide (DMSO), and acetonitrile (AN), which exhibit good electrochemical reduction resistance to Mg metal anode and can stabilize Mg^{2+} by coordinating with electron-donating ether oxygen (Figure 5.3b) [13]. In addition, ethers generally have a low viscosity and high dielectric constant, which can promote the dissociation of electrolyte salts and increase the ionic conductivity of multivalent ions. However, their relatively strong solvating ability leads to the formation of a stable solvated structure (such as $Mg(DME)_3{}^{2+}$), which is not beneficial for improving the interfacial kinetics.

Fine design of electrolyte salts is crucial for developing high-performance multivalent-ion batteries because the anions of these salts generally participate in the electrochemical formation of SEIs (Figure 5.3c). The electrolyte design should meet several requirements: (i) High solubility in solvents to improve the mobility of multivalent ions; (ii) wide electrochemical stability windows to be compatible with electrode materials; (iii) high thermal stability and processability to enhance safety and reduce electrolyte cost; and (iv) the ability of anions to form SEI/cathode electrolyte interphase (CEI)-forming ability to prevent the formation of a passivation layer and to enable the smooth transport of multivalent ions.

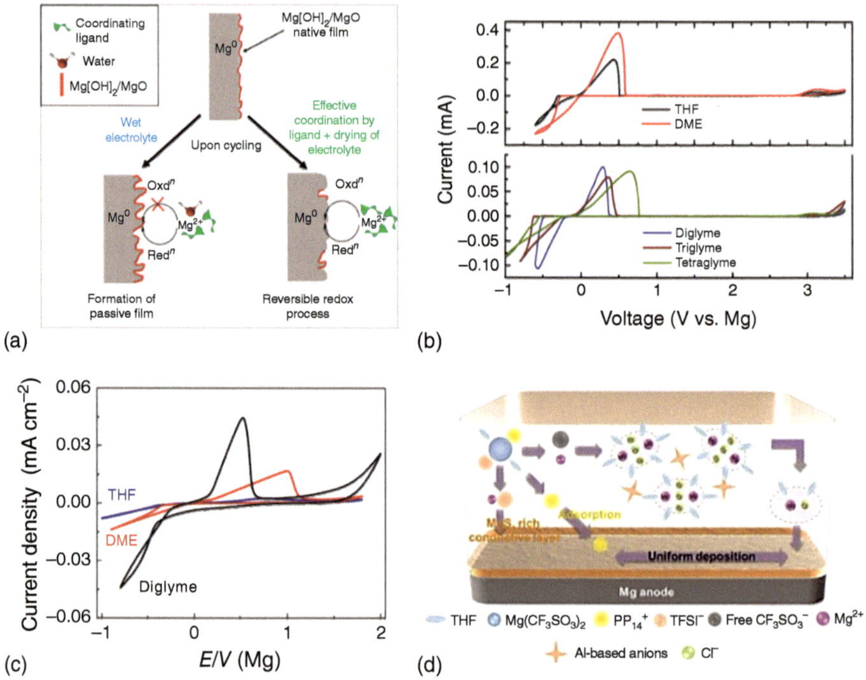

Figure 5.3 (a) Schematic diagram of the working mechanism of Mg anode. Source: Reproduced with permission [33]. Copyright 2019, Wiley-VCH. (b) Cyclic voltammograms of Mg electrolytes synthesized by reacting 0.4 M $MgCl_2$ with a desired amount of $AlEtCl_2$ in different solvents as indicated. Source: Reproduced with permission [34]. Copyright 2015, The Royal Society of Chemistry. (c) Cyclic voltammograms recorded on a Pt electrode in Cl-free 0.01 M $Mg(BH_4)_2$ in DGM, DME, and THF. Source: Reproduced with permission [35]. Copyright 2013 Springer Nature. (d) Bulk phase model of the MAM-IL (8 : 1) electrolyte. Source: Reproduced with permission [36]. Copyright 2023, Elsevier.

These anions are generally composed of high-valence central elements (e.g. B, P, S, and Cl) and their bonding groups (e.g. =F and =O) [13]. However, some of these elements are prone to generate chemical components (such as MgF_2 and CaO) of the passivation layer via interfacial parasitic reactions between the metal anode and electrolyte. It is clear that the selection of electrolyte salts requires insight into their electrochemical nature in solvents.

In addition, electrolytes, typically consisting of solvents and salts (usually Mg salts), enable the reversible cycling of these multivalent-ion batteries. However, their comprehensive, performance, such as the delivered capacity, rate capability, and long-term cycling stability, is insufficient for practical applications. Electrolyte additives have been regarded as critical components to enhance the electrochemical stability of electrolyte and electrodes, to improve interfacial diffusion kinetics, and to facilitate the transport of multivalent ions [38–40]. For example, in order to avoid the formation of a passivation layer on the Mg metal anode, different protective layers have been constructed using electrolyte additives such as $PP_{14}TFSI$, bismuth trifluoromethanesulfonate ($Bi[OTf]_3$), and $GeCl_4$ (Figure 5.3d) [36, 39, 40].

5.2 Intercalation Chemistry of Multivalent-Ion Batteries

Compared to liquid electrolyte environment, solid hosts in cathode materials pose greater restrictions on the multivalent-ion diffusion because the high charge density of multivalent ions cannot be shielded by an assisting solvent sheath, and thus the strong interactions between multivalent ions and the host crystal structure hinder the ion hopping among lattice positions [29]. Therefore, developing advanced electrode materials is a research hotspot for constructing high-performance multivalent-ion batteries [29, 32]. Although there has been some progress in the design of cathode materials allowing for multivalent-ion diffusion, it remains challenging for the cathode host structure to simultaneously allow sufficient mobility and high storage capacity for multivalent ions.

Accordingly, an in-depth understanding of the crystallographic and electronic structures of cathode compounds is extremely important. This understanding contributes to the rational optimization of the physical geometries of cathode materials for smooth multivalent-ion chemistry [19]. On this basis, crystal engineering can be implemented to fine-tune the local lattice structure of cathode materials, thus effectively lowering the energy barrier for multivalent-ion migration. For instance, manganese dioxide (MnO_2) has different oxidation states and abundant polymorphic structures, due to the six-coordination structure of Mn by O, where adjacent octahedron can be stacked with different edge-shared or corner-shared configurations [41]. The diversity of crystal structure supports various crystal engineering approaches for the regulation of electrochemical Mg^{2+} storage properties. Hollandite α-MnO_2, possessing large 1D migration channels, has enough space to allow Mg^{2+} transport [42]. However, the reversible and robust electrochemical Mg^{2+} storage behavior of α-MnO_2 was confined to a low magnesiation level ($Mg/Mn \leq 0.25$) [30]. Otherwise, the migration channels are prone to collapsing, and the crystal material tends to become amorphous, induced by the deep magnesiation of hollandite α-MnO_2. Theoretical calculations suggest that the strong affinity between Mg^{2+} and O^{2-} preferably drives the electrochemical formation of oxides (MgO and MnO) rather than Mg^{2+}-intercalated Mg_xMnO_2 compounds [30].

5.2.1 Diffusion Channel Engineering

It is vital to rationally design the crystal structure of cathode materials to form favorable migration channels. As mentioned above, in transition metal oxides, O atoms surrounding cations build up the migration channels for active ions. When oxide frameworks are changed to sulfide or selenide structures, a favorable migration environment with a weaker polarizability contributes to a lower migration energy barrier and facilitates the transport of multivalent ions. For example, in vanadium dichalcogenides with a layered structure, the transport of Mg^{2+} species from a thermodynamically stable octahedral site to the next equivalent one must cross an intermediate tetrahedral site [14]. Theoretical calculations suggest that the highest energy barrier of Mg^{2+} species along the migration path in VO_2 is up to 1032 meV. The value is much higher than the threshold of the energy barrier (525–650 meV) required for moderate Mg^{2+} diffusion kinetics. In contrast, vanadium disulfide (VS_2) has a reduced value of 593 meV, which decreases to 346 meV in its selenide framework [14, 43]. The kinetic superiority of VSe_2 results in distinct voltage plateaus in the reversible Mg^{2+} intercalation/de-intercalation process.

Generally, a four-coordinated structure is desired for the cathode materials in LIBs due to the presence of favorable Li^+ diffusion channels, while some multivalent ions, such as Mg^{2+}, favor a six coordination in the host structure [19]. Promising cathode materials with olivine or layered structures for LIBs have stably six-coordinated insertion sites. To accomplish migration between adjacent octahedral sites, active ions have to go through a four-coordinated tetrahedral site that is facile for Li^+. However, the tetrahedral site is unfavorable for Mg^{2+} diffusion, accompanied by a high migration energy barrier. Different from the olivine and layered structures, the spinel structure (for example, Mn_2O_4, Co_2O_4, and Ni_2O_4) can provide stable tetrahedral sites for Mg^{2+} insertion, bridged by face-shared octahedral intermediate sites, which present a relatively smaller Mg^{2+} diffusion energy barrier (650–850 meV), close to the threshold of the energy barrier for moderate Mg^{2+} diffusion kinetics. Such Mg^{2+} storage processes in spinel oxides can theoretically deliver a relatively high voltage (~3.0 V vs. Mg/Mg^{2+}) and a considerably high specific capacity (above 200 mAh g^{-1}) [44, 45]. However, so far, the electrochemical Mg^{2+} storage performance of spinel oxides is practically limited by the high energy barrier of Mg^{2+} diffusion and the formation of rocksalt structures [46]. Additionally, the polymorphs of V_2O_5 enable a large tunable space to accommodate volume changes induced by the de-/intercalation of multivalent cations such as Ca^{2+} ions and Al^{3+} ions [47, 48]. Density functional theory (DFT) calculations suggest that different V_2O_5 polymorphs exhibit distinct cycling stability and diffusion energy barriers for Ca^{2+} intercalation. δ-V_2O_5 has a significantly lower energy barrier for Ca^{2+} diffusion than α-V_2O_5, even though the latter presents a better structure stability during the Ca^{2+} ion intercalation process. The kinetic superiority of δ-V_2O_5 can be attributed to the smaller ion-coordination number fluctuation during the Ca^{2+} diffusion process [49].

In addition, weakening the interaction between multivalent ions and host lattice, which lowers the diffusion energy barrier, and increasing the diffusion kinetics have also been a feasible method to enhance the migration kinetics of multivalent ions and the structural stability of cathode materials. For example, molybdenum disulfide (MoS_2) has a layered structure, where the tunable interlayer spacing allows for the maneuverability of the interaction between Mg^{2+} and host lattice, due to the weak van der Waals interplay forces between adjacent layers. Theoretical calculations suggest that increasing the interlayer spacing of MoS_2 from 0.62 nm (pristine value) to 0.772 nm could significantly enhance the Mg^{2+} diffusion kinetics comparable to that of Li^+ in pristine MoS_2 [50]. Moreover, it has also been demonstrated that rational defect engineering of the known structures can boost the diffusion kinetics of active multivalent ions. In addition, a defect-free anatase TiO_2 only delivered a specific capacity of 25 mAh g^{-1} (0.037 Mg^{2+} per formula unit) for electrochemical Mg^{2+} storage, while the introduction of titanium vacancies through aliovalent doping enhanced the electrochemical de-/magnesiation activity of TiO_2, resulting in a greatly improved specific capacity up to 165 mAh g^{-1} and superior rate capability [51]. Furthermore, the defect engineering may also provide materials with an extra driving force for thermodynamically favorable diffusion of active ions. It is proven that the existence of anionic vacancies facilitates Mg^{2+} diffusion in TiO_2. Cation-deficient anatase TiO_2 also showed a similar improvement for reversible Al^{3+} intercalation.

5.2.2 Delocalizing Electronic Structure

The multivalency of active cations results in their high charge densities, which on the one hand cause a strong electrostatic interaction between active cations and the host lattice, on the other hand give rise to the formation of electrochemical redox products with high thermodynamic stability [30]. It is clear that, compared to the monovalent Li^+, the multivalent ions exert a stronger influence on the local electronic structure surrounding the ions. As a result, the multivalent nature imposes significant limitation on the transport kinetics of these active ions. In addition to engineering the host structure, regulating the local charge allocation is of significant importance to weaken their impact on the local electronic structure. Thermodynamically, delocalized electronic structures can result in the existence of multiple redox centers with similar valence bands, which is beneficial to enhance the electrochemical kinetics of Mg^{2+} ions in electrode materials [19]. Chevrel phase Mo_6X_8 (X represents S, Se, etc.) clusters have been used as cathode materials with the good capability of reversible Mg^{2+} storage. For example, Aurbach et al. [52] developed the first Mg^{2+} cell prototype based on a Mo_6S_8 cathode, which delivered good cycling reversibility (Figure 5.4a). The superior Mg^{2+} transport kinetics of the Chevrel phase can be ascribed to its unique Mo_6S_8-block cluster structure that consists of S_8 cubes individually encaging a Mo_6 octahedron (Figure 5.4b). Six Mo atoms in the octahedron share the double charge of Mg^{2+} ions at outer ring sites, owing to the delocalized electronic structure caused by metallic Mo=Mo bonding, which leads to minimal change in the oxidation state of Mo atoms [55, 56]. In addition, there are extra Mg^{2+} storage sites in the inner ring with a charge shift toward S atoms, even though this Mg^{2+} storage process is generally subjected to intrinsically sluggish kinetics [53]. The Chevrel phase Mo_6S_8 cathode delivered a reversible room-temperature capacity of only approximately 70 mAh g^{-1} at 1.1 V [52]. The replacement of Se for S in Mo_6Se_8 cathode could weaken the Mg^{2+} cathode interaction and regain the capacity from anionic redox [57]. Nevertheless, electrochemical energy storage processes stemming from metallic bonds usually result in a relatively low potential. Analogous Chevrel phases ($CaMo_6X_8$, X = S, Se, Te) are also theoretically available for Ca^{2+} and Zn^{2+} intercalation [16, 54, 58]. $CaMo_6S_8$ cathode was theoretically predicted to deliver a favorable voltage of 1.4 V (vs. Ca/Ca^{2+}); however, the diffusion barrier for Ca^{2+} was significantly higher than that of Mg^{2+} in the cathode. In contrast, the selenide replacement greatly reduced the diffusion barrier from 780 to 520 meV (Figure 5.4c).

In addition, there is another method to produce a delocalized electronic structure where the multicenter redox reactions are associated with the simultaneous anionic

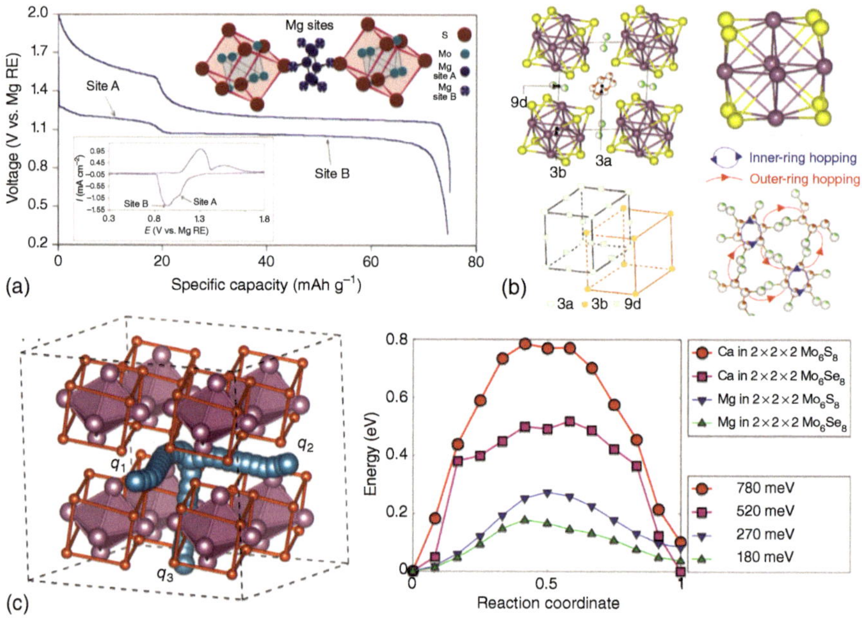

Figure 5.4 (a) Typical electrochemical behavior and the basic structure of the $Mg_xMo_3S_4$ cathodes, $0 < x < 1$, corresponding to a maximal charge capacity of $122\ mAh\ g^{-1}$. The electrolyte was 0.25 M $Mg(AlCl_2BuEt)_2$ in THF. A chronopotentiogram (main figure; voltage vs. capacity at a constant current of $0.3\ mA\ cm^{-2}$) and a cyclic voltammogram (inset; $0.05\ mV\ s^{-1}$) of steady-state Mg insertion–deinsertion cycles are shown. Source: Reproduced with permission [52]. Copyright 2000, Springer Nature. (b) Crystalline structure and sublattices in Chevrel phase Mo_6S_8, Mo_6S_8 superanion and positions of highly symmetric 3a, 3b, and 9d sites (viewed along the [211] direction). Mo is colored light purple, S yellow, the inner site orange and white, and the outer site green and white. Outer-ring and inner-ring hopping between partially occupied inner and outer sites. Source: Reproduced with permission [53]. Copyright 2017, American Chemical Society. (c) The three diffusion pathways considered in a 2 × 2 × 2 supercell, and calculated diffusion barriers for Ca^{2+} in Mo_6S_8 (red circles), Mo_6Se_8 (magenta squares), Mg^{2+} in Mo_6S_8 (blue inverted triangles), and Mg^{2+} in Mo_6S_8 (green upright triangles). The barriers are anisotropic due to the distorted geometry. Only the energies of the pathway corresponding to the lowest diffusion barrier and the corresponding activation energies are shown. Source: Reproduced with permission [54]. Copyright 2015, Elsevier.

and cationic redox at a similar redox potential. As a result, the charges of multivalent carriers could be uniformly distributed among multiple redox centers. For example, transition metal chalcogenide compounds can induce the simultaneous cationic and anionic redox when there is an overlapped valence orbitals of their transition metal with the chalcogenide [19, 59]. Additionally, compared to other classical ionic transition metal compounds, higher covalency inside these transition metal chalcogenide compounds can be obtained for their bonding with multivalent carriers. Therefore, there is a weaker interplay environment between multivalent carriers and host structure in cathode materials to cause a faster diffusion kinetics. It is noteworthy that the replacement of Se for S (such as VSe_2 and $TiSe_2$) can produce larger overlapped areas of valence orbitals for Se and transition metal bands [43]. For example, the intercalation of Mg^{2+} ions into a $TiSe_2$ cathode can deliver a relatively high reversible capacity of c. 120 mAh g^{-1} and robust cycling stability, accompanied only by a slight change in lattice structure (the variation of *d*-spacing ratio before and after Mg^{2+} intercalation is less than 1%) [59]. A similar phenomenon could be observed for the Mg^{2+} storage in a VSe_2 cathode. It is clear that selenide compounds present promising superiority of diffusion kinetics for multivalent carriers over their sulfide counterparts.

5.2.3 Properly Shielding Charges of Multivalent Carriers

High charge densities of multivalent carriers play a crucial role in the strong interaction between multivalent carriers and cathode hosts. Proper shielding of the charges of multivalent carriers is another approach to reduce the interaction. The approach relies on the intercalation chemistry of anion or solvent-assistant intercalants coupling with a multivalent carrier, which is beneficial to reduce the charge densities due to the decreased net charge or/and the increased total volume of the intercalants such as $Mg_xCl_y^+$ clusters, $AlCl_4^-$ clusters, and solvent-coordinated Mg^{2+} species [19, 60–62]. The addition of Cl^- ions into Mg^{2+} electrolytes can boost the dissociation of Mg salts via constructing electrochemically active monovalent $Mg_xCl_y^+$ clusters with an increased volume, which results in an obviously lowered charge density. As a result, much faster diffusion kinetics can be thermodynamically obtained for these multimeric $Mg_xCl_y^+$ clusters. Furthermore, the intercalation chemistry of $MgCl^+$ species can limit the energetically unfavorable Mg=Cl dissociation and thus enhance the interfacial kinetics. For example, Yoo et al. reported that the replacement of Mg^{2+} ion with $MgCl^+$ species dramatically lowered the diffusion energy barrier of active ions from 0.51 to 0.18 eV in an interlayer-expanded TiS_2 material, corresponding to a fivefold higher diffusivity of $MgCl^+$ species (Figure 5.5a,b) [63]. Notably, there are several hurdles for the intercalation chemistry of $Mg_xCl_y^+$ clusters [19]. One is the expanded structure of electrode materials that is necessary to enable the storage of $Mg_xCl_y^+$ clusters without severe steric effect. The second hurdle stems from the low energy capacity of $Mg_xCl_y^+$ intercalation. Moreover, the corrosive disadvantage of Cl^- ions raises concern on electrochemical compatibility with electrode materials.

As for solvent-coordinated multivalent ions, highly polarizing groups of solvent molecules (for example, H_2O) enable strong coordination to form solvated multivalent species, where their multiple charges can be effectively shielded by the solvation shell, weakening the electrostatic interactions between multivalent carrier and the host lattice. For instance, the H_2O solvent is beneficial to improve the electrochemical properties of MnO_2 cathodes in Mg^{2+}-based electrolytes (Figure 5.5c), which could deliver a reversible capacity of above 230 mAh g^{-1} at 2.8 V (vs. Mg^{2+}/Mg) [64]. However, there are some formidable issues remaining to be surmounted. (i) A narrow electrochemical stable voltage window of the H_2O solvent limits its electrochemical compatibility with high-voltage cathode materials. (ii) The H_2O solvent presents poor electrochemical compatibility with Mg metal anode. (iii) The as-produced protons accompanying the hydrolysis of H_2O solvent may participate in the electrochemical processes of Mg^{2+} storage. Many efforts have

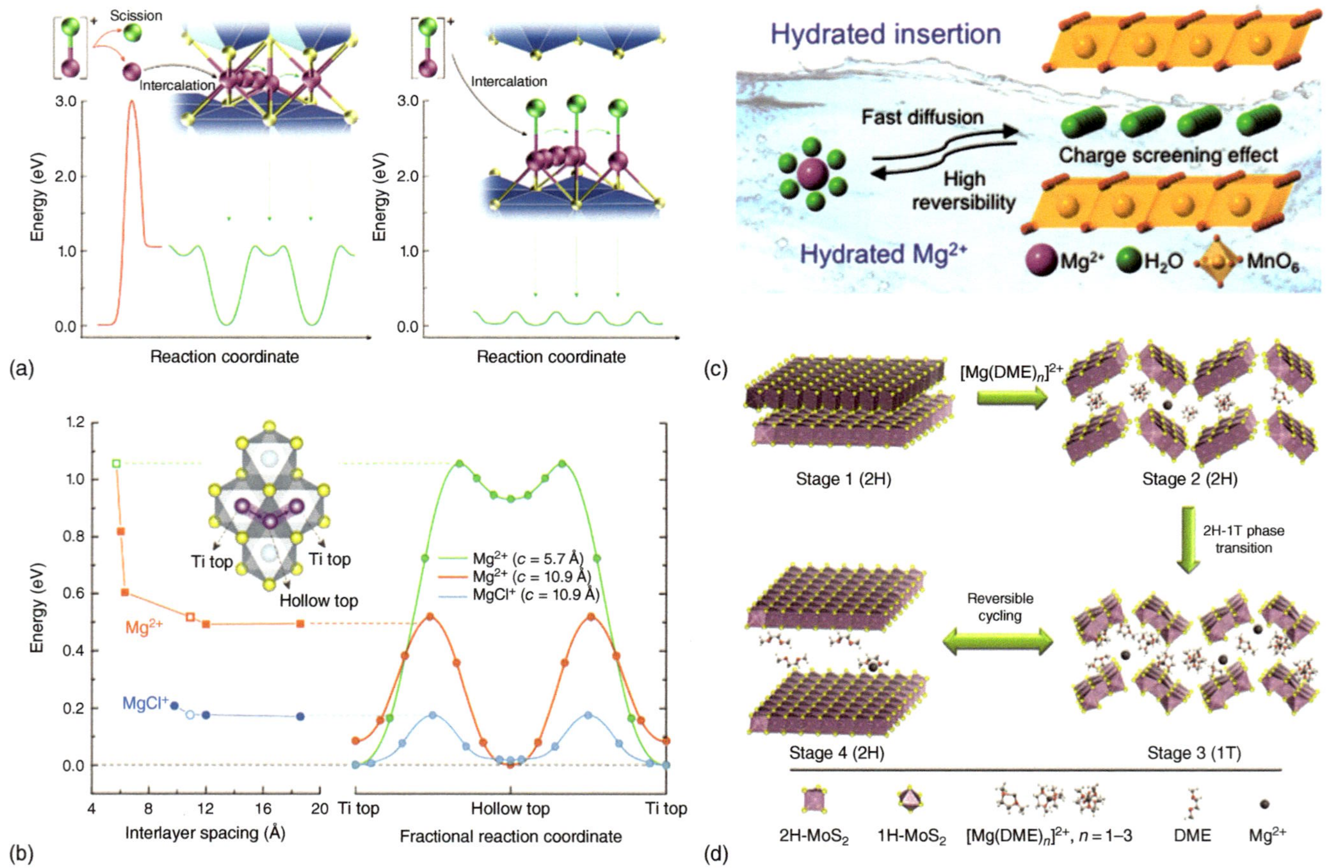

Figure 5.5 (a) Energy diagrams for the intercalation and diffusion of Mg^{2+} and $MgCl^+$. (b) First-principles calculations for the diffusion of Mg ions in TiS_2: Energy barriers for the migration of Mg^{2+} and $MgCl^+$ as a function of the interlayer distance of TiS_2 at the dilute limit. The diffusion path from a Ti top to another Ti top site via the adjacent Hollow top site is shown in the inset. Source: Reproduced with permission [63]. Copyright 2000, Springer Nature. (c) Schematic illustration of hydrated Mg^{2+} boosting the Mg^{2+}-storage ability of birnessite MnO_2. Source: Reproduced with permission [64]. Copyright 2015, American Chemical Society. (d) Schematic illustration of the Mg storage mechanism in MoS_2 structures with MgBOR/DME electrolyte. Source: Reproduced with permission [65]. Copyright 2018, Springer Nature.

been made to alleviate these issues. For example, crystal water could be introduced as pillar in the lattice structure to provide favorable electrochemical performance of hydrated cathodes (such as $Mg_{0.3}V_2O_5{\cdot}1.1\ H_2O$) for Mg^{2+} storage [66]. Similarly, organic solvent molecules with highly polarizing groups are applicable to form solvent-coordinated multivalent carriers. For instance, Li et al. [65] reported an intercalation behavior of $Mg(DME)_3{}^{2+}$ carriers into a 2D interlayer-expanded MoS_2 structure, even though there was a phase transition occurring from stable 2H-MoS_2 to metallic 1T-MoS_2, accompanying with structural distortion (Figure 5.5d). Hou et al. [31] demonstrated the reversible intercalation of solvated Mg^{2+} into a layered $Mg_{0.15}MnO_2$, where less compactly coordinated Mg^{2+} could be produced by solvating methoxyethyl amines with tailored chain length, enabling tunable configuration and thus reducing energy barrier for the intercalation of solvated Mg^{2+} into the layered $Mg_{0.15}MnO_2$. In the case of Ca-ion batteries, Bervas et al. [67] prepared (VO_x, PC) nanocomposites of crystalline V_2O_5 xerogels and PC by a prolonged soaking procedure and subsequent heat treatment, which could deliver Ca^{2+}-intercalated capacities of above 270 mAh g^{-1}. In contrast, only 20 mAh g^{-1} could be obtained for dried V_2O_5 xerogels without PC, implying that the co-intercalation of PC solvent boosts the Ca^{2+}-storage ability of the cathode materials. This phenomenon could be explained by the partially shielded electrostatic interactions between Ca^{2+} and the host lattice.

5.3 Interfacial Chemistry of Multivalent-Ion Batteries

Differing from the bulk diffusion kinetics of multivalent carriers within cathode materials, their interfacial kinetics involve solvation/desolvation process, interfacial reactions, interfacial charge migration, and evolution of chemical composition. It is noteworthy that the interfacial mass transfer between electrode and electrolyte has to surmount high energy barrier involving desolvation behavior of multivalent ions. Moreover, the sluggish interfacial transfer kinetics not only impedes their intercalation ability of multivalent ions into cathode materials but also results in an increased electrochemical overpotential, thus lowering the energy storage efficiencies. For example, theoretical calculations suggested that the dissociation energy of the Mg=Cl bonds is much higher than the energy barrier for Mg^{2+} diffusion in layered TiS_2. The desolvation energies of Mg^{2+} ions in different organic solvents are onefold higher than that of Li^+ counterpart [63, 68]. These analyses suggest that interfacial kinetics rather than solid diffusion in multivalent-ion batteries might play a dominant role in electrochemical energy storage kinetics.

For example, thus far, the Chevrel phase Mo_6S_8 cathode delivered a superiority of Mg^{2+}-intercalation reversibility, so much attention have been paid on the interfacial kinetics and electrochemical diffusion mechanism during the magnesiation of Mo_6S_8 cathode. Previous studies showed that the catalytic effect originating from the exposed Mo atoms of its (100) surface could reduce the activation energy from 3 to 0.2 eV, and boost the breaking of the Mg=Cl bonds (Figure 5.6a) [69]. Moreover, Cl-free ethereal solvent molecules with weakly coordinated ability also enabled the

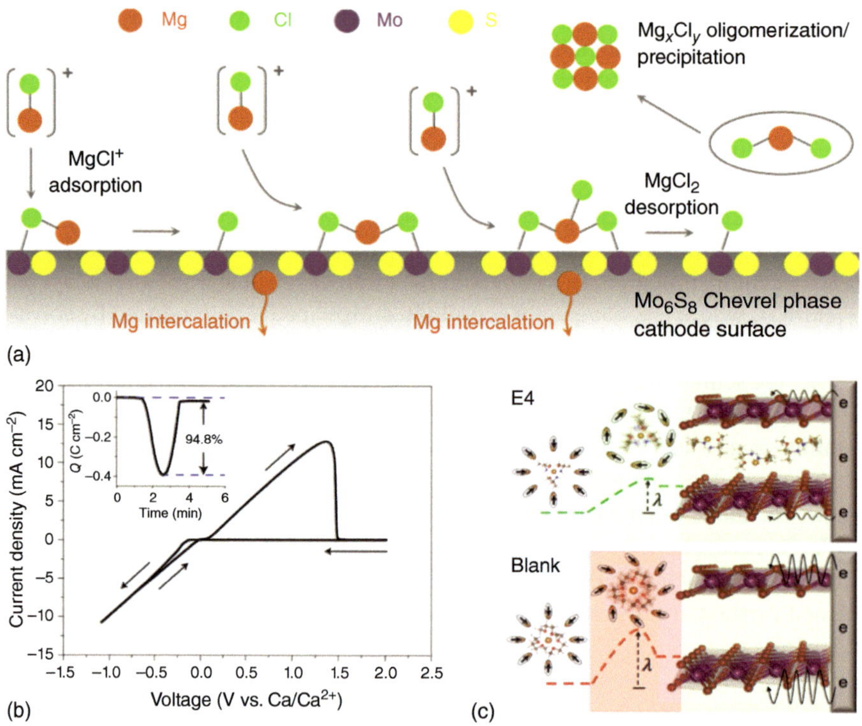

Figure 5.6 (a) Schematic catalytic effect of Mo atoms on the Chevrel phase (100) surface to reduce the activation energy of Mg—Cl bond and boost the interfacial kinetics of Mg^{2+} ions. Source: Reproduced with permission [69]. Copyright 2015, American Chemical Society. (b) Electrochemical calcium plating/stripping test of calcium plating/stripping in 1.5 M $Ca(BH_4)_2$ in THF. Source: Reproduced with permission [24]. Copyright 2017, Springer Nature. (c) The scheme of concerted ion and electron transfer in the cathode host limited by the solvation sheath reorganization. Source: Reproduced with permission [31]. Copyright 2021, American Association for the Advancement of Science.

reversible intercalation of Mg^{2+} ions in the Mo_6S_8 cathode. Nevertheless, it is necessary to get further insight into the interfacial kinetics and electrochemical diffusion mechanism in these Cl-free electrolytes. In addition, the Chevrel phase Mo_6S_8 cathode takes the advantage of relatively high ionic conductivity, which allows a good penetration ability of Mg^{2+} into its bulk [70]. In contrast, on the surface region of other cathode materials, the accumulation of Mg^{2+} content causes unfavorable parasite reactions, which results in the distortion of lattice structure and the formation of an amorphous layer. Consequently, the further electrochemical reactions of Mg^{2+} ions intercalation into these cathode materials are hindered, and only a low specific capacity can be delivered. The insufficient penetration depth makes a size effect prevalent in the Mg^{2+}-intercalated cathode materials [71, 72]. These understandings promote the design and development of cathode materials for Mg^{2+} storage. However, the reduced size also causes larger surface area and more parasite reactions at the interface between cathode and electrolyte. As a result, limiting the interfacial side reactions and improving the interfacial kinetics are important for developing high-performance multivalent-ion batteries.

As mentioned above, proper cathode materials and compatible electrolytes are crucial for the effective construction of multivalent-ion batteries. In the case of Mg metal anodes, the existence of a passivation layer limits the reversible Mg plating/stripping. Although the electrolytes containing Cl-based Lewis acidity are beneficial to remove the passivation layer such as MgO and $Mg(OH)_2$, the contrary characteristics between these nucleophilic electrolytes and electrophilic cathodes (such as oxides and redox-active organics) pose restriction on their chemical compatibility [19]. Accordingly, developing nonnucleophilic electrolytes to improve electro-/chemical stability is highly desired. $Mg(BH_4)_2$ with a suitable anion was firstly reported for a Cl-free Mg salt [73], but its insufficient reduction stability (~2.0 V vs. Mg^{2+}/Mg) impedes further application. Moreover, bulky monovalent anions have also been designed to weakly coordinate with highly polarized Mg^{2+}, relying on a delocalized electron effect. In order to match high-voltage cathode, some other anions (such as carborane anions and fluorinated alkoxyborates) with weakly coordinating abilities and a relatively wide electrochemical voltage window (~4.0 V vs. Mg^{2+}/Mg) have been reported to exhibit high-efficiency Mg plating/stripping reversibility [74, 75]. There is no doubt that electrolyte solvents are also of significant importance for interfacial compatibility in batteries. Among conventional electrolyte solvents for LIBs, ethers are the only feasible type of solvents that enable reversible Mg plating/stripping, but their application is still limited by their poor oxidation stability (~3.5 V vs. Mg^{2+}/Mg). Although the oxidation stability of ether-based electrolytes can be enhanced by coordination with active cations, some concerns still remain regarding the long-term cycling stability.

It has also been demonstrated that Ca^{2+} plating/stripping on Ca metal anode is generally irreversible in conventional carbonate electrolytes at room temperature, due to the blocking effect of the SEI layer. At elevated temperatures of 75–100 °C, the reversible Ca^{2+} plating/stripping could be observed in $Ca(ClO_4)_2$ and $Ca(BF_4)_2$-based electrolytes [23]. Analyses suggested that many defects were formed in the passivation layer formed at elevated temperatures, enabling reversible Ca plating/stripping, which first validated the reversibility of Ca plating/stripping at moderate temperatures. However, the high operating temperature potentially accelerates battery degradation and lowers the energy efficiency of Ca-ion batteries. Bruce and coworkers [24] constructed an SEI layer composing of a small amount of CaH_2 evenly distributed Ca via applying an electrolyte dissolving $Ca(BH_4)_2$ in THF solvent (Figure 5.6b), which enabled sufficient room-temperature Ca^{2+} diffusion, and the formation of CaH_2 on the Ca metal surface limited the further parasitic reactions at the interface between Ca metal anode and the electrolyte.

Fine designing protective layer on the electrode materials is another strategy to improve the interfacial compatibility. For example, in LIBs, differing from the *in situ* formed CEI via the decomposition of partial electrolyte; however, this CEI layer in Mg-ion batteries becomes a passivation barrier for Mg^{2+} diffusion into cathode. How to construct a properly functioning CEI layer for multivalent-ion batteries remains challenging. In addition, the protective layer can be constructed by artificially enwrapping electrode particles with a well-functioning coating. Ideally, the artificially protective layer requires a strong ability to conduct multivalent ions and good chemical inertness to the electrolyte and presents a good electrochemical compatibility with electrode materials [19]. Considering the high charge density of multivalent metal ions, fine controls of chemical component and physical structure are crucial for the stability of their transport across interface. For instance, DFT calculations predicted that $Mg(PO_3)_2$ and MgP_4O_{11} were potential components for protecting Mg-based cathodes [76]. Developing protective layers with proper vacancies and/or porous structure enables the effective diffusion of multivalent metal ions.

For the interfacial kinetics of multivalent metal ions, the desolvation process also plays a key role. Generally, the solvation structure of electrolytes involves the competitive coordination of solvents and anions with multivalent metal ions. The electrolyte design must simultaneously take into consideration the desolvation energy of active cations and the dissociation energy of anions and cations, ensuring the sufficient ion diffusion in the electrolyte bulk and enhancing the interfacial kinetics relying on the facile desolvation of active cations. Currently, most studies on designing the solvation structure focus on improving the reversible metal plating/stripping on the metal anode. Hou et al. [31] realized divalent metal (Mg and Ca) ion batteries with fast interfacial charge transfer kinetics via reorganizing solvation sheath of electrolytes individually based on a methoxyethyl-amine chelant solvent (Figure 5.6c). Compared to traditional ether solvents, these chelants present much higher affinity for Mg^{2+}, and the reorganization of the chelant-rich solvation sheaths can get rid of the energetically unfavorable desolvation process. As a result, the electrolyte design limited the concomitant parasitic reactions, reduced the interfacial impedance, and improved interfacial kinetics for both the anode and cathode. The solvation regulation of electrolytes is also applied to control the intercalation chemistry of multivalent ions into cathode. For example, compared to the high desolvation barrier of the formed $Mg(DME)_3{}^{2+}$ ions in the DME solvent, AN solvent ($Mg(ClO_4)_2$/AN) with a relatively weaker coordination energy enabled a good reversibility of Mg^{2+} intercalation into a V_2O_5 cathode [77]. Additionally, electrolyte additives are also important components for the regulation of the solvation structure by competitive coordination with ions and/or competitive decomposition on the surface of anode and cathode.

5.4 Concluding Remarks

On the path to the high-performance multivalent-ion batteries, the fine design of key materials and interfaces plays a dominant role. It is important for the development of advanced electrolytes to consider the following aspects: (i) sufficient ionic conductivity and electrochemical stability to increase the diffusion kinetics of multivalent ions in the electrolyte bulk and reduce interfacial parasitic reactions, (ii) an in-depth understanding of the formation mechanism of the passivation layer at the interface between electrolyte and anode to guide the rational design of electrolyte composition and artificial interphase layer, (iii) the exploitation of electrolyte additives for existing electrolyte systems to enhance the compatibility between electrolyte and electrode materials and thus improve the electrochemical performance of multivalent-ion batteries, and (iv) enhancing electrochemical compatibility between the electrolyte and electrode materials to prevent degradation of these battery components and increase the stability and reliability of multivalent-ion batteries.

As for the cathode materials, kinetic hindrance remains one of the main bottlenecks for multivalent-ion battery systems, due to the strong interaction between the host lattice with relatively narrow spacing and the highly polarizing charge carriers with high charge density. What is worse, thermodynamically superior conversion reactions during the ion intercalation process are induced by the strong ion-cathode interaction, which can damage the host structure and increase the complexity of multivalent-ion chemistry in cathode materials. To overcome these inherent challenges of the multivalent charge carriers, the design of active cathode materials can learn from research strategies of conventional monovalent-ion storage materials to gain further insight into the kinetic limitations and the host structural instability of multivalent-ion intercalation/de-intercalation reactions.

In the case of interface construction, a full understanding of the formation mechanism of interfacial layers and the diffusion mechanism of charge carriers, including multivalent ions, is of significant importance for improving the plating/stripping reversibility and enhancing interfacial diffusion kinetics of multivalent ions across interfaces. In addition, the interfacial reactions involve complex parasitic reactions, finely deciphering the electrochemical reaction process relies on the development of advanced characterization techniques, such as *in situ/operando* imaging technologies, and the establishment of theoretical models for analyzing the interfacial reaction processes.

References

1 Zhang, L., Wang, H., Zhang, X., and Tang, Y. (2021). A review of emerging dual-ion batteries: fundamentals and recent advances. *Advanced Functional Materials* 31 (20): 2010958.

2 Zhou, X., Liu, Q., Jiang, C. et al. (2020). Strategies towards low-cost dual-ion batteries with high performance. *Angewandte Chemie International Edition* 59 (10): 3802–3832.

3 Liu, Q.R., Wang, H.T., Jiang, C.L., and Tang, Y.B. (2019). Multi-ion strategies towards emerging rechargeable batteries with high performance. *Energy Storage Materials* 23: 566–586.

4 Tu, J., Song, W.L., Lei, H.P. et al. (2021). Nonaqueous rechargeable aluminum batteries: progresses, challenges, and perspectives. *Chemical Reviews* 121 (8): 4903–4961.

5 Huang, Z., Du, X.F., Ma, M.B. et al. (2023). Organic cathode materials for rechargeable aluminum-ion batteries. *ChemSusChem* 16 (9): e202202358.

6 Pan, Q., Gong, D., and Tang, Y. (2020). Recent progress and perspective on electrolytes for sodium/potassium-based devices. *Energy Storage Materials* 31: 328–343.

7 Xie, J. and Zhang, Q. (2019). Recent progress in multivalent metal (Mg, Zn, Ca, and Al) and metal-ion rechargeable batteries with organic materials as promising electrodes. *Small* 15 (15): 1805061.

8 Muldoon, J., Bucur, C.B., and Gregory, T. (2014). Quest for nonaqueous multivalent secondary batteries: magnesium and beyond. *Chemical Reviews* 114 (23): 11683–11720.

9 Lei, X., Zheng, Y.P., Zhang, F. et al. (2020). Highly stable magnesium-ion-based dual-ion batteries based on insoluble small-molecule organic anode material. *Energy Storage Materials* 30: 34–41.

10 Qin, K.Q., Huang, J.H., Holguin, K., and Luo, C. (2020). Recent advances in developing organic electrode materials for multivalent rechargeable batteries. *Energy & Environmental Science* 13 (11): 3950–3992.

11 Lakhnot, A.S., Bhimani, K., Mahajani, V. et al. (2022). Reversible and rapid calcium intercalation into molybdenum vanadium oxides. *Proceedings of the National Academy of Sciences of the United States of America* 119 (30): e2205762119.

12 Yao, Z.P., Hegde, V.I., Aspuru-Guzik, A., and Wolverton, C. (2019). Discovery of calcium-metal alloy anodes for reversible Ca-ion batteries. *Advanced Energy Materials* 9 (9): 1802994.

13 Man, Y.H., Jaumaux, P., Xu, Y.F. et al. (2023). Research development on electrolytes for magnesium-ion batteries. *Science Bulletin* 68 (16): 1819–1842.

14 Rong, Z.Q., Malik, R., Canepa, P. et al. (2015). Materials design rules for multivalent ion mobility in intercalation structures. *Chemistry of Materials* 27 (17): 6016–6021.

15 Wang, S.B., Ran, Q., Yao, R.Q. et al. (2020). Lamella-nanostructured eutectic zinc-aluminum alloys as reversible and dendrite-free anodes for aqueous rechargeable batteries. *Nature Communications* 11 (1): 1634.

16 Gummow, R.J., Vamvounis, G., Kannan, M.B., and He, Y.H. (2018). Calcium-ion batteries: current state-of-the-art and future perspectives. *Advanced Materials* 30 (39): 1801702.

17 Wang, G.X., Wang, Z.C., Shi, H.C. et al. (2022). Progress and perspective on rechargeable magnesium-ion batteries. *Science China Chemistry* 65: https://doi.org/10.1007/s11426-022-1454-0.

18 Nie, C.H., Wang, G.L., Wang, D.D. et al. (2023). Recent progress on Zn anodes for advanced aqueous zinc-ion batteries. *Advanced Energy Materials* 13 (28): 2300606.

19 Li, Z.Y., Häcker, J., Fichtner, M., and Zhao-Karger, Z. (2023). Cathode materials and chemistries for magnesium batteries: challenges and opportunities. *Advanced Energy Materials* 13 (27): 2300682.

20 Liang, Z. and Ban, C. (2021). Strategies to enable reversible magnesium electrochemistry: from electrolytes to artificial solid–electrolyte interphases. *Angewandte Chemie International Edition* 60 (20): 11036–11047.

21 Wu, S., Zhang, F., and Tang, Y.B. (2018). A novel calcium-ion battery based on dual-carbon configuration with high working voltage and long cycling life. *Advanced Science* 5 (8): 1701082.

22 Ji, B.F., He, H.Y., Yao, W.J., and Tang, Y.B. (2021). Recent advances and perspectives on calcium-ion storage: key materials and devices. *Advanced Materials* 33 (2): 2005501.

23 Ponrouch, A., Frontera, C., Bardé, F., and Palacín, M.R. (2016). Towards a calcium-based rechargeable battery. *Nature Materials* 15 (2): 169–172.

24 Wang, D., Gao, X.W., Chen, Y.H. et al. (2018). Plating and stripping calcium in an organic electrolyte. *Nature Materials* 17 (1): 16–20.

25 Liu, S.L., Zhang, R.Z., Mao, J.F. et al. (2022). From room temperature to harsh temperature applications: fundamentals and perspectives on electrolytes in zinc metal batteries. *Science Advances* 8 (12): eabn5097.

26 Jia, B.E., Thang, A.Q., Yan, C.S. et al. (2022). Rechargeable aqueous aluminum-ion battery: progress and outlook. *Small* 18 (43): e2107773.

27 Wei, T.T., Peng, P.P., Qi, S.Y. et al. (2021). Advancement of technology towards high-performance non-aqueous aluminum-ion batteries. *Journal of Energy Chemistry* 57: 169–188.

28 Ma, D.W., Yuan, D., de León, C.P. et al. (2023). Current progress and future perspectives of electrolytes for rechargeable aluminum-ion batteries. *Energy & Environmental Materials* 6 (1): e12301.

29 Canepa, P., Gautam, G.S., Hannah, D.C. et al. (2017). Odyssey of multivalent cathode materials: open questions and future challenges. *Chemical Reviews* 117 (5): 4287–4341.

30 Ling, C., Zhang, R.G., Arthur, T.S., and Mizuno, F. (2015). How general is the conversion reaction in Mg battery cathode: a case study of the magnesiation of α-MnO_2. *Chemistry of Materials* 27 (16): 5799–5807.

31 Hou, S., Ji, X., Gaskell, K. et al. (2021). Solvation sheath reorganization enables divalent metal batteries with fast interfacial charge transfer kinetics. *Science* 374 (6564): 172–178.

32 Liang, Y.L., Dong, H., Aurbach, D., and Yao, Y. (2020). Current status and future directions of multivalent metal-ion batteries. *Nature Energy* 5 (9): 646–656.

33 Ma, Z., MacFarlane, D.R., and Ka, M. (2019). Mg cathode materials and electrolytes for rechargeable Mg batteries: a review. *Batteries & Supercaps* 2 (2): 115–127.

34 Cheng, Y., Stolley, R.M., Han, K.S. et al. (2015). Highly active electrolytes for rechargeable Mg batteries based on a $[Mg_2(\mu\text{-}Cl)_2]^{2+}$ cation complex in dimethoxyethane. *Physical Chemistry Chemical Physics* 17 (20): 13307–13314.

35 Shao, Y.Y., Liu, T.B., Li, G.S. et al. (2013). Coordination chemistry in magnesium battery electrolytes: how ligands affect their performance. *Scientific Reports* 3 (1): 3130.

36 Zhang, D., Duan, S., Liu, X. et al. (2023). Deeping insight of $Mg(CF_3SO_3)_2$ and comprehensive modified electrolyte with ionic liquid enabling high-performance magnesium batteries. *Nano Energy* 109: 108257.

37 Zhang, J.L., Liu, J., Wang, M. et al. (2023). The origin of anode-electrolyte interfacial passivation in rechargeable mg-metal batteries. *Energy & Environmental Science* 16 (3): 1111–1124.

38 Li, Y.H., Yao, H., Liu, X.J. et al. (2023). Roles of electrolyte additive in Zn chemistry. *Nano Research* 16 (7): 9179–9194.

39 Meng, Z., Li, Z.Y., Wang, L.P. et al. (2021). Surface engineering of a Mg electrode via a new additive to reduce overpotential. *ACS Applied Materials & Interfaces* 13 (31): 37044–37051.

40 Zhang, J.H., Guan, X.Z., Lv, R.J. et al. (2020). Rechargeable Mg metal batteries enabled by a protection layer formed *in vivo*. *Energy Storage Materials* 26: 408–413.

41 Kitchaev, D.A., Dacek, S.T., Sun, W., and Ceder, G. (2017). Thermodynamics of phase selection in MnO_2 framework structures through alkali intercalation and hydration. *Journal of the American Chemical Society* 139 (7): 2672–2681.

42 Zhang, R.G., Yu, X.Q., Nam, K.W. et al. (2012). α-MnO_2 as a cathode material for rechargeable Mg batteries. *Electrochemistry Communications* 23: 110–113.

43 Mao, M.L., Ji, X., Hou, S. et al. (2019). Tuning anionic chemistry to improve kinetics of Mg intercalation. *Chemistry of Materials* 31 (9): 3183–3191.

44 Liu, M., Rong, Z.Q., Malik, R. et al. (2015). Spinel compounds as multivalent battery cathodes: a systematic evaluation based on *ab initio* calculations. *Energy & Environmental Science* 8 (3): 964–974.

45 Okamoto, S., Ichitsubo, T., Kawaguchi, T. et al. (2015). Intercalation and push-out process with spinel-to-rocksalt transition on Mg insertion into spinel oxides in magnesium batteries. *Advanced Science* 2 (8): 1500072.

46 Truong, Q.D., Kobayashi, H., and Honma, I. (2019). Rapid synthesis of $MgCo_2O_4$ and $Mg_{2/3}Ni_{4/3}O_2$ nanocrystals in supercritical fluid for Mg-ion batteries. *RSC Advances* 9 (63): 36717–36725.

47 Wang, H.L., Bi, X.X., Bai, Y. et al. (2017). Open-structured $V_2O_5{\cdot}nH_2O$ nanoflakes as highly reversible cathode material for monovalent and multivalent intercalation batteries. *Advanced Energy Materials* 7 (14): 1602720.

48 Gautam, G.S., Canepa, P., Abdellahi, A. et al. (2015). The intercalation phase diagram of Mg in V_2O_5 from first-principles. *Chemistry of Materials* 27 (10): 3733–3742.

49 Wang, D., Liu, H., Elliott, J.D. et al. (2016). Robust vanadium pentoxide electrodes for sodium and calcium ion batteries: thermodynamic and diffusion mechanical insights. *Journal of Materials Chemistry A* 4 (32): 12516–12525.

50 Liang, Y.L., Yoo, H.D., Li, Y.F. et al. (2015). Interlayer-expanded molybdenum disulfide nanocomposites for electrochemical magnesium storage. *Nano Letters* 15 (3): 2194–2202.

51 Koketsu, T., Ma, J.W., Morgan, B.J. et al. (2017). Reversible magnesium and aluminium ions insertion in cation-deficient anatase TiO_2. *Nature Materials* 16 (11): 1142–1148.

52 Aurbach, D., Lu, Z., Schechter, A. et al. (2000). Prototype systems for rechargeable magnesium batteries. *Nature* 407 (6805): 724–727.

53 Ling, C. and Suto, K. (2017). Thermodynamic origin of irreversible magnesium trapping in Chevrel phase Mo_6S_8: importance of magnesium and vacancy ordering. *Chemistry of Materials* 29 (8): 3731–3739.

54 Smeu, M., Hossain, M.S., Wang, Z. et al. (2016). Theoretical investigation of Chevrel phase materials for cathodes accommodating Ca^{2+} ions. *Journal of Power Sources* 306: 431–436.

55 Levi, E., Lancry, E., Mitelman, A. et al. (2006). Phase diagram of Mg insertion into Chevrel phases, $Mg_xMo_6T_8$ (T = S, Se). 1. Crystal structure of the sulfides. *Chemistry of Materials* 18 (23): 5492–5503.

56 Richard, J., Benayad, A., Colin, J.F., and Martinet, S. (2017). Charge transfer mechanism into the Chevrel phase Mo_6S_8 during Mg intercalation. *Journal of Physical Chemistry C* 121 (32): 17096–17103.

57 Mitelman, A., Levi, E., Lancry, E., and Aurbach, D. (2007). On the Mg trapping mechanism in electrodes comprising Chevrel phases. *ECS Transactions* 3 (27): 109.

58 Juran, T.R. and Smeu, M. (2017). Hybrid density functional theory modeling of Ca, Zn, and Al ion batteries using the Chevrel phase Mo_6S_8 cathode. *Physical Chemistry Chemical Physics* 19 (31): 20684–20690.

59 Gu, Y., Katsura, Y., Yoshino, T. et al. (2015). Rechargeable magnesium-ion battery based on a $TiSe_2$-cathode with *d-p* orbital hybridized electronic structure. *Scientific Reports* 5 (1): 12486.

60 Chen, H., Xu, H., Wang, S. et al. (2017). Ultrafast all-climate aluminum-graphene battery with quarter-million cycle life. *Science Advances* 3 (12): eaao7233.

61 Salama, M., Shterenberg, I., Shimon, L.J.W. et al. (2017). Structural analysis of magnesium chloride complexes in dimethoxyethane solutions in the context of Mg batteries research. *The Journal of Physical Chemistry C* 121 (45): 24909–24918.

62 Canepa, P., Jayaraman, S., Cheng, L. et al. (2015). Elucidating the structure of the magnesium aluminum chloride complex electrolyte for magnesium-ion batteries. *Energy & Environmental Science* 8 (12): 3718–3730.

63 Yoo, H.D., Liang, Y., Dong, H. et al. (2017). Fast kinetics of magnesium monochloride cations in interlayer-expanded titanium disulfide for magnesium rechargeable batteries. *Nature Communications* 8 (1): 339.

64 Nam, K.W., Kim, S., Lee, S. et al. (2015). The high performance of crystal water containing manganese Birnessite cathodes for magnesium batteries. *Nano Letters* 15 (6): 4071–4079.

65 Li, Z.Y., Mu, X.K., Zhao-Karger, Z. et al. (2018). Fast kinetics of multivalent intercalation chemistry enabled by solvated magnesium-ions into self-established metallic layered materials. *Nature Communications* 9 (1): 5115.

66 Xu, Y.N., Deng, X.W., Li, Q.D. et al. (2019). Vanadium oxide pillared by interlayer Mg^{2+} ions and water as ultralong-life cathodes for magnesium-ion batteries. *Chem* 5 (5): 1194–1209.

67 Bervas, M., Klein, L.C., and Amatucci, G.G. (2005). Vanadium oxide-propylene carbonate composite as a host for the intercalation of polyvalent cations. *Solid State Ionics* 176 (37–38): 2735–2747.

68 Okoshi, M., Yamada, Y., Yamada, A., and Nakai, H. (2013). Theoretical analysis on de-solvation of lithium, sodium, and magnesium cations to organic electrolyte solvents. *Journal of the Electrochemical Society* 160 (11): A2160–A2165.

69 Wan, L.W.F., Perdue, B.R., Apblett, C.A., and Prendergast, D. (2015). Mg Desolvation and intercalation mechanism at the Mo_6S_8 Chevrel phase surface. *Chemistry of Materials* 27 (17): 5932–5940.

70 Saha, P., Jampani, P.H., Datta, M.K. et al. (2014). A convenient approach to Mo_6S_8 Chevrel phase cathode for rechargeable magnesium battery. *Journal of the Electrochemical Society* 161 (4): A593–A598.

71 Hu, L.H., Jokisaari, J.R., Kwon, B.J. et al. (2020). High capacity for Mg^{2+} deintercalation in spinel vanadium oxide nanocrystals. *ACS Energy Letters* 5 (8): 2721–2727.

72 Blanc, L.E., Sun, X.Q., Shyamsunder, A. et al. (2020). Direct nano-synthesis methods notably benefit Mg-battery cathode performance. *Small Methods* 4 (5): 2000029.

73 Mohtadi, R., Matsui, M., Arthur, T.S., and Hwang, S.J. (2012). Magnesium borohydride: from hydrogen storage to magnesium battery. *Angewandte Chemie-International Edition* 51 (39): 9780–9783.

74 Zhao-Karger, Z., Bardaji, M.E.G., Fuhr, O., and Fichtner, M. (2017). A new class of non-corrosive, highly efficient electrolytes for rechargeable magnesium batteries. *Journal of Materials Chemistry A* 5 (22): 10815–10820.

75 Tutusaus, O., Mohtadi, R., Arthur, T.S. et al. (2015). An efficient halogen-free electrolyte for use in rechargeable magnesium batteries. *Angewandte Chemie-International Edition* 54 (27): 7900–7904.

76 Chen, T.N., Ceder, G., Gautam, G.S., and Canepa, P. (2019). Evaluation of Mg compounds as coating materials in Mg batteries. *Frontiers in Chemistry* 7: 24.

77 Attias, R., Salama, M., Hirsch, B. et al. (2018). Solvent effects on the reversible intercalation of magnesium-ions into V_2O_5 electrodes. *ChemElectroChem* 5 (22): 3514–3524.

6

Dual-Ion Batteries: Materials and Mechanisms

Luojiang Zhang[1] *and Yongbing Tang*[1,2]

[1] *Advanced Energy Storage Technology Research Center, Shenzhen Institute of Advanced Technology, Chinese Academy of Sciences, 1068 Xueyuan Avenue, Shenzhen University Town, Shenzhen 518055, China*
[2] *University of Chinese Academy of Sciences, School of Chemical Engineering, No.19A Yuquan Road, Shijingshan District, Beijing 100049, China*

6.1 Introduction

As the current best-known energy storage system/rechargeable batteries, LIBs have been widely applied in human life, including portable electronics, power tools, electrical vehicles, and grid energy storage, owing to their high energy density, long lifespan, and no memory effect. In 2019, the Nobel Prize in Chemistry was awarded to John B. Goodenough, M. Stanley Whittingham, and Akira Yoshino for the development of LIBs. Although technological innovation leads to new chemistries being used over time, the current commercial LIBs commonly utilize cathode materials containing limited and unevenly distributed resources such as Li and Co. Therefore, it is urgent to develop novel low-cost but highly efficient energy storage systems, apart from the recycling/reuse of LIBs and exploiting other available cathode materials.

As early as 1938, Rüdorff and Hofmann first discovered the anion (HSO_4^-) intercalation phenomenon into graphite when they studied a cell with graphite as the electrodes and concentrated H_2SO_4 as the electrolyte [1]. The research makes it possible for energy storage systems to use cathode material hosting anions and anode material hosting cations. However, the cell based on cation and anion "dual-intercalation" was not designed until the 1990s by McCullough et al. in a patent [2]. In 1994, Carlin et al. reported dual-intercalating molten electrolyte batteries with graphite as both cathode and anode materials, in which a series of room-temperature or low-melting single molten salt electrolytes composed of substituted imidazolium cations and inorganic/organic anions were studied [3]. Then in 2012, Winter and coworkers first renamed the "dual-carbon" or "dual-graphite cell" to "dual-ion cell," and the concept has been in use ever since [4]. Afterward, research of DIBs has sprung up, not only from lithium-based DIBs (LDIBs) to other alkaline metals (e.g. Na and K) [5–10] and even alkaline earth metals (e.g. Mg and Ca)-based DIBs [11–13] but also

Electrochemical Energy Storage Devices: Non-Conventional Technologies and Materials,
First Edition. Edited by Yongbing Tang and Luojiang Zhang.

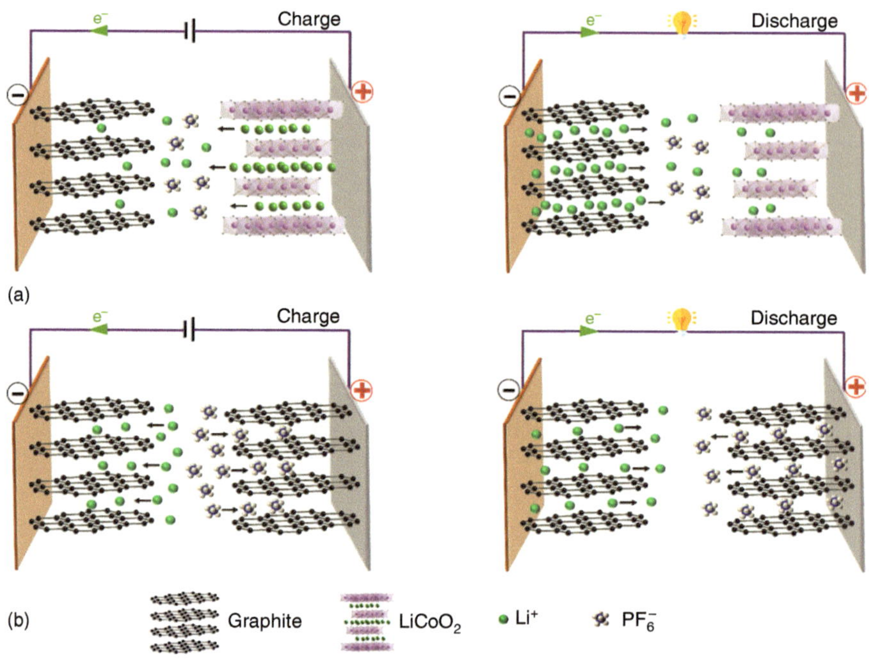

Figure 6.1 Schematic illustration of the operational working mechanisms of different batteries. (a) Conventional LIB using $LiCoO_2$ and graphite as cathode and anode, respectively, and (b) typical DIB using graphite as both cathode and anode.

exploiting numerous appropriate electrode materials hosting anions/cations as well as electrolytes. In addition, other dual-ion systems such as reverse DIBs have also been developed.

The rise and development of DIBs should be associated with the storage mechanism. Figure 6.1 illustrates the charge–discharge mechanism of LIB and DIB. In a traditional LIB (Figure 6.1a), the charge–discharge process is described as a classical "rocking-chair" model, in which the Li^+ ions stored in cathode material (e.g. $LiCoO_2$) migrate back and forth between the cathode and anode (e.g. graphite). The electrolyte in the battery only serves as a "vehicle" to transfer Li^+ ions (on the premise of not considering the consumption for the formation of solid electrolyte interphase (SEI) film). The reactions are as follows:

$$\text{Cathode:}\ \ LiCoO_2 \leftrightarrow Li_{1-n}CoO_2 + nLi^+ + ne^- \tag{6.1}$$

$$\text{Anode:}\ \ C + nLi^+ + ne^- \leftrightarrow Li_nC \tag{6.2}$$

$$\text{Overall reaction:}\ \ LiCoO_2 + C \leftrightarrow Li_{1-n}CoO_2 + Li_nC \tag{6.3}$$

While for DIB (Figure 6.1b), the cations (e.g. Li^+) and anions (e.g. PF_6^-) act as active species (or charge carriers) that simultaneously insert into anode (e.g. graphite) and cathode (e.g. graphite), respectively, during the charging process.

During the discharging process, the inserted Li^+ and PF_6^- ions would be released back to electrolyte from the corresponding electrode materials. The reactions can be described as follows:

$$\text{Cathode:}\ \ C_C + nPF_6^- \leftrightarrow (PF_6)_nC_C + ne^- \tag{6.4}$$

$$\text{Anode:}\ \ C_A + nLi^+ + ne^- \leftrightarrow Li_nC_A \tag{6.5}$$

$$\text{Overall reaction:}\ \ C_C + nPF_6^- + C_A + nLi^+ \leftrightarrow (PF_6)_nC_C + Li_nC_A \tag{6.6}$$

where C_C and C_A represent graphite cathode and graphite anode, respectively. It is easy to see that the active cations and anions participated in the whole reactions originate from the electrolyte only. Because of this, relatively more electrolytes should be commonly needed in DIBs than in LIBs, ensuring the sufficient amount of charge carriers.

The working potential (ΔV) of the aforementioned DIB can be expressed in Eq. (6.7) [14]:

$$ne\Delta V = n\left(\mu_{Li} - \mu'_{Li}\right) + n\left(\mu_{PF_6} - \mu'_{PF_6}\right) \tag{6.7}$$

where μ_{Li} and μ_{PF_6} are the chemical potentials of Li^+ and PF_6^- in electrolyte, respectively; μ'_{Li} and μ'_{PF_6} stand for the chemical potentials of Li^+ and PF_6^- insert into C_A and C_C, respectively. It is worth noting that μ_{Li} and μ_{PF_6} depend on the Li^+ and PF_6^- concentration in electrolyte, respectively, while μ'_{Li} and μ'_{PF_6} are determined as a function of inserted Li^+ and PF_6^- content in C_A and C_C, respectively. According to the Nernst equation, μ_{Li} and μ_{PF_6} can be given as follows [15, 16]:

$$\mu_{Li} = \mu^0_{Li} + \text{kTln}[Li^+] \tag{6.8}$$

$$\mu_{PF_6} = \mu^0_{PF_6} + \text{kTln}\left[PF_6^-\right] \tag{6.9}$$

where μ^0_{Li} and $\mu^0_{PF_6}$ represent the chemical potentials of Li^+ and PF_6^- in 1 M solution, respectively, which are directly determined by the electrolyte solvent; $[Li^+]$ and $[PF_6^-]$ are the Li^+ and PF_6^- concentration in electrolyte, respectively. By combing Eqs. (6.7)–(6.9), the working potential of the DIB can be shown as follows:

$$e\Delta V = \mu^0_{Li} + \mu^0_{PF_6} + \text{kTln}[Li^+] + \text{kTln}\left[PF_6^-\right] - \mu_{Li} - \mu_{PF_6} \tag{6.10}$$

As a result, it can be concluded that the types of solvents and ions as well as electrolyte concentration play great roles in the working potential of the DIBs, demonstrating the active nature of electrolytes in DIBs.

On this basis, it is known that the charge of electrolyte should match that of cathode or anode. That is [17],

$$Q_E = Q_A = Q_C = Q \tag{6.11}$$

where Q_E, Q_A, Q_C, and Q correspond to the charges of electrolyte, anode, cathode, and the DIB cell, respectively.

According to the equation $Q = mc$, one can obtain [17, 18]

$$c = \frac{Q}{m} = \frac{Q_A}{m_A + m_C + m_E} = \frac{m_A c_A}{m_A + m_C + m_E} = \frac{m_A c_A}{m_A + m_C + \frac{Q_E}{c_E}} = \frac{m_A c_A}{m_A + m_C + \frac{m_A c_A}{c_E}} \tag{6.12}$$

where m, m_A, m_C, and m_E are the mass of DIB cell, anode, cathode, and electrolyte, respectively, in the unit of gram; c, c_A, and c_E represent the capacities of the DIB cell, anode, and electrolyte, respectively, in the unit of mAh g^{-1}.

For the electrolyte, the c_E can be calculated as follows [17]:

$$c_E = \frac{Fx(M_d - M_c)}{10^3 \rho} \tag{6.13}$$

where F is the Faraday constant (26.8×10^3 mAh mol^{-1}), x is the charge of the active species in the electrolyte ($x = 1$ for the aforementioned DIB system), M_d and M_c stand for the molarities of the electrolyte in the full discharged and charged state, respectively, in the unit of mol l^{-1}, and ρ is the density of the electrolyte (g ml^{-1}).

In this case, the capacity of a DIB cell based on electrode materials and electrolytes can be expressed as follows:

$$c = \frac{m_A c_A}{m_A + m_C + \frac{m_A c_A}{c_E}} = \frac{m_A c_A}{m_A + m_C + \frac{10^3 \rho m_A c_A}{Fx(M_d - M_c)}} \tag{6.14}$$

The gravimetric energy density (E) of a DIB cell is

$$E = cV \tag{6.15}$$

where V is the discharge medium voltage of the DIB cell. In consideration of current collectors, separators, and packaging, the energy density of a full cell could be reduced by ~30–45% [19].

The intercalation of anions into graphite cathodes always enables DIBs with cutoff voltages higher than 5.0 V, which are much higher than those of traditional LIBs. This remarkable feature presents a high energy density as well as several challenges for DIBs. First, other advanced cathode materials with satisfactory intercalation capacity and structural stability should be developed as the graphite materials can be easily exfoliated and destroyed during the continuous anion intercalation process. Second, anode materials need to be considered to match the reaction kinetics with the cathode materials. Third, electrolytes with an extended electrochemical stability window (ESW) must be required to meet the demand of anion intercalation at high potential.

6.2 Cathode Materials

Cathode in DIBs should function as a host material for reversible anion intercalation/deintercalation and is generally considered a critical limiting factor in determining the electrochemical performance of DIBs.

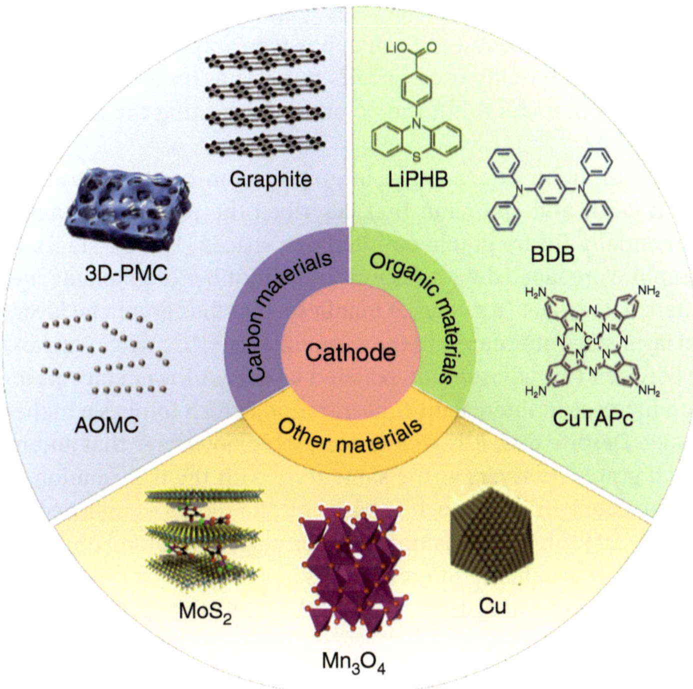

Figure 6.2 The classification of typical cathode materials for DIBs. Source: Image of 3D-PMC: Reproduced with permission Zhang et al. [20]. Copyright 2019, Wiley. Image of AOMC: Reproduced with permission Wang et al. [21]. Copyright 2019, Elsevier. Image of LiPHB: Reproduced with permission Rajesh et al. [22]. Copyright 2020, Wiley-VCH. Image of BDB: Reproduced with permission Glatz et al. [23]. Copyright 2019, American Chemical Society. Image of CuTAPc: Reproduced with permission Wang et al. [24]. Copyright 2019, Wiley-VCH. Image of MoS_2: Reproduced with permission Li et al. [25]. Copyright 2020, Elsevier. Image of Mn_3O_4: Reproduced with permission [26]. Copyright 2019, Wiley-VCH. Image of Cu: Reproduced with permission Yu et al. [27]. Copyright 2022, Wiley-VCH.

The key characteristics of cathode materials contain (i) rational and stable structures for effective anion intercalation/deintercalation during long-term continuous charge–discharge processes; (ii) appropriate intercalation potential of anions within the ESW of electrolytes; and (iii) accommodating anions as much as possible to achieve relative high capacity. The following part will introduce the currently developed cathode materials including carbon materials, organic materials, and other cathode materials (Figure 6.2).

6.2.1 Carbon Cathode Materials

6.2.1.1 Graphite Materials

Graphite is a kind of layered structured carbon material with sp^2 hybridized carbon atoms tightly bonded in standard hexagonal rings within a single sheet, possessing the attracted properties such as weak interlayer force (a bond energy of 16.7 kJ mol^{-1}) [28], high electron mobility (~200,000 $cm^2\ V^{-1}\ S^{-1}$) [29], and low cost.

Specifically, the redox-amphoteric character enables graphite to participate in the reaction not only as an electron acceptor to form donor-type graphite intercalation compounds (GICs) that are commonly used in LIBs but also as an electron donor to form acceptor-type GICs that can fulfill the potential for hosting the anions in DIBs [17].

The pioneering work on anion intercalation in graphite can be traced back to year 1938, in which Rüdorff and Hofmann first described the process as staging that HSO_4^- ions sequentially fill up graphene interlayer spaces [1]. The stages of intercalation into graphite are labeled with respect to the number of graphene layers between each intercalated layer (e.g. stage 4 manifests that four graphene layers and one intercalated layer stage alternately in the acceptor-type GICs; stage 1 means one graphene layer between two adjacent intercalated layers). A lower stage number corresponds to an increase in intercalant concentration, which implies a higher charge storage capacity. Despite this, it is widely acceptable nowadays that anions intercalate between all graphene layers at the same time with the deformation of layers around the intercalated anions, which is based on the model established by Daumas and Herold in 1969 [30]. The transition between stages via lateral diffusion as well as the coexistence of domains at different stages in a single crystallite can be allowed. In GICs, the periodic repeat distance (I_c) and the intercalant gallery height (d_i) obey the following relationship:

$$I_c = d_i + 3.35\,\text{Å} \cdot (n-1) = \Delta d + 3.35\,\text{Å} \cdot n = l \cdot d_{\text{obs}} \tag{6.16}$$

where n is the well-defined stages, Δd is the gallery expansion, l is the index of (00l) planes oriented in the stacking direction, and d_{obs} is the observed value of the spacing between two adjacent planes.

The merits of low cost, relatively high capacity, and cycling stability enable graphite as the most widely studied cathode material in DIBs. The intercalated anions have been extended from original HSO_4^- to BF_4^- [31], PF_6^- [32], ClO_4^- [33], $CF_3SO_3^-$/OTf$^-$ [23], fluorosulfonyl-(trifluoromethanesulfonyl) imide (FTFSI$^-$) [34], bis(fluorosulfonyl) imide/amide (FSI$^-$/FSA$^-$) [35], bis(trifluoromethanesulfonyl) imide (TFSI$^-$) [36], and so on [37]. The TFSI$^-$ intercalation behaviors were investigated in various graphite materials with different particle size distributions, specific surface areas, and particle morphologies by Winter and coworkers [36]. Several investigations could be obtained: (i) Different from the graphite anodes in LIBs, the graphite cathode in DIBs exhibited no correlation between the initial Coulombic efficiency (CE) and the particle size and/or specific surface area. (ii) The insertion of TFSI$^-$ anions was strongly affected by the amount of the "nonbasal plane" surface area of graphite due to the fact that anion intercalation only takes place via edge or "defect" surface sites. (iii) The anion insertion, rather than the anion deinsertion, was thought to be the kinetic limiting factor for the rate performance of DIB system. The ion mobility increased as the testing temperature rose to 60 °C, which resulted in the improvement of capacity, rate performance, as well as cycling stability. It was also confirmed from another work that kish graphite flakes with the highest structural perfection (degree of graphitization and crystal perfection) were able to exhibit better electrochemical performance than other kinds of graphite materials including natural graphite flakes, highly oriented pyrolytic graphite, potato-shaped graphite particles, and acetylene black.

The continuous intercalation/deintercalation of anions with large size could induce unexpected volume/thickness changes in the graphite cathodes, leading to rapid capacity fading as well as low CE during cycles. It was confirmed that the graphite cathode initially presented an irreversible thickness expansion of ~60% with the insertion of PF_6^-, and the thickness remained <20% in the following cycles [38]. On considering this, it is expected that surface modification on graphite cathode could be an effective strategy to maintain the structural stability, thus improving the cycling performance. Materials serving as protect layers in high-voltage cathodes in alkaline-metal ion batteries may be also applied in DIBs to form a robust and compatible cathode electrolyte interface (CEI) that can protect the graphite cathode from structural collapse. $Li_4Ti_5O_{12}$ (LTO) is such an example. The protection function of LTO has been demonstrated in various LIB cathode materials, including $LiCoO_2$ and $LiNi_{0.5}Co_{0.2}Mn_{0.3}O_2$. In DIBs, an *in situ* generated CEI layer on LTO surface can stabilize the electrode/electrolyte interface, maintaining the structural integrity of mesocarbon microbead (MCMB) cathode during the insertion/deinsertion processes of PF_6^- ions [39]. As expected, the LTO-CEI-modified MCMB could operate stably for 2000 cycles with a capacity retention of 85.1%. In addition to this, it has been also discovered that the rational designed SEI film can also be useful to improve the structural stability of graphite cathode via the formation of a robust graphite/electrolyte interface. After discharging/charging at 100 mA g^{-1} in the potential window of 0.3–2.0 V (vs. Li^+/Li) for five cycles, the graphite surface was modified with an artificial SEI protector, which was revealed to establish the optimal anion-transport path as well as facilitate the electron transfer [40]. As a result, such SEI-modified graphite exhibited a cycling stability with 96% capacity retention after 500 cycles under 200 mA g^{-1} at the upper cutoff voltage of 5.0 V and 98% capacity retention after 2000 cycles at the upper cutoff voltage of 4.9 V, which outperformed the pristine graphite electrode. However, in consideration of limited theoretical capacity of graphite cathode ($C_{16}PF_6$: ~140 mAh g^{-1}), most reported graphite cathodes delivered capacities in the range of 60–120 mAh g^{-1}. Therefore, it is highly desirable for researchers to develop other advanced cathode materials.

6.2.1.2 Other Carbon Materials

High conductive carbon materials are the preferred opinions for cathode materials in DIBs other than graphite. Meanwhile, it is worth noting that the high specific surface area makes carbon materials possible to store energy from anion adsorption. This phenomenon is seen in a 3D porous microcrystalline carbon (3D-PMC) material with a specific surface area of ~180 $m^2\,g^{-1}$ [20]. The as-assembled 3D-PMC cathode-based DIB exhibited a pair of redox peaks in the cyclic voltammetry (CV). The corresponding galvanostatic charge–discharge curves manifested as the combination of linear PF_6^- adsorption behavior existed in activated carbon cathode (the lower charge–discharge plateaus) and staged PF_6^- intercalation behavior in graphite cathode (the relative higher charge–discharge plateaus). Benefiting from this, the DIB possessed a high capacity of 168.0 mAh g^{-1} at 0.3 A g^{-1}, and stable cycling over 2000 cycles with a capacity retention of ~70% at 1 A g^{-1}. Apart from this, the simultaneous accommodation of cations and anions during charging and discharging processes can be another effective way to boost the performance of carbon cathode materials. The dual energy storage mechanism was realized in the amorphous ordered mesoporous carbon (AOMC) with a high Brunauer–Emmett–Teller (BET) surface area of 1070 m^2 g^{-1} and an average pore size of 6.5 nm [21]. When in a sodium-based DIB (SDIB) system with $NaPF_6$ as the electrolyte salt, the electrochemical reaction was investigated as follows:

In the discharge process of the first cycle:

$$\text{Anode:}\ \ \text{Na} - \text{e}^- \rightarrow \text{Na}^+ \tag{6.17}$$

$$\text{Cathode:}\ \ \text{C} + x\text{Na}^+ + x\text{e}^- \rightarrow \text{Na}_x\text{C} \tag{6.18}$$

$$\text{Overall reaction:}\ \ \text{C} + x\text{Na} \rightarrow \text{Na}_x\text{C} \tag{6.19}$$

In the charge–discharge processes of the second cycle:

$$\text{Anode:}\ \ \text{Na}^+ + \text{e}^- \leftrightarrow \text{Na} \tag{6.20}$$

$$\text{Cathode:}\ \ \text{Na}_x\text{C} - x\text{e}^- \leftrightarrow x\text{Na}^+ + \text{C} \tag{6.21}$$

$$\text{C} + y\text{PF}_6^- - y\text{e}^- \leftrightarrow \text{C(PF}_6)_y \tag{6.22}$$

$$\text{Overall reaction:}\ \ (x - y)\text{Na} + y\text{NaPF}_6 + 2\text{C} \rightarrow \text{NaC}_x + \text{C(PF}_6)_y \tag{6.23}$$

As a result, the AOMC delivered an enhanced capacity of 139 mAh g^{-1} at 2 A g^{-1}, which can maintain 136 mAh g^{-1} over 800 cycles.

In short, carbon materials especially graphite materials showcase the unique ability of anion intercalation/deintercalation at a high working potential. Future research can be focused on the improvement of capacity as well as cycling stability. Some strategies such as structural optimization and introducing dual or multiple energy storage mechanisms could be available. Besides, it should be necessary to develop adaptable electrolyte with high oxidation tolerance.

6.2.2 Organic Cathode Materials

Organic materials are considered another type of cathode material to host anions due to their composition designability, structural tunability, as well as environmental friendliness. According to the difference of active centers, organic cathode materials can be generally divided into two categories. One is the *p*-type or bipolar-type organics with nitrogen-containing materials as the representatives. The *N* active centers can first extract electrons to form a positively charged state, which then attracts anions to ensure the material electroneutrality after charging. Lithium 4-(10*H*-phenothiazin-10-yl) benzoate (LiPHB) with a nonpolymeric phenothiazine-based structure was synthesized to serve as an efficient high-voltage

p-type cathode material [22]. Anions like PF_6^-, $TFSI^-$, and ClO_4^- have been demonstrated to react with the N active center via a one-electron reaction. It was suggested that LiPHB was partially soluble at oxidized states resulting in a kind of particle erosion and forming new small surface particles after reduction or a surface layer (CEI) on the surface of the composite. For ClO_4^- insertion/deinsertion process in the potential range from 2.8 to 4.2 V, an excellent cycling stability over 500 cycles at 0.2C can be achieved, although the reversible discharge capacity was only 86 mAh g^{-1}. The capacity of nonpolymeric organics could be further enhanced by introducing more N active centers. Take 1,4-bis(diphenylamino)benzene (BDB) as an example, the two tertiary nitrogens can reversibly oxidize/reduce in two steps accompanied by chemical evolution between BDB, BDB^+, and BDB^{2+} species [23]. As a result, the material delivered a capacity of 125 mAh g^{-1} at an average voltage of 1.25 V in a dual-ion aqueous zinc battery with 1 m ZnOTf and 19 m LiTFSI in water as the electrolyte. The two-electron redox reaction was also seen in the material 4,4′-(phenaine-5,10-diyl)debenzoate dilithium salt (PZDB-Li_2) that presented a capacity of around 100 mAh g^{-1} at 123.4 mA g^{-1} [41]. Besides, it is worth noting that the bipolar feature also endowed PZDB-Li_2 with oxidizability as an electron acceptor to attract with cations (e.g. Li^+), thus extending the applications in all-organic symmetric DIBs. The cell exhibited a discharge voltage of c. 2.5 V, which was higher than most binary transition metal oxide-based symmetric cells (<2 V in most cases). It achieved an energy density of 127 Wh kg^{-1} and a power density of 617 W kg^{-1}. Apart from the nonpolymeric organics, p-type conducting polymers also show their potentials in DIB cathode materials. For example, poly(N-vinylcarbazole) (PVK) was proposed to undergo one-electron redox reaction along with PF_6^- insertion/desinertion in a potassium-based dual-ion battery (PDIB), which displayed a reversible capacity of 114 mAh g^{-1} at 50 mA g^{-1} and 61 mAh g^{-1} at 2000 mA g^{-1} for 500 cycles [42].

Metal organic frameworks (MOFs) are another type of organic cathode material, in which the transitional metal ions and/or organic ligands can function as active centers to incorporate the anions during their oxidation processes. A mixed-valence MOF, $Co_2^{II}Co^{III}(DFc)_2(OH)_3{\cdot}H_2O$ ($Co(DFc)_x$), with 1,1′-dicarboxyferrocene (DFc) was prepared to serve as the cathode material for LDIBs [43]. The redox reaction of Fe^{2+}/Fe^{3+} and Co^{2+}/Co^{3+} couples during the charge–discharge processes would trigger the intercalation/deintercalation of PF_6^- anions into/from $Co(DFc)_x$. The results indicate that $Co(DFc)_x$ delivered an operating potential of 3.63 V (vs. Li^+/Li), a high discharging capacity of 170 mAh g^{-1} at 50 mA g^{-1}, and excellent long-cycling stability with an average capacity of 74.9 mAh g^{-1} for 8000 cycles at 2000 mA g^{-1}. While in another work, a bipolar-type MOF, Cu tetraaminephthalocyanine (CuTAPc), was reported as electrode material in LDIBs with phthalocyanine as the organic ligand [24]. The centrosymmetric planar 18 π-electron aromatic macrocycles with four nitrogen-linked isoindole units enable CuTAPc to be oxidized into dicationic species ($CuTAPc^{2+}$, 16 π-electrons), which can then interact with anions like PF_6^-, resulting in a high discharge capacity of 236 mAh g^{-1} at 50 mA g^{-1}. The all-organic symmetric DIBs using CuTAPc as both the cathode and anode materials displayed a specific energy of 239 Wh kg^{-1} and a specific power of 11.5 kW kg^{-1}.

Overall, organic cathode materials with high capacity can be rationally designed by introducing more active centers. However, some concerns like the relatively lower operating potential than graphite materials, and the possible dissolution in electrolyte cannot be neglected. In addition, it is important to produce high-quality organic materials using the proper synthetic method.

6.2.3 Other Cathode Materials

Similar to graphite, layered transitional metal disulfides (TMDs) could potentially offer capacity to host anions. MoS_2, the commonly used anode material in alkaline-metal ion batteries, has been proven to successfully store difluoro(oxalato)borate ($DFOB^-$) anions in the potential range from 3.0 to 5.0 V (vs. Li^+/Li) due to the large interlayer spacing [25]. The weakly bonded MoS_2-$DFOB^-$ could enable the MoS_2 intercalation compounds formed during the charge process to reversibly release the $DFOB^-$ anions during the discharge process. As a result, the material displayed a capacity of 118 mAh g^{-1} at a current density of 0.5 mA cm^{-2}. Beyond that, it should also be feasible to utilize transitional metal oxides (TMOs) as the cathode materials for DIBs due to the variability of chemical valence of transitional metal ions. Mn_3O_4 has been demonstrated to store/release NO_3^- in aqueous electrolyte through creating amorphous nanoregions along with the rise and fall of valence state of Mn [26].

Despite the progress, the aforementioned carbon and organic cathode materials always suffer from limited theoretical capacities. In contrast, materials based on conversion reactions can promise higher specific capacities. Metals are the prime examples as they have the tendency to lose electrons to form metal salts. Previous work reported the reversible anodic behaviors of copper metal, in both aqueous and organic electrolytes [44]. In the aqueous electrolyte of 4.5 m K_2CO_3 and 9 m KOH (m: mol kg^{-1}), Cu was found to host CO_3^{2-} and OH^- to form $Cu_2CO_3(OH)_2$, and the capacity can achieve 555 mAh g^{-1} at 500 mA g^{-1} in the potential range from −0.1 to 1.0 V with activated carbon as the anode. While in a $TFSI^-$ anion-based organic electrolyte system, the Cu cathode exhibited a high capacity of 762 mAh g^{-1} at 42.2 mA g^{-1} at a high average discharge potential of 3.2 V (vs. Li^+/Li) [27]. More complicated conversion reactions occurred as the oxidized Cu^{2+} existed as both the $Cu(TFSI)_2$ catholyte and solid after charging:

$$\mathrm{Cu(s) + 2TFSI^-(sol) - 2e^- \rightarrow 0.24Cu(TFSI)_2(sol) + 0.76Cu(TFSI)_2(s)} \quad (6.24)$$

During the discharge process, the oxidized Cu^{2+} would change back to Cu metal within two steps:

$$\mathrm{0.76Cu(TFSI)_2(s) + 0.24Cu^{2+}(sol) + e^- \rightarrow \mathit{x}CuTFSI(s) + (1-\mathit{x})Cu^+(sol) + (1.52-\mathit{x})TFSI^-(sol)}\ (1 \geq x \geq 0) \quad (6.25)$$

$$\mathrm{\mathit{x}CuTFSI(s) + (1-\mathit{x})Cu^+(sol) + e^- \rightarrow Cu(s) + \mathit{x}TFSI^-(sol)}\ (1 \geq x \geq 0) \quad (6.26)$$

To sum up, the development of other cathode materials including TMDs, TMOs, and metals would further extend the application of DIB systems. However, there is a bottleneck to the improvement of the cycling stability.

6.3 Anode Materials

In DIBs, the anode materials play the roles of storing/releasing cations (e.g. Li^+, Na^+, and K^+), which is similar to those in rocking-chair batteries. Therefore, the anode materials in rocking-chair batteries could have the potential to be used in DIBs.

However, it is worth noting that the design of anode materials in DIBs should match the anion-storing kinetics in the cathode side as well as the high-voltage requirements in the electrolyte. The following part will introduce the currently developed anode materials including metallic materials, intercalation materials, alloying materials, conversion materials, and other materials (Figure 6.3).

6.3.1 Metallic Anode Materials

The metallic anode materials (e.g. Li, Na, and K) store/release cations based on the plating/stripping working mechanism during the charging and discharging processes. Specifically, this kind of material always exhibits the lowest redox potential, which would endow the corresponding DIBs with the highest working voltage. However, two critical problems limit the development of the metallic anode materials: (i) the high activity of the metals usually reacts spontaneously and rapidly with electrolyte to form unstable SEI film, leading to unsatisfactory CE and cycle life. (ii) The crack of mechanically fragile SEI film usually causes the growth of dendrite, resulting in serious security issues such as short circuit, fire, and explosion. Therefore, research work for metallic cathode materials primarily focuses on their stabilization.

Surface modification has been considered one of the effective strategies to stabilize the metallic anode. The first method is introducing a conductive carbon layer between anode and electrolyte. To protect the Li anode, a carbon nanofiber (CNF) film was utilized based on its high conductivity, large specific surface area, and excellent structural stability [45]. The results confirmed that the CNFs film can reduce the polarization of Li anode (symmetric CNF-Li cell worked stable within the anodic/cathodic potential of ±180 mV), regulate the Li deposition sites and slow down the dendrite formation (dendrite-free morphology was observed after

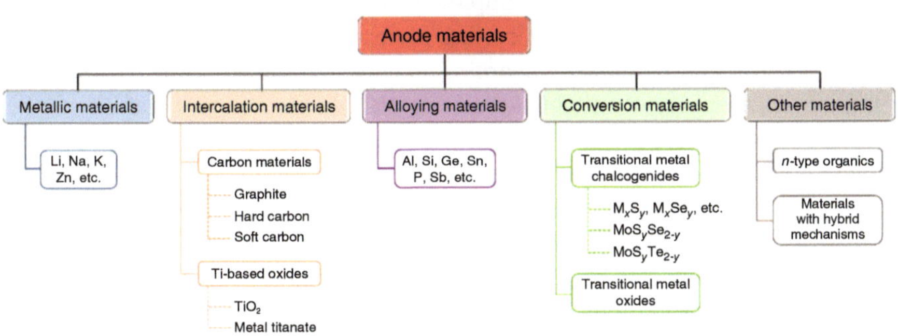

Figure 6.3 The classification of anode materials in DIBs.

200 cycles), and suppress the durative decomposition of electrolyte. As a result, the as-assemble CNFs-Li//graphite DIB delivered a high capacity retention of 86.4% over 2000 cycles at $1\,A\,g^{-1}$ with an average CE of 98.5%. The second method is constructing an artificial SEI layer on metallic anode. In a research study, a robust and conductive SEI film composed of LiN_xO_y, Li_2S_x, and $Li_2S_xO_y$ was deposited on Li anode surface by cycling a symmetric Li cell in the electrolyte of LiTFSI, $LiNO_3$, and Li_2S_8 dissolved in 1,3-dioxolane (DOL) and 1,2-dimethoxyethane (DME) [46]. It was proved that the artificial heterogeneous SEI (*h*-SEI) can effectively prevent the dendrite growth, improve Li anode/electrolyte interfaces, and decrease the electrochemical impedance, enabling the SEI-Li//graphite DIB with enhanced cycling stability (a capacity retention of up to 89.6% after 500 cycles at $1\,A\,g^{-1}$).

Besides, the selection of milder working temperature is also beneficial to the elimination of metal dendrites formation as the spread of the metal can be effectively prevented due to the high surface tension of liquid anode. This strategy is better for sodium and potassium metallic anode-based DIBs due to the lower melting points of Na (97.8 °C), K (63.7 °C), and NaK alloys (<25 °C for 30–85 mol% K) [47]. A Na symmetric cell with $Na_{0.56}K_{0.44}$FSA molten salt as the electrolyte designed by Xu et al. was reported to operate durably for hundred cycles at current densities of 0.5 and $1\,mA\,cm^{-2}$ at 90 °C. Small overpotential lower than 100 mV as well as uniform and intact SEI film can be observed.

The design of effective, metallophilic layer artificial SEI on the metallic anode materials as well as construction of special system with milder working temperature can be the powerful synergistic strategies to improve the anode/electrolyte interface and suppress the dendrites growth. Significantly, the utilization of appropriate electrolytes and additives should also greatly affect the stability of the anode, which will be discussed in the next part. Furthermore, the safety needed to be assessed before the practical application.

6.3.2 Intercalation Anode Materials

Carbon materials are regarded as the most typical intercalation anode materials in DIBs due to their good stability and high electrical conductivity [48, 49]. As the representative one, graphite with low cost has been widely used to store/release cations (e.g. Li^+ and K^+), commonly in dual-graphite batteries. However, it suffers from the relative low capacity as well as poor cycling performance ascribed to the large volume change during the continuous charging and discharging processes. In addition, the graphite material is not suitable for storing Na^+ due to the poor stability of the Na-based GIC in the electrolyte with high working voltage. Therefore, it is required to explore other carbon anode materials for DIBs.

Compared with graphite, hard carbon can achieve a higher capacity of $>279\,mAh\,g^{-1}$ [50], suggesting the probability of acting as anode material in SDIBs. A pine needle-derived carbon (PNC) anode was reported to construct SDIB by coupling with graphite cathode [51]. The SDIB delivered a capacity of $140\,mAh\,g^{-1}$ at $500\,mA\,g^{-1}$ in the voltage range of 0.8–4.7 V. The capacity can remain at $127\,mAh\,g^{-1}$ after 1000 cycles with a retention rate of 87.2%. However, hard carbon with tortuous atomic configuration, defects, and microvoids always exhibits a discharge plateau at a low potential of <0.1 V (vs. Na^+/Na), probably bringing out poor CE and safety issues. In contrast, soft carbon (SC) contains a semigraphitic structure with relatively fewer defects and higher crystallinity, showing single sloping region in the charge–discharge profiles, and enabling high conductivity and excellent rate performance [52]. A SC with an interlayer distance of around 0.39 nm synthesized by thermal polymerization of perylene-3,4,9,10-tetracarboxylic acid dianhydride was demonstrated to deliver capacities of 186 and 96 mAh g^{-1} at 100 and 2000 mA g^{-1}, respectively [53]. The expansion and shrinkage of interlayer distance during the charging and discharging processes implied the reversible intercalation and deintercalation of Na^+. After combining with graphite cathode, the SDIB possessed a capacity of 103 mAh g^{-1} at 200 mA g^{-1} with the cutoff voltage from 2.0 to 4.7 V, and long-term cycling stability with a capacity retention rate of 81.8% after 800 cycles at 1000 mA g^{-1}. From the above results, it is easy to find that hard carbon and soft carbon have their own strong points and weaknesses in terms of capacity, discharge plateau, rate performance, and cycling stability. Doping with heteroatoms and nanostructure design could be efficient methods to further improve their electrochemical performance.

Apart from the carbon materials, the Ti-based oxides (e.g. TiO_2 [54] and $Li_4Ti_5O_{12}$ [55]) have also been utilized as intercalation-type anode materials for DIBs due to their safe operating potential as well as excellent cycling stability. A $Li_4Ti_5O_{12}$/carbon composite with *in situ* implanted carbon nanofilms and 3D porous structure (LTO@3DC) was developed by Tang and coworkers to serve as the anode material in LDIB using graphite as the cathode [56]. The LTO@3DC anode showed a capacity of 148 mAh g^{-1} at 87.5 mA g^{-1} and maintained a capacity of 156 mAh g^{-1} after 500 cycles, indicating the reversible intercalation/deintercalation of Li^+. Correspondingly, the as-assembled LDIB exhibited a capacity of ~110 mAh g^{-1} at 200 mA g^{-1}, and long cycling stability with a capacity retention of ~100% after 700 cycles in the working voltage of 1.0–3.8 V.

Despite the efforts on the development of intercalation anode materials, the low capacity caused by the limited active sites is still the main bottleneck problem to clear up.

6.3.3 Alloying Anode Materials

In comparison with the intercalation anode materials, the alloying anode materials (e.g. Al [57], Si [58], Ge [59], Sn [6], P [60], and Sb [8]) usually display extremely high capacity due to the outstanding intrinsic ability to store specific cations. Especially, the alloying-type metallic materials (such as Al foil and Sn foil) have the potential to be directly the anode in DIBs [61]. However, the large volume changes (e.g. LiAl: ~97% [62], $Li_{15}Si_4$: ~270% [63], $Li_{4.4}Ge$: ~370% [64], $Na_{15}Sn_4$: ~420% [65], Na_3P: ~331% [66], and K_3Sb: ~400% [67]) during the alloying/dealloying processes always cause serious crack and pulverization of the anode, leading to large irreversible capacities and rapid capacity fading. In the first report on DIB using Al foil as the anode, the reversible Al-Li alloying and dealloying occurred at the flat plateau of 0.22 and 0.52 V (vs. Li^+/Li) along with the formation/breakdown of LiAl alloy, respectively [68]. The Al//graphite DIB only worked for 200 cycles with a capacity of 92 mAh g^{-1} maintained at 200 mA g^{-1}.

Feasible strategies should be adopted to modify the electrode for the improvement of cycling stability. The construction of carbon-based nanoarchitectures is one of the most common approaches. With the coating of an amorphous carbon layer, a modified Al nanosphere anode (nAl@C) showed significantly outstanding cycling stability with a capacity of 88 mAh g^{-1} with 94.6% capacity retention after 1000 cycles at a current density of 1500 mA g^{-1} in a nAl@C//graphite DIB [69]. The amorphous carbon layer was demonstrated to not only buffers the mechanical strain induced by the volume change but also facilitates the electron and ion transport. Note that the rigid contacts between the active materials and metal foil current collectors would result in significant interfacial stress, leading to the structural degradation and resultant fast capacity decay, particularly at high current densities. The rational design of flexible interface could be useful to solve the problem. Take Si anode as an example, a flexible interface was established by using a conductive soft nylon substrate with a Cu–Ni buffer layer [58]. The notable elastic deformation of nylon can effectively regulate the alloying stress of Si anode, giving rise to stress relaxation. As a result, a Si//graphite LDIB exhibited enhanced cycling stability with a high capacity retention of 97% over 2000 cycles at 1000 mA g^{-1}, and high flexibility with a high capacity retention of ~84% after 1500 bends and a low self-discharging voltage loss of 0.0015% per bend after 10,000 bends.

Compared with the single-component metal anode, the alloys of no less than two kinds of metals would show the synergistic effect to react with cations at different potentials, thus resulting in the enhancement of cycling performance. A free-standing alloy anode with SnSb nanodots encapsulated in free-standing porous carbon nanowires (SnSb@PCNWs) can deliver a reversible capacity of 408.5 mAh g^{-1} at 300 mA g^{-1} and 86.8% capacity retention after 200 cycles [8]. The dealloying processes of K_3Sb and KSn can be clearly observed from the anodic scan of CV curve with the presence of two oxidation peaks located at around 0.52 and 1.3 V (vs. K^+/K). As a result, the assembled SnSb@PCNMs//graphite PDIB displayed good cycling performance with a capacity of 128.8 mAh g^{-1} retained at 1000 mA g^{-1} after 300 cycles.

The focus of alloying anode materials will be still on the structural stability during the alloying/dealloying processes. The formation of a stable SEI film on the anode with an easy and efficient method is required.

6.3.4 Conversion Anode Materials

The conversion anode materials (dominated by TMOs/chalcogenides, etc.) have also come into the spotlight due to the merits of high capacity usually from multielectron conversion reactions and relatively lower volume expansion when compared with the alloying-type anode materials.

Owing to the distinctive electronic structures and superior electrochemical behaviors, 2D transitional metal dichalcogenides (TMDs) have received tremendous research interest for sodium storage. MoS_2, with a tunnel spacing of ~0.62 nm and a theoretical capacity of 670 mAh g^{-1}, is the typical representative. However, some critical issues like the intrinsic poor conductivity and possibly severe aggregation during charging/discharging processes need to be handled. Nanostructure configuration combined with conductive carbon materials has commonly been proposed as the most effective strategy. For example, in a MoS_2/C nanocomposite, the hierarchical penne-like structure with an expanded interlayer distance of 0.98 nm and the homogenously distributed carbon endowed the material with a reversible capacity of 315 mAh g^{-1} after 200 cycles at 201 mA g^{-1} [70]. By coupling with graphite cathode, the SDIB delivered a capacity of 65 mAh g^{-1} at 200 mA g^{-1}.

The anion doping is also a promising method to modify the physicochemical and electrochemical properties of nanomaterials. The selenium or tellurium doping element with lower electronegativity and larger atomic radius has been demonstrated to enlarge the interlayer spacing of MoS_2, increase the conductivity, and provide more active sites, therefore boosting the electrochemical performance. As a result, a carbon-coated nanowire-in-nanotube tellurium doped MoS_2 material ($MoS_{1.5}Te_{0.5}$@C) with an interlayer spacing of 0.75 nm expressed reversible capacities of 462.9 and 352.3 mAh g^{-1} at 0.1 and 5 A g^{-1}, respectively, and displayed a capacity of 355.4 mAh g^{-1} after 500 cycles at 1 A g^{-1} with a retention of 93.3% [71]. On this basis, the as-assembled $MoS_{1.5}Te_{0.5}$@C//graphite SDIB maintained a capacity of 145.7 mAh g^{-1} over 1500 cycles at 1 A g^{-1}.

Despite the progress of conversion anode materials, the relatively higher working potential is the critical shortcoming that would greatly limit the energy density of as-assembled DIBs. It is challenging but valuable for researchers to exploit novel conversion-type anode materials with suitable cation uptake potential.

6.3.5 Other Anode Materials

Other kinds of materials apart from the aforementioned anodes have also been developed under the unremitting efforts of experts. One is the *n*-type organics containing —C=O— functional group(s), which can coordinate with cations during the charge process. 3D porous dipotassium terephthalate (pK_2TP) has been verified to reversibly store K^+ due to the existence of carbonyl as well as feasible ion diffusion pathways [72]. The consequent pK_2TP//expanded graphite (EG) K-based dual ion battery (KDIB) can display a capacity of 68 mAh g^{-1} at 200 mA g^{-1} and work stably with no capacity decay over 2000 cycles at 2 A g^{-1}. Significantly, the insolubility in electrolyte as well as the structural stability of organic anode materials should be guaranteed. In a magnesium-based DIB (MDIB) with $Mg(TFSI)_2/Pyr_{14}TFSI$ as organic electrolyte, the solubility test and morphology comparison after cycling confirmed that 3,4,9,10-perylenetetracarboxylic diimide (PTCDI) exhibited better stability than 3,4,9,10-perylenetetracarboxylic dianhydride (PTCDA) [12].

Beyond the organic anodes, materials with hybrid energy storage mechanisms also show their superiorities in electrochemical performance enhancement. Owing to the involvement of N-doped carbon matrix (N-C) as well as the conversion and alloying synergistic reactions, the SnP_2O_7@N-C anode delivered a high specific capacity (~400 mAh g^{-1} at 0.1 A g^{-1}) and good cycling stability (a capacity retention of 92% after 1200 cycles under 1.5 A g^{-1}), which endowed the SnP_2O_7@N-C//KS6 SDIB with long-term cycling life with a capacity retention of ~96% after 1000 cycles at 2 A g^{-1} [73]. This phenomenon is commonly observed for perovskite materials including $NaTaO_3$, $KNi_{0.1}Co_{0.9}F_3$, and $KZnF_3$.

In summary, the experience in the anodes of rocking-chair batteries can promote the development of anodes in DIBs. Apart from the concerns about the electrochemical properties and safety, the match between the anode and cathode as well as the high-voltage electrolyte should be studied further.

6.4 Electrolytes

As the active species in DIBs, electrolytes greatly affect the electrochemical performance (capacity, energy density, CE, cycling stability, etc.). An ideal electrolyte in DIBs must call for some requirements: (i) wide ESW to support the anion intercalation in cathode at a high potential over ~5 V; (ii) enabling reversible electrochemical reactions of cations at anode side and anions at cathode side, respectively; (iii) fast anion/cation flow ensuring high ionic conductivity to achieve good rate performance; and (iv) low cost, good safety, and high compatibility with the inactive components. This section will introduce four kinds of electrolytes in DIBs: organic electrolytes, ionic liquid electrolytes, aqueous electrolytes, and gel polymer electrolytes (GPEs) (Figure 6.4).

6.4.1 Organic Electrolytes

The booming development of LIBs motivates researchers to investigate the electrochemical behaviors of organic electrolytes in DIBs. According to Eq. (6.14), it can be observed that the capacity of a DIB cell is quite associated with the electrolyte characters.

In terms of electrolyte solvents, a lower energy of the highest occupied molecular orbital (HOMO) than the electrochemical potential of the cathode and a higher energy of the lowest unoccupied molecular orbital (LUMO) should be necessary. Only in this case can the electrolyte be prevented to be oxidized or reduced at the proper working potential. The carbonate esters, the most commonly used organic solvents in LIBs, have also attracted wide attention in DIBs due to their proper HOMO/LUMO values as well as mature market environment. Among them, the cyclic carbonates (e.g. ethylene carbonate (EC) and propylene carbonate (PC)) with high dielectric constant (ε) are capable of dissolving enough salts; while the linear carbonates (e.g. ethylmethyl carbonate (EMC), dimethyl carbonate (DMC), and diethyl carbonate (DEC)) with low viscosity (η) are favorable to ion diffusion [75]. Therefore, the combination of cyclic carbonate(s) and linear carbonate(s) can be more applicable for solvents in DIBs. Besides, the employment of other kinds of co-solvents could also be feasible in DIBs, especially in the case of multivalent electrolytes. It is worth noting that in multivalent electrolytes, the cations could

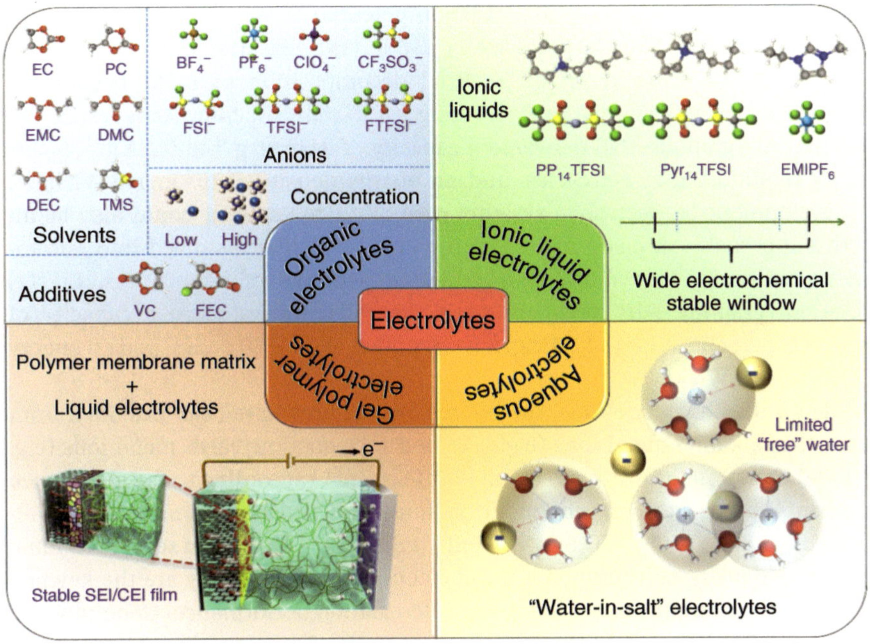

Figure 6.4 The classification of typical electrolytes for DIBs. Source: Image of gel polymer electrolyte: Reproduced with permission Xu et al. [74]. Copyright 2020, Elsevier.

be co-intercalated with anions into graphite cathodes in the form of large-sized anionic complexes due to the strong cation–anion association, leading to the limited anion-storage capacity of a graphite cathode. The utilization of co-solvents with high permittivity and oxidation tolerance is effective in suppressing this cation co-intercalation behavior. For instance, in a $Cu_{2-x}Se$//graphite MDIB, the introduction of adiponitrile (ε: 29.8, η: 6.93 cP) into $Mg(TFSI)_2$-EMC electrolyte can result in a decrease of 83% in Mg^{2+} co-intercalation and an increase of ~29.5% in capacity of the graphite cathode [76].

Despite this, the carbonate SDIBs usually work at limited oxidation potential (e.g. <4.4 V vs. Li^+/Li), which leads to the underutilized of cathode material, resulting in poor cycling stability and unsatisfactory energy density of DIBs. The employment of electrolyte additives could be a simple and practical method due to their functions of suppressing electrolyte decomposition, improving the ionic conductivity, and forming stable SEI/CEI layers. Only a proper concentration of additives is beneficial for improving electrochemical behaviors of DIBs. For example, 4 M $LiPF_6$ in EMC with 5 wt% vinylene carbonate (VC) additive served as the optimal electrolyte to endow the nAl@C anode with sufficient SEI film [69]. In another work, a stable CEI layer on graphite cathode was revealed in the electrolyte of 3 M $LiPF_6$ in EMC with 5 vol% fluoroethylene carbonate (FEC) additive [77]. Another effective way is to design concentrated electrolytes which show enhanced resistance to reduction/oxidation and decrease the usage of electrolyte. A concentrated electrolyte with 7.5 m LiFSI in 1 : 1 (v/v) EC/DMC demonstrated an improved higher oxidation potential of around 6.0 V (vs. Li^+/Li) than 0.8 *m* (~4.2 V) and 4.0 *m* (~4.9 V) LiFSI [18]. Besides, the concentrated electrolyte would bring about preferential decomposition of LiFSI because of the decreased LUMO energy level, forming a LiF-rich SEI film. As a result, the assembled Al//graphite DIB delivered a capacity of 94 mAh g^{-1} at 0.2 A g^{-1}, 96.8% capacity retention after 500 cycles, and an energy density of up to 178 Wh kg^{-1}. The development of electrolyte systems with advanced solvent could also be an effective approach. An electrolyte system with 4.0 *m* LiFSI dissolved into tetramethylene sulfone (TMS) showed an oxidation potential of ~6.44 V (vs. Li^+/Li) [78]. A Li//graphite DIB based on such electrolyte can achieve a capacity of 113.3 mAh g^{-1} at 0.2 A g^{-1}, 94.7% capacity retention after 1000 cycles, and an energy density of ~180 Wh kg^{-1}.

While for the electrolyte salts, the intercalated cations have been extended from Li^+ to other alkaline metal ions (e.g. Na^+ and K^+), alkaline earth metal ions (e.g. Mg^{2+} and Ca^{2+}), and transitional metal ions (e.g. Zn^{2+}) according to the abundance of elements as well as environmental friendliness. The anions have also been developed to BF_4^-, PF_6^-, ClO_4^-, $CF_3SO_3^-$, $FTFSI^-$, FSI^-/FSA^-, $TFSI^-$, and so on. The ionic size, charge density, solvation effect, and electrochemical stability are the key factors that need to be considered in the ion intercalation performance. Generally, the larger radii of anions would cause higher resistance for anions intercalation into cathode materials than cations intercalation into anode materials, which always leads to unsatisfactory rate capability and cycling stability. The anion-hybridization strategy has been proposed to facilitate ion diffusion and alleviate volume expansion in the cathode during cycling. Based on the electrolyte with the combination of PF_6^- anions (radius: 0.254 nm) and smaller sized BF_4^- (radius: 0.229 nm), the Al//graphite DIB demonstrated a capacity of 90.5 mAh g^{-1} at 1.5 A g^{-1}, and ~100% capacity retention after 500 cycles at 0.5 A g^{-1}, which were superior to the DIB with the single PF_6^- anion-based electrolyte (53.4 mAh g^{-1} at 1.5 A g^{-1}, and failed after 150 cycles at 0.5 A g^{-1}) [79].

Despite the great efforts in the development of organic electrolytes in DIBs, the exploration of suitable electrolyte with effective ionic kinetics is still on the way. In addition, the pursuit of security in organic electrolytes should be always the changeless tenet.

6.4.2 Ionic Liquid Electrolytes

Ionic liquids (ILs), known as liquid-state molten salts at room temperature, also show good potential in DIBs due to the merits of broad ESW, high thermal stability, nonflammability, and low volatility. Pure ILs such as 1,2-dimethyl-3-propylimidazolium chloroaluminate ($DMPI^+$) ($AlCl_4^-$) [80] have been applied as electrolytes in dual graphite cells with well-defined charge–discharge plateaus, and the reversible intercalation and deintercalation of cations and anions into/from graphite cathode and anode have also been demonstrated. However, the poor compatibility of ILs with the electrodes usually endows the cells with dismal performance. In this case, IL electrolytes are utilized by mixing ILs (e.g. 1-butyl-1-methylpiperidinium bis(trifluoromethylsulfonyl) imide ($PP_{14}TFSI$), 1-butyl-1-methylpyrrolidinium bis(trifluoromethylsulfonyl) imide ($Pyr_{14}TFSI$), and 1-ethyl-3-methylimidazolium-hexafluorophosphate ($EMIPF_6$)) with metal salts (e.g. LiTFSI) [81, 82].

Besides, the metal salts could serve as active species for extra redox reactions. In a PTCDI//graphite MDIB, the addition of $Mg(TFSI)_2$ in $Pyr_{14}TFSI$ IL electrolyte could induce the magnesiation/demagnesiation processes at the PTCDI anode beyond the insertion/deinsertion of Pyr_{14}^+ [12]. As expected, the DIB displayed a discharge capacity of 57.7 mAh g^{-1} at 200 mA g^{-1}, which was much higher than that of 23 mAh g^{-1} for pure IL system. Similar to organic electrolytes, the active species concentration as well as the usage of additives also have a great influence on the DIB performance with IL as the electrolyte. For instance, in an Al//MCMB DIB based on the electrolyte of LiTFSI in $Pyr_{14}TFSI$ IL, the introduction of FEC can clearly improve the capacity as well as the cycle stability [81]. Specifically, the capacity of DIB can enhance from 29 to 80 mAh g^{-1} at 200 mA g^{-1} as the FEC amount increases from 0 to 15 wt%. While further increase of FEC (15 wt%) would cause the fast decay of cycling stability. In addition, it was also demonstrated that a LiTFSI concentration of 1.5 M in $Pyr_{14}TFSI$ achieved the best performance based on the balance of active species content with the viscosity of electrolyte.

The IL electrolytes with high thermal stability and nonflammability can endow the as-assembled DIBs with high security. However, the inferior diffusion kinetics as well as high cost limit the practical application.

6.4.3 Aqueous Electrolytes

Compared with organic and IL electrolytes, aqueous electrolytes have been also paid attention to DIBs due to their cost and safety benefits. A 1 M $(NH_4)_2SO_4$ aqueous solution was reported to serve as the electrolyte in an aqueous ammonium DIB, in which the *n*-type polyimide anode and *p*-type radical polymer can react with NH_4^+ and SO_4^{2-} ions, respectively [83]. Based on the weight of both cathode and anode materials, the DIB delivered an energy density of 51.3 Wh kg^{-1}, a power density of 15.8 kW kg^{-1}, and ~86.4% capacity retention after 10,000 cycles at 5 A g^{-1}. Unfortunately, the DIB can only work at a maximum voltage of 1.9 V, which was ascribed from the narrow ESW of aqueous electrolytes without decomposing.

It has been revealed that the activity of water can be significantly decreased in superconcentrated aqueous electrolytes, commonly called "water-in-salt" electrolytes (WiSEs), as the little "free" water acts as parts of ion solvation shells for the active salts [84], which reduces the onset potential of oxygen evolution reaction (OER) on the cathode material. A WiSE of 20 *m* NaFSI and 0.5 *m* $Zn(TFSI)_2$ was reported to provide a much larger ESW of ~2.7 V than the diluted electrolyte containing 1 *m* NaFSI and 0.5 *m* $Zn(TFSI)_2$ (~2.3 V) [85]. The Zn//graphite DIB based on this WiSE presented a cell voltage of ~2.3 to ~2.5 V and delivered a capacity of 110 mAh g^{-1} at 200 mA g^{-1}. In addition, designing hybrid aqueous/nonaqueous WiSEs can also be acceptable. By mixing an aqueous solution of 7 *m* LiOTf and 21 *m* LiTFSI with a nonaqueous component of 9.25 *m* LiTFSI in DMC, a hybrid WiSE can be developed to apply in the Nb_2O_5//graphite DIB which can work stably in the voltage range of 0–3.2 V [86]. Meanwhile, the hybrid electrolyte also exhibited excellent combustion resistance.

To sum up, the aqueous electrolytes have demonstrated their advantages in DIBs. However, it is quite important to find a balance between the electrochemical performance and the cost and safety benefits.

6.4.4 Gel Polymer Electrolytes

Considering the co-intercalation process of solvents in cathode materials in traditional liquid electrolyte-based DIBs, the development of GPEs can be an available effective strategy to avoid it, thus optimizing the cycling stability. Generally, the GPEs can be *ex situ* or *in situ* synthesized to fulfill the requirements of oxidative stability as well as interfacial compatibility.

The *ex situ* synthesis of GPE always involves two steps: the formation of polymer membrane matrix and subsequent soaking process within conventional liquid electrolytes. For the polymer matrix, the chemical stability, mechanical stability, ionic conductivity, and crystallinity are the several essential conditions to be considered. In one report, a polymer matrix was prepared by doping poly(vinylidene fluoride-hexafluoro propylene) (PVDF-HFP) with poly(ethylene oxide) (PEO) and graphene oxide (GO) via the H—F or H—O hydrogen bond interactions according to their respective advantages: low crystallinity, high ionic conductivity and mechanical stability for PVDF-HFP, good pore-making capacity for PEO, and excellent mechanical property as well as thermal stability for GO [87]. As expected, the obtained polymer with a 3D porous network achieved a good ionic conductivity of 2.1×10^{-3} S cm^{-1} and high electrolyte (i.e. 4 M $LiPF_6$ in EMC with 2% VC) uptake of 368%. The optimized GPE enabled the Al//graphite DIB to operate stably for 2000 cycles with a capacity retention of 92% at 500 mA g^{-1} in a voltage range of 3.0–4.95 V.

While for the *in situ* prepared GPEs, the polymer matrix can be directly polymerized from the monomer in the presence of optimized liquid electrolytes, effectively regulating the anion and cation fluxes. By *in situ* polymerizing an ethoxylated pentaerythritol tetraacrylate monomer in a liquid electrolyte (0.5 M $NaPF_6$ in PC/EMC/FEC [v/v/v = 1 : 1 : 1]) with 4 wt% 1,3-propanesultone as an additive, a GPE was developed to possess a high ionic conductivity of 5.33×10^{-3} S cm^{-1} and wide ESW of up to 5.5 V (vs. Na^+/Na) and facilitate the formation of stable SEI and CEI films on the Na anode and graphite cathode, respectively [74]. As a result, the GPE-based Na//graphite SDIB delivered a capacity of 78 mAh g^{-1} after 1000 cycles with a capacity retention of 86.7% except the initial 10 activation cycles at 100 mA g^{-1}.

The development of GPEs can further expand the application scenarios of DIBs, especially for the fields of flexible energy storage devices. However, future work should focus on the improvement of interfacial compatibility between GPEs and anode/cathode materials.

6.5 Conclusion and Prospects

In conclusion, the merits of high working voltage, enhanced energy density, and cost-effectiveness can enable DIBs to serve as good candidates for next-generation energy storage. In DIBs, especially, the electrolytes would function as active species to participate in redox reactions, which is totally different from the traditional rocking-chair batteries. Until now, various kinds of cathode materials (i.e. carbon materials represented by graphite, organic materials, and other cathode materials like TMDs/oxides), anode materials (i.e. metallic materials, intercalation-type materials, alloying-type materials, conversion-type materials, and other materials such as organic materials), and electrolytes (i.e. organic electrolytes, ionic liquid electrolytes, aqueous electrolytes, and GPEs have been explored to applied in DIBs. Meanwhile, with the development of science and technology, numerous advanced characterizations have been utilized to investigate the energy storage mechanism: *in situ/ex situ* X-ray diffraction (XRD), *in situ/ex situ* Raman, *in situ/ex situ* transmission electron microscope (TEM), *in situ* electrochemical quartz crystal microbalance (EQCM), *in situ* atomic force microscope (AFM), real-time optical microscope, and so on. Despite these encouraging achievements, DIBs can still not be available for practical application currently. More efforts need to be invested to promote the further developments of DIBs:

(1) It is necessary to exploit novel host materials for reversible anion storing/releasing. Graphite materials have been currently considered the most studied cathode materials in DIBs due to their unique layered structure, low cost, and wide applicability for various anions. However, low capacity is the fatal flaw. Besides, the continuous intercalation/deintercalation of large-sized anions and possible solvent co-intercalation behavior would cause severe volume change, leading to unsatisfactory cycling stability. Exploiting a new generation of cathode material with high capacity, long cycle life, and high and stable working potential is difficult at this stage, while research and development of materials/composites with hybrid energy storage mechanisms could be an alternative strategy.
(2) Optimization design of high-voltage organic electrolyte is needed. Although IL electrolytes, aqueous electrolytes, and GPE have an advantage in safety due to their low flammability or nonflammability, the relatively mature research system as well as flexible component adjustability enable organic electrolytes to be still considered as the most promising currently available electrolytes in DIBs. The development of solvents and additives with enhanced resistance to reduction/oxidation is the key and difficult point. The energy storage mechanism especially for the anion storing/releasing behavior in the cathode side must be studied in-depth during the process of adjusting electrolyte components. Note that the SEI-forming solvents or additives can be utilized on the premise of without suppressing/blocking the anion intercalation into cathode materials. Meanwhile, the interfacial compatibility between electrodes (cathode/anode) and electrolytes should also deserve attention.

In a word, the development of DIBs is still in the initial stage. Based on the current achievements, it is encouraged to assemble advanced DIB demos including pouch cells, cylindrical cells, and square batteries. With unremitting efforts on improvements in cyclability, energy density, and safety, the DIB technology will be an option for sustainable stationary energy storage in the future.

References

1 Rüdorff, W. and Hofmann, U. (1938). Uber graphitsalze. *Zeitschrift für Anorganische und Allgemeine Chemie* 238: 1–50.

2 McCullough, F.P., Levine, C.A., and Snelgrove, R.V. (1989). Secondary battery. US Patent 4830938.

3 Carlin, R.T., Delong, H.C., Fuller, J., and Trulove, P.C. (1994). Dual intercalating molten electrolyte batteries. *Journal of the Electrochemical Society* 141 (7): L73–L76.

4 Placke, T., Bieker, P., Lux, S.F. et al. (2012). Dual-ion cells based on anion intercalation into graphite from ionic liquid-based electrolytes. *Zeitschrift für Physikalische Chemie* 226 (5–6): 391–407.

5 Liang, Z., Gong, D., Shang, J. et al. (2022). Low volume expansion carbon-coated $Fe_2P_4O_{12}$ anode material for high-performance sodium dual-ion battery. *Energy Storage Materials* 53: 331–339.

6 Sheng, M.H., Zhang, F., Ji, B.F. et al. (2017). A novel tin-graphite dual-ion battery based on sodium-ion electrolyte with high energy density. *Advanced Energy Materials* 7 (7): 1601963.

7 Ji, B., Zhang, F., Song, X., and Tang, Y. (2017). A novel potassium-ion-based dual-ion battery. *Advanced Materials* 29 (19): 1700519.

8 Wei, C.Y., Gong, D.C., Xie, D.H., and Tang, Y.B. (2021). The free-standing alloy strategy to improve the electrochemical performance of potassium-based dual-ion batteries. *ACS Energy Letters* 6 (12): 4336–4344.

9 Han, C., Wang, H., Wang, Z. et al. (2023). Solvation structure modulation of high-voltage electrolyte for high-performance K-based dual-graphite battery. *Advanced Materials* 35 (24): 2300917.

10 Zhang, X., Zhang, L., Zhang, W. et al. (2022). A fast and stable sodium-based dual-ion battery achieved by Cu_3P@P-doped carbon matrix anode. *Journal of Power Sources* 518: 230741.

11 Yang, R., Zhang, F., Lei, X. et al. (2020). Pseudocapacitive Ti-doped niobium pentoxide nanoflake structure design for a fast kinetics anode toward a high-performance Mg-ion-based dual-ion battery. *ACS Applied Materials & Interfaces* 12 (42): 47539–47547.

12 Lei, X., Zheng, Y.P., Zhang, F. et al. (2020). Highly stable magnesium-ion-based dual-ion batteries based on insoluble small-molecule organic anode material. *Energy Storage Materials* 30: 34–41.

13 Wang, M., Jiang, C.L., Zhang, S.Q. et al. (2018). Reversible calcium alloying enables a practical room-temperature rechargeable calcium-ion battery with a high discharge voltage. *Nature Chemistry* 10 (6): 667–672.

14 Seel, J.A. and Dahn, J.R. (2000). Electrochemical intercalation of PF_6^- into graphite. *Journal of the Electrochemical Society* 147 (3): 892–898.

15 Wang, M. and Tang, Y. (2018). A review on the features and progress of dual-ion batteries. *Advanced Energy Materials* 8 (19): 1703320.

16 Zhang, L., Wang, H., Zhang, X., and Tang, Y. (2021). A review of emerging dual-ion batteries: fundamentals and recent advances. *Advanced Functional Materials* 31 (20): 2010958.

17 Kravchyk, K.V. and Kovalenko, M.V. (2019). Rechargeable dual-ion batteries with graphite as a cathode: key challenges and opportunities. *Advanced Energy Materials* 9 (35): 1901749.

18 Xiang, L., Ou, X., Wang, X. et al. (2020). Highly concentrated electrolyte towards enhanced energy density and cycling life of dual-ion battery. *Angewandte Chemie International Edition* 59 (41): 17924–17930.

19 Placke, T., Heckmann, A., Schmuch, R. et al. (2018). Perspective on performance, cost, and technical challenges for practical dual-ion batteries. *Joule* 2 (12): 2528–2550.

20 Zhang, G., Ou, X.W., Cui, C.Y. et al. (2019). High-performance cathode based on self-templated 3D porous microcrystalline carbon with improved anion adsorption and intercalation. *Advanced Functional Materials* 29 (2): 1806722.

21 Wang, W., Huang, H.X., Wang, B. et al. (2019). A new dual-ion battery based on amorphous carbon. *Science Bulletin* 64 (21): 1634–1642.

22 Rajesh, M., Dolhem, F., Davoisne, C., and Becuwe, M. (2020). Reversible anion insertion in molecular phenothiazine-based redox-active positive material for organic ion batteries. *ChemSusChem* 13: 2364–2370.

23 Glatz, H., Lizundia, E., Pacifico, F., and Kundu, D. (2019). An organic cathode based dual-ion aqueous zinc battery enabled by a cellulose membrane. *ACS Applied Energy Materials* 2 (2): 1288–1294.

24 Wang, H.G., Wang, H.D., Si, Z.J. et al. (2019). A bipolar and self-polymerized phthalocyanine complex for fast and tunable energy storage in dual-ion batteries. *Angewandte Chemie International Edition* 58 (30): 10204–10208.

25 Li, C., Lao, B., Li, Z. et al. (2020). Dual-ion battery with MoS_2 cathode. *Energy Storage Materials* 32: 159–166.

26 Jiang, H., Wei, Z., Ma, L. et al. (2019). An aqueous dual-ion battery cathode of Mn_3O_4 via reversible insertion of nitrate. *Angewandte Chemie International Edition* 58: 5286–5291.

27 Yu, M., Sui, Y., Sandstrom, S.K. et al. (2022). Reversible copper cathode for non-aqueous dual-ion batteries. *Angewandte Chemie International Edition* 61 (47): e202212191.

28 Cai, W., Yao, Y., Zhu, G. et al. (2020). A review on energy chemistry of fast-charging anodes. *Chemical Society Reviews* 49 (12): 3806–3833.

29 Morozov, S.V., Novoselov, K.S., Katsnelson, M.I. et al. (2008). Giant intrinsic carrier mobilities in graphene and its bilayer. *Physical Review Letters* 100 (1): 016602.

30 Daumas, N. and Herold, A. (1969). Relations between phase concept and reaction mechanics in graphite insertion compounds. *Comptes Rendus Hebdomadaires Des Seances De L Academie Sciences Serie C* 268 (878): 373–380.
31 Zhang, L., Li, J., Huang, Y. et al. (2019). Synergetic effect of ethyl methyl carbonate and trimethyl phosphate on BF_4^- intercalation into a graphite electrode. *Langmuir* 35 (11): 3972–3979.
32 Ou, X., Li, J., Tong, X. et al. (2020). Highly concentrated and nonflammable electrolyte for high energy density K-based dual-ion battery. *ACS Applied Energy Materials* 3 (10): 10202–10208.
33 Zhang, E.J., Cao, W., Wang, B. et al. (2018). A novel aluminum dual-ion battery. *Energy Storage Materials* 11: 91–99.
34 Meister, P., Schmuelling, G., Winter, M., and Placke, T. (2016). New insights into the uptake/release of $FTFSI^-$ anions into graphite by means of in situ powder X-ray diffraction. *Electrochemistry Communications* 71: 52–55.
35 Li, X., Ou, X., and Tang, Y. (2020). 6.0 V High-voltage and concentrated electrolyte toward high energy density K-based dual-graphite battery. *Advanced Energy Materials* 10 (41): 2002567.
36 Placke, T., Rothermel, S., Fromm, O. et al. (2013). Influence of graphite characteristics on the electrochemical intercalation of bis(trifluoromethanesulfonyl) imide anions into a graphite-based cathode. *Journal of the Electrochemical Society* 160: A1979–A1991.
37 Guo, Q., Kim, K.-I., Li, S. et al. (2021). Reversible insertion of I–Cl interhalogen in a graphite cathode for aqueous dual-ion batteries. *ACS Energy Letters* 6 (2): 459–467.
38 Li, N., Xin, Y., Chen, H. et al. (2019). Thickness evolution of graphite-based cathodes in the dual ion batteries via in operando optical observation. *Journal of Energy Chemistry* 29: 122–128.
39 Han, X.Q., Xu, G.J., Zhang, Z.H. et al. (2019). An in situ interface reinforcement strategy achieving long cycle performance of dual-ion batteries. *Advanced Energy Materials* 9 (16): 1804022.
40 Li, W.H., Ning, Q.L., Xi, X.T. et al. (2019). Highly improved cycling stability of anion de-/intercalation in the graphite cathode for dual-ion batteries. *Advanced Materials* 31 (4): 1804766.
41 Dai, G.L., He, Y., Niu, Z.H. et al. (2019). A dual-ion organic symmetric battery constructed from phenazine based artificial bipolar molecules. *Angewandte Chemie International Edition* 58 (29): 9902–9906.
42 Li, C., Xue, J., Huang, A. et al. (2019). Poly(N-vinylcarbazole) as an advanced organic Cathode for potassium-ion-based dual-ion battery. *Electrochimica Acta* 297: 850–855.
43 Li, C., Yang, H., Xie, J. et al. (2020). Ferrocene-based mixed-valence metal–organic framework as an efficient and stable cathode for lithium-ion-based dual-ion battery. *ACS Applied Materials & Interfaces* 12 (29): 32719–32725.
44 Gallagher, T.C., Wu, C.-Y., Lucero, M. et al. (2022). From copper to basic copper carbonate: a reversible conversion cathode in aqueous anion batteries. *Angewandte Chemie International Edition* 61 (31): e202203837.

45 Xi, X.T., Feng, X., Nie, X.J. et al. (2019). Dendrite-free deposition on lithium anode toward long-life and high-stable Li//graphite dual-ion battery. *Chemical Communications* 55 (58): 8406–8409.

46 Xi, X.T., Li, W.H., Hou, B.H. et al. (2019). Dendrite-free lithium anode enables the lithium//graphite dual-ion battery with much improved cyclic stability. *ACS Applied Energy Materials* 2 (1): 201–206.

47 Chen, C.-Y., Matsumoto, K., Kubota, K. et al. (2020). An energy-dense solvent-free dual-ion battery. *Advanced Functional Materials* 30 (39): 2003557.

48 Wang, M., Liu, Q.R., Wu, G.M. et al. (2023). Coral-like and binder-free carbon nanowires for potassium dual-ion batteries with superior rate capability and long-term cycling life. *Green Energy & Environment* 8 (2): 548–558.

49 Yang, K., Liu, Q.R., Zheng, Y.P. et al. (2021). Locally ordered graphitized carbon cathodes for high-capacity dual-ion batteries. *Angewandte Chemie International Edition* 60 (12): 6326–6332.

50 Qiu, S., Xiao, L., Sushko, M.L. et al. (2017). Manipulating adsorption–insertion mechanisms in nanostructured carbon materials for high-efficiency sodium ion storage. *Advanced Energy Materials* 7 (17): 1700403.

51 Wang, X.H., Zheng, C., Qi, L., and Wang, H.Y. (2017). Carbon derived from pine needles as a Na^+-storage electrode material in dual-ion batteries. *Global Challenges* 1 (7): 1700055.

52 Jian, Z., Bommier, C., Luo, L. et al. (2017). Insights on the mechanism of Na-ion storage in soft carbon anode. *Chemistry of Materials* 29 (5): 2314–2320.

53 Fan, L., Liu, Q., Chen, S. et al. (2017). Soft carbon as anode for high-performance sodium-based dual ion full battery. *Advanced Energy Materials* 7 (14): 1602778.

54 Wang, X.H., Qi, L., and Wang, H.Y. (2019). Anatase TiO_2 as a Na^+-storage anode active material for dual-ion batteries. *ACS Applied Materials & Interfaces* 11 (33): 30453–30459.

55 Nozu, R., Suzuki, E., Kimura, O. et al. (2020). Dual-ion battery using graphitic carbon and $Li_4Ti_5O_{12}$: suppression of gas formation and increased cyclability. *Electrochimica Acta* 332: 135238.

56 Yu, A., Gong, D.C., Zhang, M., and Tang, Y.B. (2020). In-situ implanted carbon nanofilms into lithium titanate with 3D porous structure as fast kinetics anode for high-performance dual-ion battery. *Chemical Engineering Journal* 401: 125834.

57 Tong, X.F., Zhang, F., Ji, B.F. et al. (2016). Carbon-coated porous aluminum foil anode for high-rate, long-term cycling stability, and high energy density dual-ion batteries. *Advanced Materials* 28 (45): 9979–9985.

58 Jiang, C.L., Xiang, L., Miao, S.J. et al. (2020). Flexible interface design for stress regulation of a silicon anode toward highly stable dual-ion batteries. *Advanced Materials* 32 (17): 1908470.

59 Zhou, J., Zhou, Y., Zhang, X. et al. (2020). Germanium-based high-performance dual-ion batteries. *Nanoscale* 12 (1): 79–84.

60 Yu, D., Cheng, L., Chen, M. et al. (2019). High-performance phosphorus-graphite dual-ion battery. *ACS Applied Materials & Interfaces* 11 (49): 45755–45762.

61 Wang, H., Wang, C., Zhang, F. et al. (2022). Homogeneous alloying reaction via self-assembly strategy for high-areal-density dual-ion batteries. *Chemical Engineering Journal* 449: 137708.

62 Wang, M., Zhang, F., Lee, C.-S., and Tang, Y. (2017). Low-cost metallic anode materials for high performance rechargeable batteries. *Advanced Energy Materials* 7 (23): 1700536.

63 Wang, Q., Meng, T., Li, Y. et al. (2021). Consecutive chemical bonds reconstructing surface structure of silicon anode for high-performance lithium-ion battery. *Energy Storage Materials* 39: 354–364.

64 Chan, C.K., Zhang, X.F., and Cui, Y. (2008). High capacity Li ion battery anodes using Ge nanowires. *Nano Letters* 8 (1): 307–309.

65 Wu, X., Lan, X., Hu, R. et al. (2022). Tin-based anode materials for stable sodium storage: progress and perspective. *Advanced Materials* 34 (7): 2106895.

66 Capone, I., Aspinall, J., Darnbrough, E. et al. (2020). Electrochemo-mechanical properties of red phosphorus anodes in lithium, sodium, and potassium ion batteries. *Matter* 3 (6): 2012–2028.

67 Gao, H., Guo, X., Wang, S. et al. (2020). Antimony-based nanomaterials for high-performance potassium-ion batteries. *EcoMat* 2 (2): e12027.

68 Zhang, X.L., Tang, Y.B., Zhang, F., and Lee, C.S. (2016). A novel aluminum-graphite dual-ion battery. *Advanced Energy Materials* 6 (11): 1502588.

69 Tong, X.F., Zhang, F., Chen, G.H. et al. (2018). Core-shell aluminum@carbon nanospheres for dual-ion batteries with excellent cycling performance under high rates. *Advanced Energy Materials* 8 (6): 1701967.

70 Zhu, H.L., Zhang, F., Li, J.R., and Tang, Y.B. (2018). Penne-like MoS_2/carbon nanocomposite as anode for sodium-ion-based dual-ion battery. *Small* 14 (13): 1703951.

71 Liu, Y.J., Hu, X., Li, J.W. et al. (2022). Carbon-coated $MoS_{1.5}Te_{0.5}$ nanocables for efficient sodium-ion storage in non-aqueous dual-ion batteries. *Nature Communications* 13 (1): 663.

72 Yu, A., Pan, Q.G., Zhang, M. et al. (2020). Fast rate and long life potassium-ion based dual-ion battery through 3D porous organic negative electrode. *Advanced Functional Materials* 30 (24): 2001440.

73 Mu, S., Liu, Q., Kidkhunthod, P. et al. (2021). Molecular grafting towards high-fraction active nanodots implanted in N-doped carbon for sodium dual-ion batteries. *National Science Review* 8: nwaa178.

74 Xu, X., Lin, K., Zhou, D. et al. (2020). Quasi-solid-state dual-ion sodium metal batteries for low-cost energy storage. *Chem* 6 (4): 902–918.

75 Quartarone, E. and Mustarelli, P. (2011). Electrolytes for solid-state lithium rechargeable batteries: recent advances and perspectives. *Chemical Society Reviews* 40 (5): 2525–2540.

76 Yang, Y., Wang, J., Du, X. et al. (2023). Cation co-intercalation with anions: the origin of low capacities of graphite cathodes in multivalent electrolytes. *Journal of the American Chemical Society* 145 (22): 12093–12104.

77 Wang, Y., Zhang, Y.J., Duan, Q.H. et al. (2020). Engineering cathode-electrolyte interface of graphite to enable ultra long-cycle and high-power dual-ion batteries. *Journal of Power Sources* 471: 228466.

78 Tong, X., Ou, X., Wu, N. et al. (2021). High oxidation potential ≈6.0 V of concentrated electrolyte toward high-performance dual-ion battery. *Advanced Energy Materials* 11 (25): 2100151.

79 Shin, K., Zhang, F., Ou, X.W. et al. (2019). Hybridizing anions towards fast diffusion kinetics for tri-ion batteries with significantly improved rate capability and cycling life. *Journal of Materials Chemistry A* 7 (18): 10930–10935.

80 Lv, Z., Han, M., Sun, J. et al. (2019). A high discharge voltage dual-ion rechargeable battery using pure $(DMPI^+)(AlCl_4^-)$ ionic liquid electrolyte. *Journal of Power Sources* 418: 233–240.

81 Zhang, F., Ji, B., Tong, X. et al. (2016). A dual-ion battery constructed with aluminum foil anode and mesocarbon microbead cathode via an alloying/intercalation process in an ionic liquid electrolyte. *Advanced Materials Interfaces* 3 (23): 1600605.

82 Shi, X., Zhang, W., Wang, J. et al. (2016). $(EMIm)^+(PF_6)^-$ Ionic liquid unlocks optimum energy/power density for architecture of nanocarbon-based dual-ion battery. *Advanced Energy Materials* 6 (24): 1601378.

83 Zhang, Y.D., An, Y.F., Yin, B. et al. (2019). A novel aqueous ammonium dual-ion battery based on organic polymers. *Journal of Materials Chemistry A* 7 (18): 11314–11320.

84 Suo, L., Borodin, O., Gao, T. et al. (2015). "Water-in-Salt" electrolyte enables high-voltage aqueous lithium-ion chemistries. *Science* 350 (6263): 938–943.

85 Rodríguez-Pérez, I.A., Zhang, L., Wrogemann, J.M. et al. (2020). Enabling natural graphite in high-voltage aqueous graphite || Zn metal dual-ion batteries. *Advanced Energy Materials* 10 (41): 2001256.

86 Zhu, J.J., Xu, Y.T., Fu, Y.J. et al. (2020). Hybrid aqueous/nonaqueous water-in-bisalt electrolyte enables safe dual ion batteries. *Small* 16 (17): 1905838.

87 Chen, G., Zhang, F., Zhou, Z. et al. (2018). A flexible dual-ion battery based on PVDF-HFP-modified gel polymer electrolyte with excellent cycling performance and superior rate capability. *Advanced Energy Materials* 8 (25): 1801219.

7

M-N-C Catalysts for Fuel Cells

Xiao Zhao

Jilin University, School of Materials Science and Engineering, Key Laboratory of Automobile Materials of MOE, 2699 Qianjin Street, Changchun, Jilin 130012, P. R. China

Fuel cells, a kind of electrochemical engine, convert chemical energy into electrical energy efficiently and are environmental friendly. However, platinum-based electrocatalysts cause scarcity and cost problems that must be addressed before popularizing fuel cells. In this context, noble-metal-free M-N-C electrocatalysts represent a class of nonconventional materials for next-generation fuel cells. In this chapter, M-N-C catalysts are introduced and focused on (i) the nature and engineering strategies of active sites, (ii) the scalable synthesis of active M-N-C catalysts, and (iii) the practical application of M-N-C catalysts in fuel cells.

7.1 Introduction

Platinum-based materials are state-of-the-art electrocatalysts for both anodic hydrogen oxidation reaction (HOR) and cathodic oxygen reduction reaction (ORR); however, their expensive cost and scarcity inhibit commercializing fuel cells. Particularly, the ORR kinetic is quite sluggish relative to the HOR, which results in high overpotential (over 400 mV) in fuel cells. Consequently, the Pt for ORR results in over 40% cost in proton exchange membrane fuel cells (PEMFCs, US$40 kW^{-1}) [1], against the cost target of the $30 kW^{-1} set by the U.S. Department of Energy (DOE) [2]. Meanwhile, Pt-based PEMFCs lose 10% of their performance after ≈3700 hours of operation [3]; thus, the degradation and lifetime of both electrocatalysts and membrane electrode assemblies (MEAs) need addressing (PEMFC systems require 8000 hours of continuous operation for cars, 20,000 hours for buses, and 40,000 hours for stationary applications, set by the U.S. DOE). In this context, the replacement of Pt with nonprecious-metal electrocatalysts composed of earth-abundant elements, such as Fe, Co, N, and C, is pressing for commercializing fuel cells. Those precious-metal-free electrocatalysts are usually made of metal single atoms (SAs) coordinating with nitrogen functionalities embedded in a carbon matrix (denoted as M-N-C catalysts; the M represents transition or main group

Electrochemical Energy Storage Devices: Non-Conventional Technologies and Materials, First Edition. Edited by Yongbing Tang and Luojiang Zhang.

metals). Advances have been made in implementing M-N-C catalysts in the cathode of fuel cells, offering great hope for developing next-generation cost-competitive fuel cells. For example, Zelenay and coworkers reported that via the pyrolysis of polyaniline with iron and cobalt, the most active Fe-Co-NC catalyst in fuel cells showed comparable activity with commercial Pt/C, remarkable stability (700 hours at 0.4 V), and high four-electron selectivity (hydrogen peroxide yield <1.0%) [4].

The studies of M-N-C catalysts can be traced to as early as 1964 and have received growing attention recently [5–17]. Raymond initiated the report of cobalt phthalocyanine (CoPc) as a new fuel cell cathode electrocatalyst for the ORR in alkaline solutions [18]. Shortly after that, various metal–N_4 complexes supported on carbon were reported to be active for the ORR [19, 20]; but their performance degraded quickly due to the potential-driven decomposition. Subsequently, to improve stability, the synthesis was improved via the pyrolysis of nitrogen-coordinated-metal macrocycles with high-surface-area carbon [19, 21] and further, using cheap nitrogenous polymer (e.g. polyacrylonitrile, polyaniline, and polypyrrole) with metal salts and carbon in inert [22] or NH_3 [23] atmospheres. Until now, the pyrolysis of M, N, and C-containing precursors remains a general step for the preparation of high-performance M-N-C electrocatalysts [5–7, 10, 12]. However, pyrolysis also brings difficulties in controlling microstructures and identifying real active sites. A wealth of reviews involved in M-N-C catalysts have been published [1, 5–8, 11, 12, 15–17, 24, 25].

In this chapter, the practical M-N-C electrocatalysts need five intrinsic attributes: (i) dense active sites or high site density (SD), (ii) high accesses for reactants (fully exposed active sites), (iii) highly intrinsic activity with desirable $4e^-$ selectivity, (iv) high durability (the robustness in a corrosive and oxidative environment), and (v) rational porous structures for efficient mass transports and site utilization. Further, the atomic/molecular-level understanding of M, N, and C roles in the ORR, the scalable synthesis methods, and the practical measurements in fuel cells should be emphasized.

7.2 Synthesis and Characterizations of M-N-C Catalysts

Targeting dense, accessible, active, and durable sites, two typical synthesis methods are introduced including the pyrolysis and chemical vapor deposition (CVD) methods (see typical examples in Figure 7.1). The most-used characterization tools from material microstructures to performance evaluations are introduced including spherical-aberration correction transmission electron microscopy (Cs-TEM), X-ray absorption spectroscopy (XAS), rotating disk electrode (RDE), and MEA measurements.

7.2.1 Pyrolysis

The pyrolysis of M, N, and C-containing precursors is a general step to synthesize M-N-C catalysts. The synthesis-structure-property relations can be tuned by

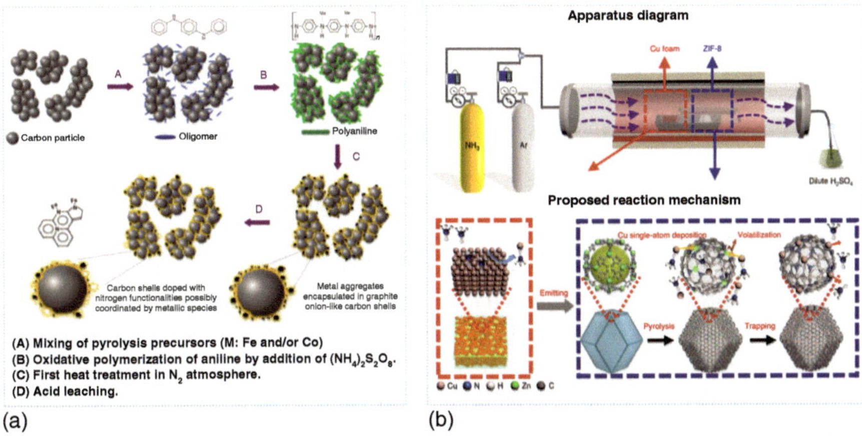

Figure 7.1 Schematic for representative synthesis methods. (a) Pyrolysis. Source: Reproduced with permission [4]. Copyright 2011, American Association for the Advancement of Science. (b) CVD. Source: Reproduced with permission [26]. Copyright 2018, Springer Nature.

changing precursors (their compositions, types, molar ratios, and nanostructures), pyrolysis conditions (temperature, duration, pressure, and atmosphere), and post-treatments (acid corrosion and second pyrolysis). Jaouen et al. [27] revealed that the structural transformation in pyrolysis generally followed the insertion of nitrogen into carbon, microporous formation through a reaction between carbon and ammonia, and the formation of active sites in micropores by a reaction of iron with ammonia. Regarding the increase in SD, they highlighted the micropores that control the macroscopic activity when nitrogen was supplied sufficiently, indicating that active sites were generated in microspores. Thus, the carbon precursors with numerous micropores (e.g. Black Pearls@2000 carbon) benefit the formation of active sites. Indeed, Dodelet and coworkers reported a great increase in SD by using a mixture of Black Pearls@2000 carbon, phenanthroline, and ferrous acetate that were ball-milled and then pyrolyzed twice (first in argon, then in ammonia) [28]. The active sites were claimed to comprise iron cations coordinated with four pyridinic functionalities in the interstices of graphitic sheets within the micropores [28]. Analysis of 15 commercial carbon blacks showed that the pristine carbon with the largest disordered phase, the smallest particle diameter, and the largest size of the graphene layers maximized micropores via a reaction of disordered carbon with ammonia, promoting the formation of Fe-N-C active sites in micropores [23]. Using microporous carbon and/or creating micropores in the pyrolysis seems to improve SD (volumetric activity) and, therefore, the power density of fuel cells [4, 6, 23, 27, 28]. Therefore, optimizing porous structure benefits not only mass transfer but also an increase in SD for M-N-C catalysts.

Following the micropores functionality in forming M-N-C active sites, the metal-organic frameworks (MOFs), also called porous coordination polymers, are promising precursors due to the MOF-based precursors with (i) atomically dispersed M, N, and C elements in integrated precursors, (ii) hosting capacity for M, N-containing precursors, (iii) ultralarge Brunauer–Emmett–Teller (BET) surface areas, (iv) rich porous structures, and (v) templating function in forming hierarchical structures, which together help to generate dense accessible M–N_x sites. Indeed, growing studies [15, 29–48] have used various MOFs, for example, zeolitic-imidazolate frameworks (ZIFs) [30, 31, 34, 35, 41–45, 47], NH_2-MIL-101(Al) [39], and MIL-88B-NH_3 [29] to synthesize M-N-C catalysts. Xu and coworkers, for the first time, reported using MOFs as a template for synthesizing porous carbon with an ultrahigh specific surface area (SSA) (BET, 2872 $m^2\ g^{-1}$) via pyrolysis of MOF-5 filled with furfuryl alcohol [49]. The core–shell structured MOF-based precursors, for example, ZIF-8@ZIF-67 and ZIF-8@ploymers, were designed to prepare complex hierarchical structures via one-step pyrolysis. Chen et al. used the ZIF-8/Fe@polymer composite as precursors via the Kirkendall effect enabling the formation of N, P, and S-doped hollow carbon [50]. Beyond MOFs, the metals coordinated with organic ligands are also promising pyrolysis precursors [51, 52]. For example, the meso/macroporous M-N-C catalysts obtained from pyrolysis of the Fe–diaminonaphthalene complex showed outstanding activity and stability toward the ORR. Generally, those reported M-N-C catalysts based on the designed M–N–C precursors such as MOFs, metal-organic complexes, and metal–polymer coordination showed promising performance for ORR in a half-cell [15, 29–47].

7.2.2 CVD

As fuel cell reactions occur at electrochemical triphase interfaces that should be accessible simultaneously for electrons, hydrated protons, and gases, this requires active sites exposed on the surface rather than buried in the interior of carbon particles. Thus, the fully exposed or highly accessible active sites are always pursued for M-N-C catalysts. With this consideration, the CVD method was proposed. In the CVD process, gaseous metal species are transported by the carrier gas and captured by defects and/or functional groups on carbon, followed by which, M-N sites are *in situ* formed on the carbon surface. As such, the CVD generates fully exposed active sites, a remarkable merit for fuel cell reactions over conventional pyrolysis methods. Li et al. reported synthesizing CuN_4 sites from Cu foam and using an ammonia carrier gas. Specifically, ammonia coordinated with copper atoms of Cu foam to form volatile $Cu(NH_3)_x$ species via strong Lewis acid–base interactions; subsequently carried by ammonia gas, $Cu(NH_3)_x$ species were trapped by defects in downstream NC supports to form isolated CuN_4 sites [26]. Such a method is versatile for the synthesis of Ni, Co, Pt, Au, and Pd SAs using their bulk metals as precursors, respectively, and for the mass production of M-N-C catalysts [26, 53, 54]. Jiao et al. synthesized Fe-N-C by flowing iron chloride vapor over a Zn-N-C substrate at 750 °C, leading to high-temperature trans-metalation of ZnN_4 sites into FeN_4 sites. Such FeN_4C catalyst has an active SD of 1.92×10^{20} sites per gram with 100% site utilization and delivered an ORR activity of 33 mA cm^{-2} at 0.90 V (IR-corrected; I, current; R, resistance) in an H_2–O_2 PEMFC at 1.0 bar and 80 °C [32].

Different from the deposition of gaseous metal species on carbon precursors, Wu and coworkers [55] reported a reversed deposition of gaseous 2-methylimidazole onto Fe-doped ZnO substrates, followed by an *in situ* thermal activation, effectively generating higher Fe SAs density relative to the catalysts synthesized from wet chemistry methods. The CVD shows the potential not only for forming accessible active sites but also for improving the durability of active sites. For example, a thin and porous NC film deposited via CVD on the surface of the Fe-N-C catalyst has drastically improved its stability [56]. Similar to the CVD method, a "plasma bombing" strategy was reported, in which metal ions were excited and stripped by plasma from bulk metals and subsequently trapped and anchored by defective and nitrogenous sites of the supports. This method enabled Fe-N-C with a high metal loading of 8.5 wt% and high ORR performance [47].

7.2.3 Cs-TEM

The state-of-the-art Cs-TEM can directly image atoms, clusters, and nanoparticles (NPs) in real space at a sub-angstrom resolution, enabling solid confirmation of metal SAs embedded in the carbon matrix. If adopting the high-angle annular dark-field (HAADF) detector, scanning-TEM (STEM) can create Z-contrast images (Z refers to the atomic number), where metal SAs appear as bright dots as the signal intensities of metals larger than carbon. Meanwhile, coupled with electron energy loss spectroscopy (EELS), Cs-TEM can provide chemical and structural information around metal SAs, showing a remarkable advantage over other characterizations. Despite these significant merits, the Cs-TEM still faces several concerns. The Cs-TEM provides partial rather than holistic information about specimens. Thus, XAS is necessary for confirming SAs entirely. The Cs-TEM works by the projection of 3D specimens on a 2D screen or photographic plate, making understanding the stereochemistry of active sites difficult. Notably, the electron radiation could damage specimens, probably causing the artificial formation of metal SAs. With these limits, studying M-N-C needs complementary techniques for a full understanding of structure–property relationships.

7.2.4 XAS

The XAS can act as an atomic probe, comprising two regimes, namely, X-ray absorption near-edge structure (XANES) and extended X-ray absorption fine structure (EXAFS) [57, 58]. XANES is sensitive to the formal oxidation state and coordination chemistry of absorbing atoms, while the EXAFS differs local coordination environment of absorbing atoms including the distances, coordination number, and neighbors [57]. The XAS measurements need synchrotrons to provide an intense and energy-tunable source of X-rays and are capable of essentially probing every element and also noncrystalline and highly disordered materials, including solutions. In practice, the XAS coupled with electrochemical measurements has been developed to provide *in situ/operando* electrochemical and structural information [59]. As for M-N-C catalysts, XAS is a powerful tool for precisely understanding the electronic and geometric information of active sites. Especially, XAS becomes necessary for confirming metal SAs and their local coordination at a bulk level. However, the XAS is not a surface-sensitive technique; and other methods, for example, electrochemical and X-ray photoelectron spectroscopy (XPS), are needed to obtain surface and interface information.

7.2.5 RDE and MEA

The RDE and MEA are the most used methods to characterize the ORR performance of M-N-C catalysts. The RDE is an easily accessible tool for quickly preselecting catalysts, whereas the MEA is more complex but represents the real condition. Lazaridis et al. compared RDE and MEA setups and analyzed their fundamental differences (see detailed comparisons in Figure 7.2) [60]. Currently, a consensus has been reached that the electrocatalytic performance tested by the RDE cannot be directly translated into the MEA performance [60–62], which originates mainly from local mass transfer resistance in the catalyst layer of MEAs. Thus, the ultimate demonstration of excellent electrocatalytic performance must be conducted by the MEAs.

The standardized protocols are accessible, thereby enabling the cross-validation of high-performance catalysts from different labs. The protocols have been

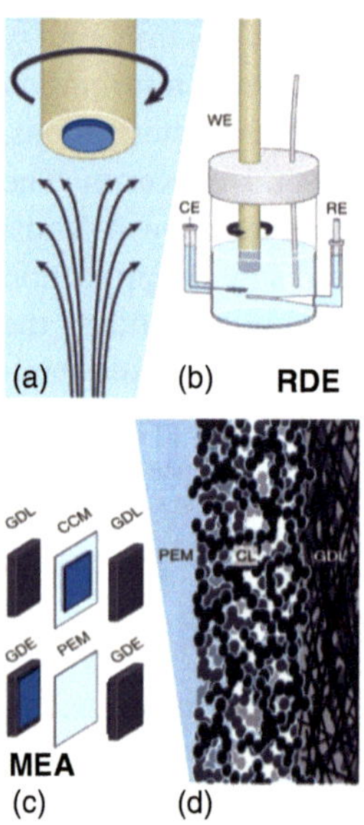

RDE	Setup	MEA
Aqueous electrolyte	Environment	H^+-conducting ionomer
~1–10 mg (required) ~0.002–0.02 mg_{PGM} cm_{geo}^{-2}	Catalyst amount	~0.2–2 g (required) ~0.02–0.5 mg_{PGM} cm_{geo}^{-2}
Minutes to hours	Experiment duration	Days to weeks
10–80 °C (mostly room temperature)	Temperature	~40–95 °C
~1 mmol $l_{electrolyte}^{-1}$	Gaseous reactant solubility	~1–3 mmol $l_{ionomer}^{-1}$
Liquid diffusion boundary layer	Dominant mass transport regime	Gas phase diffusion
<10 mA cm^{-2}_{geo}	Mass transport-limited current density	>4 A cm_{geo}^{-2}
<10 mA cm^{-2}_{PGM}	Maximum specific current density	~100 mA cm_{PGM}^{-2}
~10^5 mol $_{H+}$ mol_{PGM}^{-1} ~10^6 $l_{electrolyte}$ mol_{PGM}^{-1}	Electrolyte/catalyst ratio	~0.1–1 mol_{H+} mol_{PGM}^{-1} ~1–10 $l_{ionomer}$ mol_{PGM}^{-1}

Figure 7.2 Comparison of the RDE and five-layer MEA setups. (a) Local environment of an RDE working electrode (WE). (b) Schematic of a three-electrode RDE setup including a counter electrode (CE) and a reference electrode (RE). (c) Five-layer MEA setups consist either of a catalyst-coated membrane (CCM) sandwiched between two gas diffusion layers (GDLs) or a PEM between two gas diffusion electrodes (GDEs). (d) Local environment in an MEA, with the porous catalyst layer (CL) consisting of catalyst particles (black) and ionomer (light blue). The table summarizes typical specifications for the RDE and MEA setups (exceptions might apply). Source: Reproduced with permission [60]. Copyright 2022, Springer Nature.

recommended by the U.S. DOE and the Fuel Cell Council of Japan (FCCJ). The DOE sets a group of performance targets for key components of fuel cells and recommends the best testing practices for both PGM and PGM-free electrocatalysts. Notably, the durability evaluation should be done in the presence of oxygen because much more performance degradation occurs under an oxygen atmosphere compared to a nitrogen atmosphere [62].

7.3 Fe-N-C-Based Catalysts

7.3.1 Understanding of the Nature of Fe-N-C Active Sites

Understanding active sites and electrocatalytic mechanisms benefits the innovative synthesis of high-performance Fe-N-C catalysts. However, such understanding remains difficult as the activities of Fe-N-C catalysts are related simultaneously with nitrogen coordination, iron spin states, F^{3+}/Fe^{2+} redox potentials, carbon matrix, *in situ* generated oxygen adsorbates, and so on. However, the pyrolysis synthesis often results in complex sites that consist of FeN_x species, encapsulated iron, Fe_3C NPs, and so on. The interrelated factors challenge an accurate and precise understanding of structure–performance relationships, making the debate on the nature of active sites exist for a long time.

FeN_4 sites hosted at the edges of graphene layers and/or in exposed carbon planes are claimed mostly as active sites for ORR; however, the precise FeN_4 structure remains elusive. Using [56] Fe Mössbauer spectroscopy, Kramm et al. showed three doublets assigned to FeN_4 sites with Fe^{2+} ions in a low (D1), intermediate (D2), or high (D3) spin state, and two other doublets assigned to a single Fe species (D4 and D5) consisting of surface oxidized nitride NPs. Only D1 and D3 are active for acid ORR, in which D3 features a protonated neighbor basic nitrogen and is by far the most active site [63]. The *in situ* reduced Fe^{2+}-N_4 moiety from high-spin O_x-Fe^{3+}-N_4 contributes to most of the ORR activity [31]. While Yang et al. stated that the ORR preferentially occurred on intermediate spin state Fe^{3+} of FeN_4 with one e_g electron ($t_{2g}4e_g1$) readily penetrating the antibonding π-orbital of oxygen, rendering high ORR activity [64].

Assuming the end-on M-O_2 interaction primarily involves σ-type bonding between O_2 molecular orbitals and the e_g-orbital ($d_z{}^2$) of the transition metal ions, the ORR onset potential can be controlled by the e_g-orbital ($d_z{}^2$) energy level and closely correlated with the metal ion's redox potential [65]. A downshift of e_g-orbitals leads to improved ORR onset potentials and the weakened adsorption energy of ORR intermediates on metal ions, accompanied by the anodic shift of metal ions' redox potential. Downshifting e_g-orbitals of metal ions may be reached via the electron-withdrawing ligands and carbon matrixes. The regulation of nitrogen coordination types, like pyridinic-, pyrrolic-, and graphitic-nitrogen is a common method to improve the ORR activity of Fe-N-C catalysts [63, 66–70]. Both FeN_4C_{10} (pyridinic) [68] and FeN_4C_{12} (pyrrolic) coordination [67, 69, 71] were described as active configurations for ORR, where the subscript on C referred to the smallest number of carbon atoms connecting the four nitrogen atoms. Cao et al. considered that the Fe-pyrrolic-N structure contributed to the high ORR activity as the p-band center of carbon in Fe@pyrrolic-N was higher than that in Fe@pyridinic-N, making it more suitable for adsorbing ORR intermediates [72]. The electron-withdrawing or electron-donating properties of carbon planes can be tuned by delocalized π-electrons [65] and/or heteroatom functionalities [73], which in turn regulate the electronic structures of Fe via a long-range interaction. For example, the covalent integration of FeN_x into the π-conjugated carbon basal plane caused a dramatic anodic shift of ~600–900 mV in the Fe ion's redox potential [65]. An oxidized sulfur functionality via withdrawing electrons from carbon planes decreases the d-band center of Fe ions, thereby facilitating ORR on FeN_4 sites [73]. Liu et al. argued that the structural descriptors, like Fe=O bond length and the d-band center gap of Fe spin states, the magnetic moment of Fe sites (M_{Fe}) and $*O_2$ (M_{*O2}) can describe activity trend in ORR activity of FeN_4 sites; however, they are essentially related with the height of Fe 3d projected orbital in the Z direction. Thus, the hybridization among Fe $3d_z{}^2$, $3d_{yz}$ ($3d_{xz}$), and O_2 π^* orbital is the origin of ORR activity on the Fe=N_4 sites [74].

On the other hand, non-FeN_4 sites were also reported, typically, the iron NPs encapsulated in graphitic layers (denoted as Fe_{NP}/CNTs) [75–77] and the iron carbide NPs encased by graphitic layers (Fe_3C_{NP}/C) [78]. For Fe_{NP}/CNT catalyst, the electron transfer from Fe NPs to the CNTs led to a decreased local work function on the carbon surface, which enabled the outmost graphitic layer to active for the ORR in both acidic [75] and alkaline solution [76]. The N doped into the graphitic lattice for Fe_{NP}/NCNTs enables an enhanced ORR activity over Fe_{NP}/CNTs [75]. A similar mechanism was also proposed for Fe_3C_{NP}/C in which the encased carbide NPs activate the surrounding graphitic layers, making the outer surface of the carbon layer active toward the acidic ORR [78]. However, the XAS, Cs-TEM, and ^{57}Fe Mössbauer spectroscopy characterizations confirmed that the pyrolyzed M-N-C catalysts often consisted of FeN_x species, iron, iron carbide, iron oxides, and iron sulfide NPs [67, 68, 79–82]. Thus, elucidating which are the realistic active sites and the possible synergism between components is necessary however still under debate. It is argued that CN_x substrate, Fe/Fe_3C NPs, or FeN_x themselves do not guarantee a high alkaline ORR activity; by contrast, the coexistence of high content of FeN_x and sufficient Fe/Fe_3C NPs enables high ORR activity, in which the encased Fe/Fe_3C NPs boost the activity of FeN_x sites on outer surface [81]. In line with this synergistic effect, a strong electronic interaction between iron clusters and neighboring FeN_4 was claimed to optimize the adsorption strength of ORR intermediates on FeN_4 and also shorten the bond amplitude of FeN_4 with incoherent vibrations. Such strong electronic interactions between multiphase metal species can improve the ORR performance of FeN_4 in acidic electrolytes [83]. Meanwhile, Mukerjee stated the surface-bound coexisting Fe/Fe_xO_y NPs as the bifunctional active sites in Fe-N-C catalysts for ORR in acidic media [79]. In contrast to the promoting role of Fe/Fe_3C/Fe_xO_y, Kramm claimed that selectively removing Fe NPs from the catalyst while retaining exclusive FeN_4 sites led to a tremendous increase in acidic ORR activity [84]. Similarly, Feng and coworkers reported that the removal of encapsulated metal NPs by a low-temperature NH_4Cl etching from M-N-C catalysts has resulted in much enhanced acidic ORR activity. Accordingly, they proposed the spectator role of graphene-encapsulated metal NPs and highlighted the pivotal role of FeN_x sites in Fe-N-C catalysts for the ORR [85]. Zelenay and coworkers speculated that the components of iron or iron sulfide NPs embedded or encased within few-layer graphene or graphitic shells were not expected to contribute to ORR activity [68].

Notably, the highly reactive SAs could restructure, driven by applied potential under electrochemical environments. For example, during the ORR, Fe could spontaneously adsorb *OH species that *in situ* change atomic configurations and electronic structure thereby intrinsically tuning ORR kinetics of active sites [68, 77, 86–88]. As such, advanced *in situ/operando* techniques are particularly needed to trace dynamical structural evolution and identify real active sites. Mukerjee and coworkers used *in situ* XAS and revealed that FeN_4 was changed to FeN_4OH with increasing potentials, accompanying which Fe moved away from the FeN_4-plane. The ORR activities of FeN_4 were governed essentially by the $Fe^{2+/3+}$ redox potentials. The higher $Fe^{2+/3+}$ redox potentials were correlated with enhanced ORR performance, decreased Fe=O binding energy, and less site-blocking effect [65, 77, 86]. Theoretically, Taylor and coworkers predicted that contrary to the traditional site-blocking effect, the *in situ* generation of *OH did not poison metal sites but acted as a modifying ligand that would lead to new active sites and provide higher catalytic activities [87]. Experimentally, Xing and coworkers showed an *in situ* generation of an electron-withdrawing *OH axial ligand that tuned the energy level of Ru and conferred Ru with excellent ORR activity [88]. Further, a spontaneously absorbed *OH ligand on $FeCoN_5$ was also proposed to act as an energy level modifier to empower easy intermediate desorption and enable the production of triangular Fe–Co–OH configuration to facilitate O–O bond scission [89]. Those results indicate that the *in situ* generation of axial oxygen ligand on M–N_4 is very likely to occur during the ORR and should be considered carefully for the identification of real active sites and reaction mechanisms.

7.3.2 Engineering FeN_x Active Sites

7.3.2.1 Engineering Coordination Environment of Fe Centers

Engineering near-range and/or long-range coordination environments of metal centers is the most direct and effective method to tune the electronic structure of metal centers and their ORR performance. The methods include engineering physicochemical properties of carbon matrixes [68, 90–94] and heteroatom ligands such as B [95], O/OH [87–89, 96–98], S [73, 99–101], P [102, 103], Cl [104, 105], and their combinations [105, 106].

Usually, FeN_4 moieties embedded in the defects and/or interstices of graphitic sheets are regarded as active for the ORR. Their ORR performances could be modulated theoretically by nearby microenvironments including the carbon vacancies/porosities, electron donating/withdrawing capacities, curvatures, and surface functional groups of carbon matrixes. Typically, according to theoretical calculations, the carbon vacancies around FeN_4 were capable of reducing the adsorption-free energy of *OH and alleviating demetallization, thereby enhancing ORR kinetics and stability [94]. Theoretical calculations also showed that the edge-hosted-FeN_4 moieties were more active than the bulk-hosted FeN_4 one for the ORR [68, 90, 91]. When FeN_4 sites were hosted in π-conjugated phthalocyanine-based covalent organic polymers (COPs), the delocalized π electrons from COPs changed the antibonding d-state energy level of FeN_4 moieties, resulting in a volcano plot between the ORR kinetics of FeN_4 sites and the delocalized π electron densities of COPs [107]. Expectedly, the curved carbon planes would distort the embedded FeN_4 moieties relative to the flat one, resulting in substrate-induced strain and concurrently tuned d-band centers for FeN_4 moieties [92, 93]. Indeed, the highly curved carbon surfaces via inducing 1.5% compressive strain on Fe=N bonds resulted in 31 times enhancement on strained FeN_4 sites than planar ones [92]. Generally, the oxygenated groups in oxidized carbon would weaken electrically conducting capacity relative to the intact graphitic one. However, oxygenated groups act as axial ligands of Fe centers and thus regulate the electronic state of FeN_4 moieties that are supported on oxidized carbon. For example, surface oxygenated groups produced by an O_2 plasma treatment of acetylene black induce electronic localization of Fe centers, thus improving the ORR performance of iron phthalocyanine [96]. Those examples demonstrated that the physicochemical properties of carbon matrix practically impacted the electronic structures of Fe centers and the correlated ORR kinetics thereby going beyond the carbon conventional functions of acting as electrically conducting support and assisting mass transfer.

The introduction of non-nitrogen heteroatoms into carbon is a widely adopted method for engineering Fe-N-C active sites. Those heteroatoms could improve the ORR kinetics by either directly coordinating with metal centers or tuning electron donating/withdrawing capacities of carbon matrixes, both of which redistribute the charge of metal centers [73, 99, 100, 102, 104, 105, 108]. For example, the phosphorus in FeN_3P moieties redistributed charge and changed the electronegativity of Fe centers directly [102]. Heteroatoms changing the electron donating/withdrawing capability of carbon microdomains could indirectly affect the d-state of metal centers and their electroactivity for the ORR. This represents a long-range electron transfer between heteroatoms and metal centers. As an example, the incorporation of B into a porous carbon matrix significantly lowered the energy barrier of FeN_4 for the ORR [95]. The oxidized sulfur species withdraw electrons from carbon planes enhancing ORR activity, while thiophene-like sulfurs donate electrons to carbon planes reducing ORR activity [73]. From a viewpoint of practical application, Sun and coworkers integrated Fe-N-S-C catalysts in the cathode of PEMFC that exhibited a maximal power density over 1 W cm^{-2}, comparable with one using commercial Pt/C [99]. Different from B, S, and N atoms, halogen atoms coordinate with metals usually in an axial direction. For example, chlorine atoms acted as an axial ligand to form FeN_4Cl_1 moieties that had much-enhanced activity against FeN_4 moieties [104, 105].

7.3.2.2 Improving Metal Loading and Site Density

Improving SD and accessibility of FeN_4 moieties is a direct and effective method to reduce local mass transfer resistance and improve the power density when integrating Fe-N-C catalysts into the cathode of PEMFC. However, the common pyrolysis method suffers from two challenges. First, metal SAs are highly reactive and suffer from sintering/aggregation during the pyrolysis process. Thus, the metal loading in pyrolyzed M-N-C catalysts is usually limited in a range of 1–3 wt%. Second, the pyrolysis of M-, N-, and C-precursors at high temperatures unavoidably results in the partial burying of metal SAs in the interiors of the carbon matrix, which are inaccessible for reactants. An RDE test showed only 4.5% of FeN_x sites contribute ORR current [109], which is estimated even lower in MEAs due to the thick catalyst layers [110]. Thus, the anchoring, stabilizing, and confining strategies between metal SAs and carbon matrix and simultaneously creating hierarchical porosity for full exposure of active sites are pressing. The innovative synthesis is the basis for addressing the above two challenges. A cascade anchoring strategy was proposed for the mass production of high-loading M-N-C catalysts and achieved a metal loading of up to 12.1 wt%. Specifically, the chelation of metal ions, the physical isolation of chelate complex upon high loading, and binding with N-species at elevated temperatures are key steps to achieve high-density MN_x sites [111]. By using a sacrificial Zn in the initial CN_x synthesis and then exchanging iron into this CN_x matrix, the 7 wt% Fe SAs loading was reached (see Figure 7.3). Specifically, the pyrolysis of ZIF-8 at 900 °C was conducted to obtain a Zn–NC matrix with abundant ZnN_x sites, followed by Zn leaching to produce vacancies, then Fe inserting into vacancies to obtain unactivated FeN_x sites and finally FeN_x activated at 900 °C using dicyandiamide as activating agent [34]. The SD of FeN_4 determined by *in situ* nitrite stripping and *ex situ* CO chemisorption methods was 4.7×10^{19} and 7.8×10^{19} sites g^{-1}, with a turnover frequency (TOF) of 5.4 e sites^{-1} s^{-1} at 0.80 V in a 0.5 M H_2SO_4 electrolyte [34].

A multilayer stabilization strategy was implemented to achieve a high metal SA loading (~16 wt%), in which metal precursors were embedded into perfluorotetradecanoic acid multilayers that were then coated with polypyrrole before the pyrolysis [112]. The selection of proper chelating agents, like bipyridine [113] carboxylate [44], to anchor metal ions is also effective in improving the metal SAs loading. Notably, the improvement of SD results in not only an increase in site numbers but also a decrease in the distance between SAs that probably causes an intrinsic change in the ORR activity of metal SAs.

7.3.2.3 Creating the Porosity and Improving Site Utilization

Engineering the porosity of catalysts facilitates mass transfer and accessibility of MN_x moieties; thus, increasing the utilization of each active site and/or exposing more sites is an appealing way to improve ORR performance. The SiO_2 colloids as hard templates are often used to prepare porous openly structured catalysts [110, 114–117]. For instance, a synthesis via the silica coating on ZIF-8, preheating, etching silica off, absorbing Fe and N precursors, and carbonization enabled Fe-N-C catalysts with sufficiently exposed sites and enhanced

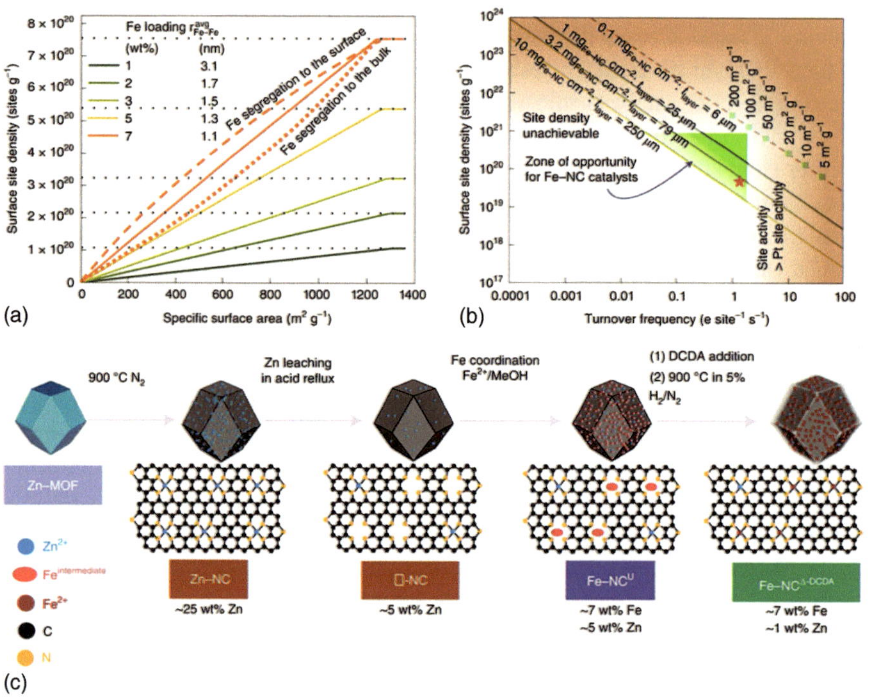

Figure 7.3 (a) Calculated surface site density (SD) as a function of different iron contents and the SSA for SA dispersed catalysts. The average iron–iron separation was constant for each iron loading and at a high SSA (~1200 $m^2\ g^{-1}$), all the iron was on the surface (horizontal dotted lines). (b) Surface SD vs. turnover frequency (TOF) plot showing the Fe–NC catalyst requirements to achieve an area-specific current density of 0.044 $A\ cm^{-2}$ at 0.90 V as a function of catalyst loading (solid lines). The star represented the SD and TOF values achieved with the Fe–NC catalyst. Shown was a comparison with the Pt/C catalyst performance requirements for the same current density at a Pt loading of 0.1 $mg\ cm^{-2}$ using the calculated surface SD of Pt nanocrystals (green squares: specific catalyst surface area and dotted line). (c) Cartoon illustrating the synthetic approach for a Fe–NC catalyst with a high Fe loading. A Zn–MOF was pyrolyzed at 900 °C in the first step to obtain a Zn–NC matrix with abundant Zn–N_x sites followed by Zn leaching (the leached matrix is labeled as □-NC) and then Fe coordination to obtain the unactivated Fe–NC^U catalyst. In the final step, Fe–NC^U was activated at 900 °C in the presence of the dicyandiamide activating agent. Source: Reproduced with permission [34]. Copyright 2022, Springer Nature.

mass transfer [110]. A silica-mediated MOF-templated approach generates a Fe-SA-decorated overhang-eave structure that possesses an edge-rich structure and three-phase boundaries to increase the utilization efficiency of each site [116]. A SiO_2-protecting-shell method enables the formation of meso/microporous Fe-N-C nanofibers, in which SiO_2-protecting shells on nanofiber precursors restrict the free migration of iron species and trap volatile gaseous species during pyrolysis [117]. Interestingly, simply regulating pyrolysis pressure results in the production of three-dimensional open-structure graphene frameworks. Controlling pyrolysis pressure enables the negative differential pressure between the inside and outside of MOFs leading to the gradual expansion and cleavage of the derived carbon layers and the final evolution toward the 3D graphene frameworks in pyrolysis [40].

7.4 Non-Fe Metal Centers

Fe-N-C-based catalysts have shown promising activity and stability for the ORR. However, the concern about their long-term durability has not been addressed sufficiently mainly due to the Fenton reactions between Fe^{2+} and H_2O_2 generating destructive free radicals that likely cause the degradation of catalysts, organic ionomers within electrodes, and polymer membranes [118]. Alternatively, non-Fe M-N-C catalysts that exhibit the mitigated detrimental effect have been explored intensively from s- to p-, f-, and d-block metals. The reported non-Fe metals include Co [36, 115, 118–128], Ni [129], Zn [130–132], Mn [133, 134], Cu [26, 93], Cr [135], Ru [88], Ir [136], Mg [137], Sn [138], Mo [139], Ce [140–142], Se [143], Pt [144], Ag [145], Zr [146], and Y [147], among which, representative examples are introduced.

The strategies for engineering non-Fe-N-C catalysts are similar to those for Fe-N-C catalysts including synthesis methods, coordination environments, porous nanostructures, site densities, and accessibilities for reactants. The pyrolysis of ZnCo MOFs at 700–1000 °C is a common method to synthesize Co-N-C catalysts [35, 36, 128]. During pyrolysis, Zn^{2+} ions served as a "fence" to isolate Co^{2+} ions and simultaneously as atomic templates to create micropores due to Zn evaporation above 800 °C. The Co/Zn ratio of ZnCo MOFs is important to maximize the density of CoN_x sites simultaneously without the formation of Co-based aggregates. Wu and coworkers reported the ZnCo MOFs-derived Co-N-C catalyst that showed a half-wave potential ($E_{1/2}$) of about 0.80 V vs RHE, comparable to Fe-based catalysts and 60 mV lower than Pt/C (60 μg Pt cm^{-2}) in acid media. Importantly, this half-cell electrocatalytic performance can be translated into practical PEMFCs [36]. Further, they utilized core–shell precursors made of surfactant F127@ZnCo MOFs that were thermally converted into graphitic layers@Co-N-C. The confinement effect of surfactant layers on ZnCo MOFs made the surface areas of up to above 900 $m^2\ g^{-1}$, capable of accommodating two times higher Co SAs against the counterpart without using surfactants. Consequently, the ORR activity of this core–shell Co-N-C catalyst reached a record ORR performance with an $E_{1/2}$ of up to 0.84 V and was much better than the conventional Co-N-C catalyst with an $E_{1/2}$ at 0.80 V in the acid medium [128]. Undoubtedly, the improvement of Co SAs density benefits the ORR activity [123, 128]. On this basis, adopting hollow NC spheres and introducing non-nitrogen heteroatoms (e.g. P and S) often enable superior ORR performance [126, 127]. The Mn–N–C catalyst is another promising non-Fe catalyst exhibiting an $E_{1/2}$ of 0.80 V comparable with Fe–N–C catalysts and encouraging performance in fuel cell tests [133], along with significantly enhanced stability in acidic media.

The main-group metal (Mg, Al, and Ca)-based M-N-C materials receive less attention as the ORR electrocatalysts due to that their ORR activity is severely hampered by the tight bonding between active centers and hydroxyl group intermediate, while Liu et al. showed that the Mg atom coordinated to two nitrogen atoms had the near-optimal adsorption strength with intermediate oxygen species by the rise of p-band center position. Experimentally, MgN_2/C exhibited an $E_{1/2}$ of 0.91 V in alkaline media, consequently turning an s/p-band metal into a highly active electrocatalyst [137]. For the challenging acid ORR, p-block SnNC-NH_3 catalysts displayed a 40–50% higher current density than FeNC-NH_3 at cell voltages below 0.7 V along with additional benefits including a $4e^-$ selectivity and a Fenton-inactive character. The stannic Sn(IV)N_x active moieties with low pyridinic N coordination numbers were argued with moderate oxygen chemisorption properties for the ORR [138].

Atomically dispersed precious metal atoms coordinated with nitrogen are also active for the ORR [88, 136, 144]. The $PtNC_2/C$ catalysts with pyridinic nitrogen were reported tolerant to carbon monoxide/methanol and highly active for the ORR [144]. Importantly, the acidic single cell with such a catalyst as a cathode delivered high performance, with power density up to 680 mW cm^{-2} at 80 °C with a low platinum loading of 0.09 mg_{Pt} cm^{-2}, corresponding to a platinum utilization of 0.13 g_{Pt} kW^{-1} in the fuel cell [144]. Xing et al. reported that the RuN_4OH active moieties were generated via spontaneous binding of OH species that served as an energy level regulator toward optimal binding energy of ORR intermediates, while the metal-free N–C and metallic Ru NPs showed negligible catalytic activity. More specifically, FeN_4 and RuN_4 were located on the left leg of the volcano plot, which meant the overly strong Fe/Ru–O binding strength restricted their intrinsic activity. The electron transfer from the Ru center to the electron-withdrawing OH ligand caused the downshift of the d-band center of Ru ions upon *OH binding, which resulted in the weakened Ru=O bond and thus the increased ORR activity for RuN_4OH moieties. Experimentally, the RuN_4OH exhibited an ORR TOF of 4.99 e^- $sites^{-1}$ s^{-1}, far exceeding the state-of-the-art FeN_xOH counterpart (0.816 e^- s^{-1} $sites^{-1}$). The RuN_4OH also possessed greatly suppressed Fenton reactivity, thus displaying much more superior stability (only 17 mV negative shift after 20,000 cycles) than the FeN_xOH counterpart (31 mV). These impressive ORR electrocatalytic performances of RuN_4OH/C have been demonstrated in a real fuel cell device [88]. Impressively, IrN_4/C exhibited orders of magnitude higher ORR activity than iridium NPs with a record-high TOF of 24.3 e^- $sites^{-1}$ s^{-1} at 0.85 V_{RHE} and mass activity of 12.2 A $mg_{Ir}{}^{-1}$, which far outperformed the previously reported RuN_4OH/C [88] and $PtNC_2/C$ [144]. The high activity of IrN_4/C was attributed to the moderate adsorption energy of reaction intermediates [136].

The *f*-block rare earth SACs were rarely reported for the ORR, while Ce SACs [140–142] are the most studied. Ce SAs supported on the P, S, and N co-doped hollow carbon substrate (Ce SAs/PSNC) possessed an $E_{1/2}$ of 0.90 V, a TOF of 52.2 s^{-1} at 0.85 V_{RHE}, and excellent stability for ORR, where the P and S dopants promoted the electroactivity of Ce SAs and improved the overall site-to-site electron transfer within the Ce SAs/PSNC [142]. The CeN_4OH/NC nanowires with a high Ce content of 8.55 wt% exhibited an ORR $E_{1/2}$ of 0.88 V in alkaline media and 0.75 V in acidic electrolytes. The CeN_4OH active moieties were produced from the *in situ* adsorption of *OH on CeN_4 moieties [140]. The group 3–4 elements (Sc, Y, La, Ti, Zr, and Hf) are Fenton-inactive but easily lose all the outermost *s* and *d* electrons, leaving behind ionic sites with nearly empty outermost orbitals that are stable but ORR-inactive due to the too strong binding with *OH intermediates. However, the ORR-inactive YN_4 could be activated by an axial ligand X through the covalency competition between Y=X and Y=OH bonds. Among YN_4, YN_4F, YN_4OH, YN_4S, YN_4Cl, furanyl YN_4O_f, thiophenyl YN_4S_t, pyridyl YN_4N_P, and benzyl YN_4C_b, YN_4Cl showed an optimal binding strength that behaves self-adaptively to facilitate ORR intermediate steps by dynamically changing its Y=Cl covalency and consequently exhibited an $E_{1/2}$ of 0.85 V in alkaline media, the same as that of the commercial Pt/C catalyst [147].

As introduced above, the metals in M-N-C ranging from s- to p-, d-, and f-blocks have been reported active for the ORR, which undoubtedly enriches our understanding and choosing possibility for real devices. Future endeavors should focus on the most promising non-Fe-N-C catalysts that meet practical requirements in MEAs.

7.5 Dual- and Multimetallic SACs

The geometry configuration of the active sites plays a decisive role in breaking O=O bonds and the resultant catalytic behavior. Thus, growing attention has been paid to bimetallic catalysts consisting of heteronuclear/homonuclear atomic pairs or the SAs neighboring the second SAs or the SAs adjacent atomic clusters. Researchers argued that bridge adsorption of an oxygen molecule would favor the dissociation mechanism via the scission of O=O bonds, whereas the side-on and end-on adsorption models prefer an association mechanism via the hydrogenation of O_2 to form *OOH intermediates. An oxygen molecule prefers adopting the side-on or end-on adsorption onto single SA sites and the bridge adsorption onto bicentric sites. Thus, the bicentric sites may promote the ORR kinetic via accelerating the breaking of O=O bonds, which represents a geometric synergistic effect between bicentric sites and has been claimed by several kinds of heteronuclear bicentric sites including Fe=Co [89, 148, 149], Zn=Co [150], Pt = N_2 = Fe [151], Fe=Mo [152], and Fe=Zn [153].

The Fe=Co atomic pairs ($FeCoN_6$) via the host–guest strategy were embedded on N-doped porous carbon and demonstrated superior ORR performance with an $E_{1/2}$ of 0.863 V_{RHE} in acid electrolyte. This $FeCoN_6$/NC-integrated fuel cell reached a kinetic current of >550 mA cm^{-2} at 0.6 V and a peak power density of >505 mW cm^{-2} at 0.42 V along with a stable operation of 100 hours at 0.6 and 1.0 A cm^{-2} in an H_2/air fuel cell. These results demonstrated the practical potential of Fe=Co dual sites in real devices. A mechanism analysis by the density functional theory (DFT) showed that the cleavage of the O=O bond easily occurred on Fe=Co dual sites due to the strong binding of O_2, resulting in sufficient activation of the O=O bond from 1.23 Å in gas phase to 1.40 Å. As such, the rate-limiting step on Fe=Co sites was the hydrogenation of the adsorbed *OH to H_2O needing an activation barrier of 0.26 eV, rather than the O=O bond cleavage [148].

In situ XAS showed a spontaneously absorbed electron-withdrawing =OH ligand on $FeCoN_5$ moieties as an energy level modifier to empower easy desorption of ORR intermediates, while the resulting triangular Fe=Co=OH coordination facilitated O=O bond scission. These attributes enabled $FeCoN_5$-OH sites with an $E_{1/2}$ of up to

0.86 V_{RHE} and a 20-fold improvement in intrinsic activity over FeN_4 sites in acidic media. The Fe=Co dual sites also had a more positive $E_{1/2}$ (0.86 V_{RHE}) against the Co_2N_5 ($E_{1/2}$ of ca. 0.79 V) and Fe_2N_5 sites ($E_{1/2}$ of ca. 0.82 V) with homonuclear atomic pairs. The practical potential of $FeCoN_5$-OH/NC catalysts in fuel cells was demonstrated with the maximum power density of 819 mW cm^{-2}, approaching that of the benchmark Pt/C cathode [89]. Similarly, the regulation of interatomic distances and thus the acceleration of O=O radical breakage without the formation of redundant *OOH intermediates were claimed for a self-adjustable N-bridged Pt = N_2 = Fe moieties [151].

Different from the covalent bonding between atomic pairs, the dual metal sites are separated in distances but may have a strong electronic correlation, resulting in enhanced ORR activity. For example, the ORR activity of FeN_4 moieties was improved by adjacent PtN_4 moieties through an electronic modulation effect, in which PtN_4 tuned the 3d electronic orbitals of FeN_4 [154]. This electronic effect was also shown effective for CoN_4/PtN_4 and MnN_4/PN_4 systems [154]. An enhanced ORR performance for the CuN_4/FeN_4 system was ascribed to that the neighboring CuN_4 decreased the magnetic moment FeN_4 [155]. FeN_x was activated by MnN_x neighbors via spin-state and electronic modulation, which led to an ideal one e_g electron filling ($t_{2g}4e_g1$) and easy penetration of electrons to the antibonding π orbital of oxygen [64]. Differently from the SA decoration, the MnN_4 sites decorated by FeMn clusters also showed enhanced ORR activity [134]. Notably, the challenge is that pyrolysis synthesis often results in the coexistence of the dual metal atomic pairs and SAs, which is hard to control and quantitatively characterize. Under this condition, some questions remain open. For example, is there any electronic modulation between dual metal atomic pairs and SAs for the ORR?

7.6 Durability of M-N-C Catalysts

The already demonstrated ORR activity in MEAs makes M-N-C catalysts promising alternatives to Pt and Pt-alloy catalysts; however, the durability concern remains. The reported degradation mechanisms include (i) micropore flooding, (ii) protonation of active sites and anion adsorption, (iii) demetallation (electrochemical leaching out of metal ions or chemical attack from reactive oxygenated species); and (iv) carbon oxidation (chemical attack from H_2O_2/radicals and/or electrochemical oxidation at high electrode potentials) [156, 157]. Notably, the dissolved metal ions can diffuse into membrane/ionomers and cause their degradation via replacing H^+ and catalyzing Fenton reactions. The carbon oxidation not only directly weakens electronic conductivity but also results in the structural destruction and leaching out of metal ions. It is possible that all these mechanisms, to different degrees, account for the performance degradation of M-N-C catalysts and the cathodic catalyst layer, depending on their microstructures and synthesis methods [156]. Regarding the durability evaluation, comparative tests emphasize that the accelerated durability test (ADT) needs a full mimicking of the working condition of M-N-C catalysts in MEAs in which the cathode works up to >1.5 V during the start-up/shutdown [157].

Typically, performing ADT in an O_2-saturated acidic electrolyte resulted in stronger degradation of the Fe-N-C structure and four times higher ORR activity loss against ADT conducted in an Ar-saturated one. The carbon corrosion, a significant loss of FeN_x sites, and the formation of Fe oxides occurred in the presence of O_2 while no carbon corrosion in Ar at a low potential [158].

Strategies for addressing degradation include adopting non-Fe metal centers, introducing radical scavengers, and coating a porous protection layer on M-N-C catalysts. Comparative experiments showed that the activity loss of Fe-N-C correlated directly with the loss of iron ions due to that H_2O_2 damaged the structural integrity of Fe-N-C catalysts [159]. As discussed above, non-Fe metal centers with a low activity to Fenton reaction have been studied intensively. For instance, the enhanced durability of Co-N-C relative to Fe-N-C was attributed to the lower activity of Co ions to Fenton reactions and the significantly reduced demetallation of Co-N-C [125]. The introduction of free radical scavengers and peroxide-decomposition catalysts in the electrodes has been used to relieve radical attacks. CeO_2 as a regenerative free radical scavenger was used to mitigate the Nafion membrane degradation previously. The Nafion membranes upon addition of 0.5–3 wt% CeO_2 exhibited over one order of magnitude reduction in fluoride emission rate with negligible loss in membrane conductivity [160]. The regenerative redox couple between Ce^{3+} and Ce^{4+} endowed CeO_2 as a radical scavenger, in which the generation of surface oxygen vacancy and the creation of catalytically active Ce^{3+} determined the scavenging efficiency of CeO_2. Naturally, the integration of radical scavengers with MN_x moieties at the molecular level decreased the H_2O_2 detrimental effect, which has been shown in the example of Ta-TiO_x/FeN_xC. The 5 nm Ta-TiO_x NPs in the vicinity of FeN_x moieties reduced the H_2O_2 yield by 51% at 0.7 V in the rotating ring disk electrode test. The MEA with Ta-TiO_x scavengers exhibited only 3% decay in current density at 0.9 $V_{iR\text{-free}}$, much less the 33% decay in an MEA without scavengers. The computable results for the radical scavenging mechanisms showed that under reaction conditions, the TaO_2–OH(110) surface bound stronger with H_2O_2, ·OH, and HOO· than Fe-N-C, thereby capable of capturing H_2O_2 and related radicals and impeding their attack to Fe-N-C active sites. The scavenging mechanism on the TaO_2–OH(110) surface was claimed via the *OH- and O*-assisted pathways. For both reaction pathways, *OH species were the initial intermediates generated from a homolytic O=O bond scission of H_2O_2. Subsequently, *OH species in the O*-assisted pathway disproportionated to form O* and H_2O*, whereas in the *OH-assisted pathway, directly form *OOH and H_2O*. An H-transfer step for both pathways completed the disproportionation reaction routes [161].

As the Fe center is located on the M–O strong binding leg of the M-N-C volcano, decreasing the adsorption energy of oxygenated intermediates on the Fe-N-C intrinsically benefits both its ORR activity and durability. For example, the carbon vacancies around MN_4 sites showed by DFT calculations reduce the adsorption-free energy of *OH and hinder the dissolution of the metal center, thereby enhancing the ORR kinetics and stability [94]. Similarly, the nitrogen-coordinated iron clusters neighboring FeN_4 active sites via a strong electronic interaction between iron clusters and satellite FeN_4 were shown to optimize the adsorption strength of oxygenated intermediates on FeN_4 and shorten the bond amplitude of FeN_4 with incoherent vibrations. Consequently, the TOF of FeN_4 was increased by about 60% and the stability was improved via inhibiting demetallation [83].

The coating of protection layers on NPs is a common method to improve the sintering resistance of NPs. Applying a very thin N-doped porous carbon layer on the Fe-N-C catalyst was shown to obtain high durability. The Fe-N-C catalyst derived from the heat treatment of Fe_2O_3@ZIF-8 was etched with NH_4Cl salts at 1100 °C (FeN_4-AC), and then a CVD process using ZIF-8 as carbon and nitrogen sources to deposit a thin porous layer of N-doped carbon onto NH_4Cl-treated Fe-N-C catalysts (FeN_4-AC-CVD). The FeN_4-AC catalyst exhibited high ORR activity ($E_{1/2} = 0.915$ V) and MEA performance (44.2 mA cm^{-2} at 0.9 $V_{iR\text{-free}}$, H_2/O_2), but suffered significant and rapid performance loss. By contrast, FeN_4-AC-CVD showed respectful activity 32.9 mA cm^{-2} at 0.9 $V_{iR\text{-free}}$ (the 26% decrease due to coating layers) but significantly enhanced durability. After 30,000 voltage cycles and a fuel cell life test (>300 hours) at 0.67 V under H_2/air conditions, the MEA using FeN_4-AC-CVD as cathode catalysts lost only 30 mV at 0.8 A cm^{-2} and 21 mW cm^{-2} (3.9%) of peak power density, meeting the DOE stability target for PEMFC cathode catalysts. Such MEA stability was comparable to a typical Pt/C cathode (0.1 mg_{Pt} cm^{-2}) after applying identical ADTs. Therefore, the combination of an NH_4Cl etching treatment to create rich carbon defects and then a CVD deposition of a thin layer of nitrogen-doped carbon on the catalyst surface is crucial for achieving promising activity and stability in the Fe-N-C catalyst. The stability improvement was ascribed to generate more graphitized and stable carbon structures adjacent to FeN_4 sites during CVD. This process also converted highly active and less stable FeN_4C_{12} (S1) to highly stable and less active FeN_4C_{10} (S2) sites. The FeN_4 demetallation process was proposed as follows. First, the protonation of two nitrogen turned FeN_4 into $FeN_2(NH)_2$ coordination, then Fe centers with adsorbed O_2 moved away from FeN_4O_2 into FeN_2O_2, and finally, FeO_2 desorbed from catalysts, among which, the nitrogen protonation was rate-determining step during a demetallation kinetic process. The nitrogen protonation for S1 and S2 sites required activation energy of 1.24 and 2.08 eV, respectively, suggesting that S2 was intrinsically stable against the demetallation kinetically [56] (Figure 7.4).

7.7 Perspective

The M-N-C catalysts undoubtedly display greatly promising activity and durability for the next generation of fuel cells as demonstrated in not only the half-cell but also real-world MEAs. The future endeavors aiming for practical applications may be as follows. (i) The experimental research for the evaluation of new M-N-C catalysts in MEAs or stacks remains few but is imperative. The application requirements for practice fuel cells may be dramatically different from the fundamental characterization using the RDE technique and highlight the integration of all necessary structural factors in a single catalyst, such as the rational pore distribution, high activity, dense active sites, and high durability, thereby calling for a systemic

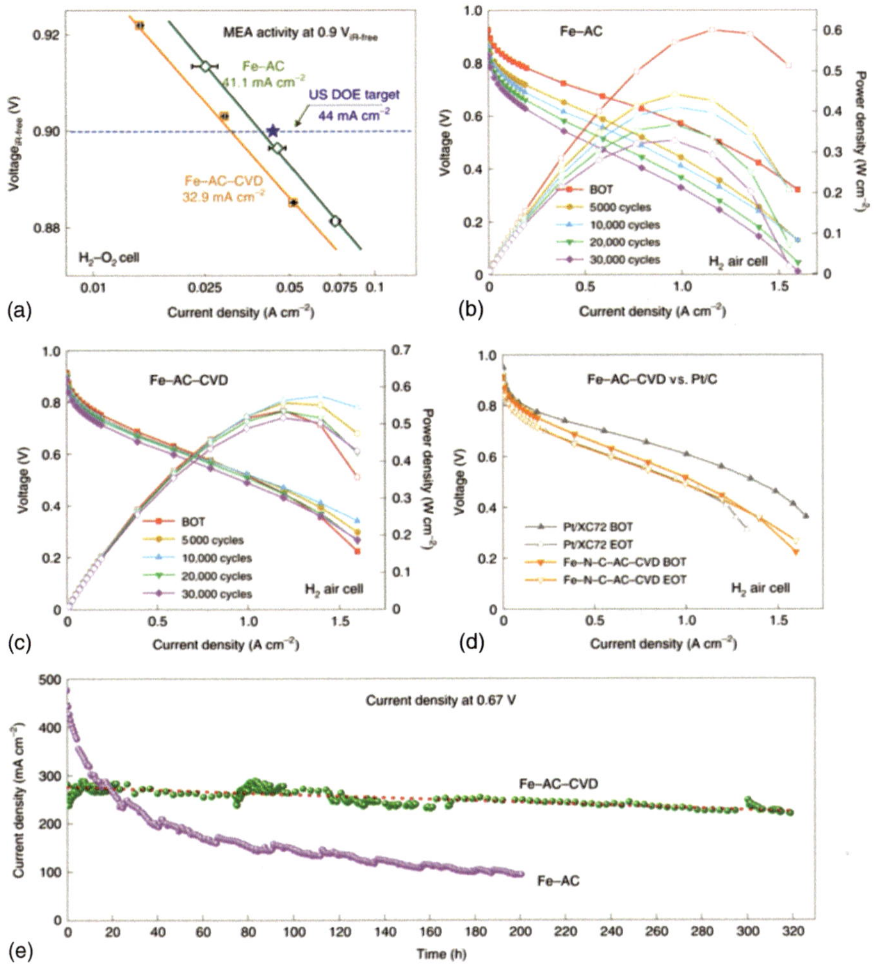

Figure 7.4 (a) Determination of the activities at 0.9 $V_{iR\text{-free}}$ in MEA at 100 kPa absolute O_2 pressure; the blue star denotes the U.S. DOE 2025 target (44 mA cm^{-2}). (b,c) H_2/air fuel cell polarization plots of the Fe–AC (b) and Fe–AC–CVD (c) cathode recorded under 100 kPa absolute air pressure at air and H_2 flow rates of 500 and 300 sccm, respectively, after various numbers of square wave ADT cycles (0.6 V to open circuit voltage [OCV], ~0.92 V) at H_2 and air flows and under ambient pressure. The sccm represents standard cubic centimeter per minute; the BOT stands for the beginning of test. (d) Comparison of H_2/air fuel cell polarization plots of the Fe–AC–CVD and a commercial Pt/XC-72 catalyst recorded at BOT and after the 30,000 cycle stability ADT (EOT: end of test). (e) Long-term fuel cell life tests for both Fe–N–C catalysts under H_2/air conditions at a constant potential of 0.67 V (100 kPa air pressure and flow rates of air 200 sccm and H_2 200 sccm). Anodes: Pt/C, 0.20 mg_{Pt} cm^{-2}, 100% RH, and 100 kPa absolute H_2 pressure. Membrane: Nafion 212. Temperature: 80 °C. MEA area: 5.0 cm^2. Error bars represent the standard deviation for three separate measurements. Source: Reproduced with permission [56]. Copyright 2022, Springer Nature.

consideration and design of M-N-C catalysts. (ii) The theory-experiment interacting design of M-N-C catalysts is desirable. Although the theory-guided design of new catalysts displays advantages over the trial-and-error mode, theoretical calculations or simulations remain a gap for capturing practical active sites and real catalytic processes as active sites often dynamically restructure during electrocatalysis. On the other hand, the *in situ/operando* spectroscopy is capable of dynamically capturing some essentially structural features of active sites and reaction intermediates. Enhancing the interaction between computation and *in situ/operando* spectroscopy should help reveal atomic/molecular active site and electrocatalytic mechanisms. (iii) Due to the complexity and diversity of structure–property relationships across catalysts and catalytic processes, computing science provides new tools like machine learning and data mining to drive the discovery of catalyst genes. These catalyst genes (features) can be validated by the experiments and in turn guide the design of new high-performance catalysts.

References

1 Wang, X.X., Swihart, M.T., and Wu, G. (2019). Achievements, challenges and perspectives on cathode catalysts in proton exchange membrane fuel cells for transportation. *Nature Catalysis* 2 (7): 578–589.

2 Thompson, S.T., Wilson, A.R., Zelenay, P. et al. (2018). ElectroCat: DOE's approach to PGM-free catalyst and electrode R&D. *Solid State Ionics* 319: 68–76.

3 Gasteiger, H.A. and Marković, N.M. (2009). Just a dream—or future reality? *Science* 324 (5923): 48–49.

4 Wu, G., More, K.L., Johnston, C.M., and Zelenay, P. (2011). High-performance electrocatalysts for oxygen reduction derived from polyaniline, iron, and cobalt. *Science* 332 (6028): 443–447.

5 Chen, Z., Higgins, D., Yu, A. et al. (2011). A review on non-precious metal electrocatalysts for PEM fuel cells. *Energy & Environmental Science* 4 (9): 3167–3192.

6 Jaouen, F., Proietti, E., Lefèvre, M. et al. (2011). Recent advances in non-precious metal catalysis for oxygen-reduction reaction in polymer electrolyte fuel cells. *Energy & Environmental Science* 4 (1): 114–130.

7 Luo, E., Chu, Y., Liu, J. et al. (2021). Pyrolyzed M–N_x catalysts for oxygen reduction reaction: progress and prospects. *Energy & Environmental Science* 14 (4): 2158–2185.

8 Wu, G. and Zelenay, P. (2013). Nanostructured nonprecious metal catalysts for oxygen reduction reaction. *Accounts of Chemical Research* 46 (8): 1878–1889.

9 Wang, X., Li, Z., Qu, Y. et al. (2019). Review of metal catalysts for oxygen reduction reaction: from nanoscale engineering to atomic design. *Chem* 5 (6): 1486–1511.

10 Shao, M., Chang, Q., Dodelet, J.-P., and Chenitz, R. (2016). Recent advances in electrocatalysts for oxygen reduction reaction. *Chemical Reviews* 116 (6): 3594–3657.

11 He, Y., Liu, S., Priest, C. et al. (2020). Atomically dispersed metal–nitrogen–carbon catalysts for fuel cells: advances in catalyst design, electrode performance, and durability improvement. *Chemical Society Reviews* 49 (11): 3484–3524.

12 Zhao, X., Yin, M., Ma, L. et al. (2011). Recent advances in catalysts for direct methanol fuel cells. *Energy & Environmental Science* 4 (8): 2736–2753.

13 Dai, L., Xue, Y., Qu, L. et al. (2015). Metal-free catalysts for oxygen reduction reaction. *Chemical Reviews* 115 (11): 4823–4892.

14 Li, Q., Cao, R., Cho, J., and Wu, G. (2014). Nanocarbon electrocatalysts for oxygen reduction in alkaline media for advanced energy conversion and storage. *Advanced Energy Materials* 4 (6): 1301415.

15 Kiani, M., Tian, X.Q., and Zhang, W. (2021). Non-precious metal electrocatalysts design for oxygen reduction reaction in polymer electrolyte membrane fuel cells: recent advances, challenges and future perspectives. *Coordination Chemistry Reviews* 441: 213954.

16 Othman, R., Dicks, A.L., and Zhu, Z. (2012). Non precious metal catalysts for the PEM fuel cell cathode. *International Journal of Hydrogen Energy* 37 (1): 357–372.

17 Deng, Y., Luo, J., Chi, B. et al. (2021). Advanced atomically dispersed metal–nitrogen–carbon catalysts toward cathodic oxygen reduction in PEM fuel cells. *Advanced Energy Materials* 11 (37): 2101222.

18 Jasinski, R. (1964). A new fuel cell cathode catalyst. *Nature* 201 (4925): 1212–1213.

19 Jahnke, H., Schonborn, M., and Zimmermann, G. (1976). Organic dyestuffs as catalysts for fuel cells. *Topics in Current Chemistry* 61: 133–181.

20 Bashyam, R. and Zelenay, P. (2006). A class of non-precious metal composite catalysts for fuel cells. *Nature* 443 (7107): 63–66.

21 Bagotzky, V.S., Tarasevich, M.R., Radyushkina, K.A. et al. (1978). Electrocatalysis of the oxygen reduction process on metal chelates in acid electrolyte. *Journal of Power Sources* 2 (3): 233–240.

22 Gupta, S., Tryk, D., Bae, I. et al. (1989). Heat-treated polyacrylonitrile-based catalysts for oxygen electroreduction. *Journal of Applied Electrochemistry* 19 (1): 19–27.

23 Charreteur, F., Jaouen, F., Ruggeri, S., and Dodelet, J.-P. (2008). Fe/N/C Non-precious catalysts for PEM fuel cells: influence of the structural parameters of pristine commercial carbon blacks on their activity for oxygen reduction. *Electrochimica Acta* 53 (6): 2925–2938.

24 Zhu, S., Ge, J., Liu, C., and Xing, W. (2019). Atomic-level dispersed catalysts for PEMFCs: progress and future prospects. *EnergyChem* 1 (3): 100018.

25 Wang, W., Jia, Q., Mukerjee, S., and Chen, S. (2019). Recent insights into the oxygen-reduction electrocatalysis of Fe/N/C materials. *ACS Catalysis* 9 (11): 10126–10141.

26 Qu, Y., Li, Z., Chen, W. et al. (2018). Direct transformation of bulk copper into copper single sites via emitting and trapping of atoms. *Nature Catalysis* 1 (10): 781–786.

27 Jaouen, F., Lefèvre, M., Dodelet, J.-P., and Cai, M. (2006). Heat-treated Fe/N/C catalysts for O_2 electroreduction: are active sites hosted in micropores? *The Journal of Physical Chemistry B* 110 (11): 5553–5558.

28 Lefevre, M., Proietti, E., Jaouen, F., and Dodelet, J.-P. (2009). Iron-based catalysts with improved oxygen reduction activity in polymer electrolyte fuel cells. *Science* 324 (5923): 71–74.

29 Zhao, S., Yin, H., Du, L. et al. (2014). Carbonized nanoscale metal–organic frameworks as high performance electrocatalyst for oxygen reduction reaction. *ACS Nano* 8 (12): 12660–12668.

30 Shang, L., Yu, H., Huang, X. et al. (2016). Well-dispersed ZIF-derived Co,N-Co-doped carbon nanoframes through mesoporous-silica-protected calcination as efficient oxygen reduction electrocatalysts. *Advanced Materials* 28 (8): 1668–1674.

31 Xiao, M., Zhu, J., Ma, L. et al. (2018). Microporous framework induced synthesis of single-atom dispersed Fe-N-C acidic ORR catalyst and its in situ reduced Fe-N4 active site identification revealed by X-ray absorption spectroscopy. *ACS Catalysis* 8 (4): 2824–2832.

32 Jiao, L., Li, J., Richard, L.L. et al. (2021). Chemical vapour deposition of Fe–N–C oxygen reduction catalysts with full utilization of dense Fe–N_4 sites. *Nature Materials* 20 (10): 1385–1391.

33 Pan, Y., Sun, K., Liu, S. et al. (2018). Core–shell ZIF-8@ZIF-67-derived CoP nanoparticle-embedded N-doped carbon nanotube hollow polyhedron for efficient overall water splitting. *Journal of the American Chemical Society* 140 (7): 2610–2618.

34 Mehmood, A., Gong, M., Jaouen, F. et al. (2022). High loading of single atomic iron sites in Fe–NC oxygen reduction catalysts for proton exchange membrane fuel cells. *Nature Catalysis* 5 (4): 311–323.

35 Yin, P., Yao, T., Wu, Y. et al. (2016). Single cobalt atoms with precise N-coordination as superior oxygen reduction reaction catalysts. *Angewandte Chemie International Edition* 55 (36): 10800–10805.

36 Wang, X.X., Cullen, D.A., Pan, Y.-T. et al. (2018). Nitrogen-coordinated single cobalt atom catalysts for oxygen reduction in proton exchange membrane fuel cells. *Advanced Materials* 30 (11): 1706758.

37 Wang, Q. and Astruc, D. (2020). State of the art and prospects in metal–organic framework (MOF)-based and MOF-derived nanocatalysis. *Chemical Reviews* 120 (2): 1438–1511.

38 Ji, S., Chen, Y., Wang, X. et al. (2020). Chemical synthesis of single atomic site catalysts. *Chemical Reviews* 120 (21): 11900–11955.

39 Xie, X., Peng, L., Yang, H. et al. (2021). MIL-101-derived mesoporous carbon supporting highly exposed Fe single-atom sites as efficient oxygen reduction reaction catalysts. *Advanced Materials* 33 (23): 2101038.

40 Zhou, H., Yang, T., Kou, Z. et al. (2020). Negative pressure pyrolysis induced highly accessible single sites dispersed on 3D graphene frameworks for enhanced oxygen reduction. *Angewandte Chemie International Edition* 59 (46): 20465–20469.

41 Zhao, D., Shui, J.-L., Grabstanowicz, L.R. et al. (2014). Highly efficient non-precious metal electrocatalysts prepared from one-pot synthesized zeolitic imidazolate frameworks. *Advanced Materials* 26 (7): 1093–1097.

42 Wang, C., Kim, J., Tang, J. et al. (2020). New strategies for novel MOF-derived carbon materials based on nanoarchitectures. *Chem* 6 (1): 19–40.

43 Proietti, E., Jaouen, F., Lefèvre, M. et al. (2011). Iron-based cathode catalyst with enhanced power density in polymer electrolyte membrane fuel cells. *Nature Communications* 2 (1): 416.

44 Li, Y., Zhang, P., Wan, L. et al. (2021). A general carboxylate-assisted approach to boost the ORR performance of ZIF-derived Fe/N/C catalysts for proton exchange membrane fuel cells. *Advanced Functional Materials* 31 (15): 2009645.

45 Li, J., Ghoshal, S., Liang, W. et al. (2016). Structural and mechanistic basis for the high activity of Fe–N–C catalysts toward oxygen reduction. *Energy & Environmental Science* 9 (7): 2418–2432.

46 Fu, S.F., Zhu, C.Z., Song, J.H. et al. (2017). Metal-organic framework-derived non-precious metal nanocatalysts for oxygen reduction reaction. *Advanced Energy Materials* 7 (19): 1700363.

47 Rao, P., Wu, D., Luo, J. et al. (2022). A plasma bombing strategy to synthesize high-loading single-atom catalysts for oxygen reduction reaction. *Cell Reports Physical Science* 3 (5): 100880.

48 Liu, X., Verma, G., Chen, Z. et al. (2022). Metal-organic framework nanocrystal-derived hollow porous materials: synthetic strategies and emerging applications. *The Innovation* 3 (5): 100281.

49 Liu, B., Shioyama, H., Akita, T., and Xu, Q. (2008). Metal-organic framework as a template for porous carbon synthesis. *Journal of the American Chemical Society* 130 (16): 5390–5391.

50 Chen, Y., Ji, S., Zhao, S. et al. (2018). Enhanced oxygen reduction with single-atomic-site iron catalysts for a zinc-air battery and hydrogen-air fuel cell. *Nature Communications* 9 (1): 5422.

51 Xiao, M., Zhu, J., Feng, L. et al. (2015). Meso/macroporous nitrogen-doped carbon architectures with iron carbide encapsulated in graphitic layers as an efficient and robust catalyst for the oxygen reduction reaction in both acidic and alkaline solutions. *Advanced Materials* 27 (15): 2521–2527.

52 Lin, L., Zhu, Q., and Xu, A.-W. (2014). Noble-metal-free Fe-N/C catalyst for highly efficient oxygen reduction reaction under both alkaline and acidic conditions. *Journal of the American Chemical Society* 136 (31): 11027–11033.

53 Qu, Y., Chen, B., Li, Z. et al. (2019). Thermal emitting strategy to synthesize atomically dispersed Pt metal sites from bulk Pt metal. *Journal of the American Chemical Society* 141 (11): 4505–4509.

54 Yang, Z.K., Zhao, C.M., Qu, Y.T. et al. (2019). Trifunctional self-supporting cobalt-embedded carbon nanotube films for ORR, OER, and HER triggered by solid diffusion from bulk metal. *Advanced Materials* 31 (12): 1808043.

55 Liu, S., Wang, M., Yang, X. et al. (2020). Chemical vapor deposition for atomically dispersed and nitrogen coordinated single metal site catalysts. *Angewandte Chemie International Edition* 59 (48): 21698–21705.

56 Liu, S., Li, C., Zachman, M.J. et al. (2022). Atomically dispersed iron sites with a nitrogen–carbon coating as highly active and durable oxygen reduction catalysts for fuel cells. *Nature Energy* 7 (7): 652–663.

57 Newville, M. (2014). Fundamentals of XAFS. *Reviews in Mineralogy and Geochemistry* 78 (1): 33–74.

58 Iwasawa, Y., Asakura, K., and Tada, M. (2016). *XAFS Techniques for Catalysts, Nanomaterials and Surfaces*, IX. New York: Springer, 556.

59 Uruga, T., Tada, M., Sekizawa, O. et al. (2019). SPring-8 BL36XU: synchrotron radiation X-ray-based multi-analytical beamline for polymer electrolyte fuel cells under operating conditions. *The Chemical Record* 19 (7): 1444–1456.

60 Lazaridis, T., Stühmeier, B.M., Gasteiger, H.A., and El-Sayed, H.A. (2022). Capabilities and limitations of rotating disk electrodes versus membrane electrode assemblies in the investigation of electrocatalysts. *Nature Catalysis* 5 (5): 363–373.

61 Fan, J., Chen, M., Zhao, Z. et al. (2021). Bridging the gap between highly active oxygen reduction reaction catalysts and effective catalyst layers for proton exchange membrane fuel cells. *Nature Energy* 6 (5): 475–486.

62 Zhang, H., Osmieri, L., Park, J.H. et al. (2022). Standardized protocols for evaluating platinum group metal-free oxygen reduction reaction electrocatalysts in polymer electrolyte fuel cells. *Nature Catalysis* 5 (5): 455–462.

63 Mineva, T., Matanovic, I., Atanassov, P. et al. (2019). Understanding active sites in pyrolyzed Fe–N–C catalysts for fuel cell cathodes by bridging density functional theory calculations and 57Fe Mössbauer spectroscopy. *ACS Catalysis* 9 (10): 9359–9371.

64 Yang, G., Zhu, J., Yuan, P. et al. (2021). Regulating Fe-spin state by atomically dispersed Mn-N in Fe-N-C catalysts with high oxygen reduction activity. *Nature Communications* 12 (1): 1734.

65 Ramaswamy, N., Tylus, U., Jia, Q., and Mukerjee, S. (2013). Activity descriptor identification for oxygen reduction on nonprecious electrocatalysts: linking surface science to coordination chemistry. *Journal of the American Chemical Society* 135 (41): 15443–15449.

66 Kramm, U.I., Herranz, J., Larouche, N. et al. (2012). Structure of the catalytic sites in Fe/N/C-catalysts for O_2^- reduction in PEM fuel cells. *Physical Chemistry Chemical Physics* 14 (33): 11673–11688.

67 Zitolo, A., Goellner, V., Armel, V. et al. (2015). Identification of catalytic sites for oxygen reduction in iron- and nitrogen-doped graphene materials. *Nature Materials* 14 (9): 937–942.

68 Chung, H.T., Cullen, D.A., Higgins, D. et al. (2017). Direct atomic-level insight into the active sites of a high-performance PGM-free ORR catalyst. *Science* 357 (6350): 479–483.

69 Ni, L., Gallenkamp, C., Wagner, S. et al. (2022). Identification of the catalytically dominant iron environment in iron- and nitrogen-doped carbon catalysts for the oxygen reduction reaction. *Journal of the American Chemical Society* 144 (37): 16827–16840.

70 Zhang, H., Hwang, S., Wang, M. et al. (2017). Single atomic iron catalysts for oxygen reduction in acidic media: particle size control and thermal activation. *Journal of the American Chemical Society* 139 (40): 14143–14149.

71 Hu, X., Chen, S., Chen, L. et al. (2022). What is the real origin of the activity of Fe–N–C electrocatalysts in the O_2 reduction reaction? Critical roles of coordinating pyrrolic N and axially adsorbing species. *Journal of the American Chemical Society* 144 (39): 18144–18152.

72 Yang, L., Cheng, D., Xu, H. et al. (2018). Unveiling the high-activity origin of single-atom iron catalysts for oxygen reduction reaction. *Proceedings of the National Academy of Sciences of the United States of America* 115 (26): 6626–6631.

73 Mun, Y., Lee, S., Kim, K. et al. (2019). Versatile strategy for tuning ORR activity of a single Fe-N4 site by controlling electron-withdrawing/donating properties of a carbon plane. *Journal of the American Chemical Society* 141 (15): 6254–6262.

74 Liu, K., Fu, J., Lin, Y. et al. (2022). Insights into the activity of single-atom Fe-N-C catalysts for oxygen reduction reaction. *Nature Communications* 13 (1): 2075.

75 Deng, D., Yu, L., Chen, X. et al. (2013). Iron encapsulated within pod-like carbon nanotubes for oxygen reduction reaction. *Angewandte Chemie International Edition* 52 (1): 371–375.

76 Chung, H.T., Won, J.H., and Zelenay, P. (2013). Active and stable carbon nanotube/nanoparticle composite electrocatalyst for oxygen reduction. *Nature Communications* 4 (1): 1922.

77 Jia, Q., Ramaswamy, N., Tylus, U. et al. (2016). Spectroscopic insights into the nature of active sites in iron–nitrogen–carbon electrocatalysts for oxygen reduction in acid. *Nano Energy* 29: 65–82.

78 Hu, Y., Jensen, J.O., Zhang, W. et al. (2014). Hollow spheres of iron carbide nanoparticles encased in graphitic layers as oxygen reduction catalysts. *Angewandte Chemie International Edition* 53 (14): 3675–3679.

79 Tylus, U., Jia, Q., Strickland, K. et al. (2014). Elucidating oxygen reduction active sites in pyrolyzed metal–nitrogen coordinated non-precious-metal electrocatalyst systems. *The Journal of Physical Chemistry C* 118 (17): 8999–9008.

80 Wu, Z.-Y., Xu, X.-X., Hu, B.-C. et al. (2015). Iron carbide nanoparticles encapsulated in mesoporous Fe-N-doped carbon nanofibers for efficient electrocatalysis. *Angewandte Chemie International Edition* 54 (28): 8179–8183.

81 Jiang, W.-J., Gu, L., Li, L. et al. (2016). Understanding the high activity of Fe–N–C electrocatalysts in oxygen reduction: Fe/Fe_3C nanoparticles boost the activity of Fe–Nx. *Journal of the American Chemical Society* 138 (10): 3570–3578.

82 Wagner, S., Auerbach, H., Tait, C.E. et al. (2019). Elucidating the structural composition of an Fe–N–C catalyst by nuclear- and electron-resonance techniques. *Angewandte Chemie International Edition* 58 (31): 10486–10492.

83 Wan, X., Liu, Q., Liu, J. et al. (2022). Iron atom–cluster interactions increase activity and improve durability in Fe–N–C fuel cells. *Nature Communications* 13 (1): 2963.

84 Kramm, U.I., Herrmann-Geppert, I., Behrends, J. et al. (2016). On an easy way to prepare metal–nitrogen doped carbon with exclusive presence of MeN_4-type sites active for the ORR. *Journal of the American Chemical Society* 138 (2): 635–640.

85 Chen, M.-X., Zhu, M., Zuo, M. et al. (2020). Identification of catalytic sites for oxygen reduction in metal/nitrogen-doped carbons with encapsulated metal nanoparticles. *Angewandte Chemie International Edition* 59 (4): 1627–1633.

86 Jia, Q., Ramaswamy, N., Hafiz, H. et al. (2015). Experimental observation of redox-induced Fe–N switching behavior as a determinant role for oxygen reduction activity. *ACS Nano* 9 (12): 12496–12505.

87 Holby, E.F. and Taylor, C.D. (2015). Activity of N-coordinated multi-metal-atom active site structures for Pt-free oxygen reduction reaction catalysis: role of *OH ligands. *Scientific Reports* 5: 9286.

88 Xiao, M.L., Gao, L.Q., Wang, Y. et al. (2019). Engineering energy level of metal center: Ru single-atom site for efficient and durable oxygen reduction catalysis. *Journal of the American Chemical Society* 141 (50): 19800–19806.

89 Xiao, M.L., Chen, Y.T., Zhu, J.B. et al. (2019). Climbing the apex of the ORR volcano plot via binuclear site construction: electronic and geometric engineering. *Journal of the American Chemical Society* 141 (44): 17763–17770.

90 Jiang, R., Li, L., Sheng, T. et al. (2018). Edge-site engineering of atomically dispersed $Fe–N_4$ by selective C–N bond cleavage for enhanced oxygen reduction reaction activities. *Journal of the American Chemical Society* 140 (37): 11594–11598.

91 Xiao, M., Xing, Z., Jin, Z. et al. (2020). Preferentially engineering FeN4 edge sites onto graphitic nanosheets for highly active and durable oxygen electrocatalysis in rechargeable Zn–air batteries. *Advanced Materials* 32 (49): 2004900.

92 Yang, J., Wang, Z., Huang, C.-X. et al. (2021). Compressive strain modulation of single iron sites on helical carbon support boosts electrocatalytic oxygen reduction. *Angewandte Chemie International Edition* 60 (42): 22722–22728.

93 Han, G., Zhang, X., Liu, W. et al. (2021). Substrate strain tunes operando geometric distortion and oxygen reduction activity of CuN_2C_2 single-atom sites. *Nature Communications* 12 (1): 6335.

94 Tian, H., Song, A., Zhang, P. et al. (2023). High durability of Fe–N–C single-atom catalysts with carbon vacancies toward the oxygen reduction reaction in alkaline media. *Advanced Materials* 35 (14): 2210714.

95 Yuan, K., Sfaelou, S., Qiu, M. et al. (2018). Synergetic contribution of boron and Fe–Nx species in porous carbons toward efficient electrocatalysts for oxygen reduction reaction. *ACS Energy Letters* 3 (1): 252–260.

96 Chen, K., Liu, K., An, P. et al. (2020). Iron phthalocyanine with coordination induced electronic localization to boost oxygen reduction reaction. *Nature Communications* 11 (1): 4173.

97 Peng, L.S., Yang, J., Yang, Y.Q. et al. (2022). Mesopore-rich Fe-N-C catalyst with FeN4-O-NC single-atom sites delivers remarkable oxygen reduction reaction performance in alkaline media. *Advanced Materials* 34 (29): 2202544.

98 Gong, L., Zhang, H., Wang, Y. et al. (2020). Bridge bonded oxygen ligands between approximated FeN4 sites confer catalysts with high ORR performance. *Angewandte Chemie International Edition* 59 (33): 13923–13928.

99 Wang, Y.-C., Lai, Y.-J., Song, L. et al. (2015). S-Doping of an Fe/N/C ORR catalyst for polymer electrolyte membrane fuel cells with high power density. *Angewandte Chemie-International Edition* 54 (34): 9907–9910.

100 Li, Q., Chen, W., Xiao, H. et al. (2018). Fe isolated single atoms on S, N codoped carbon by copolymer pyrolysis strategy for highly efficient oxygen reduction reaction. *Advanced Materials* 30 (25): 1800588.

101 Zhang, J., Zhao, Y., Chen, C. et al. (2019). Tuning the coordination environment in single-atom catalysts to achieve highly efficient oxygen reduction reactions. *Journal of the American Chemical Society* 141 (51): 20118–20126.

102 Yuan, K., Lützenkirchen-Hecht, D., Li, L. et al. (2020). Boosting oxygen reduction of single iron active sites via geometric and electronic engineering: nitrogen and phosphorus dual coordination. *Journal of the American Chemical Society* 142 (5): 2404–2412.

103 Yin, H., Yuan, P., Lu, B.-A. et al. (2021). Phosphorus-driven electron delocalization on edge-type FeN_4 active sites for oxygen reduction in acid medium. *ACS Catalysis* 11 (20): 12754–12762.

104 Hu, L., Dai, C., Chen, L. et al. (2021). Metal-triazolate-framework-derived FeN_4Cl_1 single-atom catalysts with hierarchical porosity for the oxygen reduction reaction. *Angewandte Chemie International Edition* 60 (52): 27324–27329.

105 Han, Y., Wang, Y., Xu, R. et al. (2018). Electronic structure engineering to boost oxygen reduction activity by controlling the coordination of the central metal. *Energy & Environmental Science* 11 (9): 2348–2352.

106 Su, D., Zhou, D., Wang, C., and Wang, G. (2018). Toward high performance lithium–sulfur batteries based on Li_2S cathodes and beyond: status, challenges, and perspectives. *Advanced Functional Materials* 28: 1800154–1800177.

107 Li, X. and Xiang, Z. (2022). Identifying the impact of the covalent-bonded carbon matrix to FeN_4 sites for acidic oxygen reduction. *Nature Communications* 13 (1): 57.

108 Shen, H., Gracia-Espino, E., Ma, J. et al. (2017). Synergistic effects between atomically dispersed Fe−N−C and C−S−C for the oxygen reduction reaction in acidic media. *Angewandte Chemie International Edition* 56 (44): 13800–13804.

109 Malko, D., Kucernak, A., and Lopes, T. (2016). In situ electrochemical quantification of active sites in Fe–N/C non-precious metal catalysts. *Nature Communications* 7 (1): 13285.

110 Wan, X., Liu, X., Li, Y. et al. (2019). Fe–N–C Electrocatalyst with dense active sites and efficient mass transport for high-performance proton exchange membrane fuel cells. *Nature Catalysis* 2 (3): 259–268.

111 Zhao, L., Zhang, Y., Huang, L.-B. et al. (2019). Cascade anchoring strategy for general mass production of high-loading single-atomic metal-nitrogen catalysts. *Nature Communications* 10 (1): 1278.

112 Zhou, Y., Tao, X., Chen, G. et al. (2020). Multilayer stabilization for fabricating high-loading single-atom catalysts. *Nature Communications* 11 (1): 5892.

113 Lin, Y., Liu, P., Velasco, E. et al. (2019). Fabricating single-atom catalysts from chelating metal in open frameworks. *Advanced Materials* 31 (18): 1808193.

114 Chen, Y., Li, Z., Zhu, Y. et al. (2019). Atomic Fe dispersed on N-doped carbon hollow nanospheres for high-efficiency electrocatalytic oxygen reduction. *Advanced Materials* 31 (8): 1806312.

115 Sun, T., Zhao, S., Chen, W. et al. (2018). Single-atomic cobalt sites embedded in hierarchically ordered porous nitrogen-doped carbon as a superior bifunctional electrocatalyst. *Proceedings of the National Academy of Sciences of the United States of America* 115 (50): 12692–12697.

116 Hou, C.-C., Zou, L., Sun, L. et al. (2020). Single-atom iron catalysts on overhang-eave carbon cages for high-performance oxygen reduction reaction. *Angewandte Chemie-International Edition* 59 (19): 7384–7389.

117 Hu, B.-C., Wu, Z.-Y., Chu, S.-Q. et al. (2018). SiO_2-protected shell mediated templating synthesis of Fe–N-doped carbon nanofibers and their enhanced oxygen reduction reaction performance. *Energy & Environmental Science* 11 (8): 2208–2215.

118 Wang, X.X., Prabhakaran, V., He, Y. et al. (2019). Iron-free cathode catalysts for proton-exchange-membrane fuel cells: cobalt catalysts and the peroxide mitigation approach. *Advanced Materials* 31 (31): 1805126.

119 Liang, H.-W., Wei, W., Wu, Z.-S. et al. (2013). Mesoporous metal–nitrogen-doped carbon electrocatalysts for highly efficient oxygen reduction reaction. *Journal of the American Chemical Society* 135 (43): 16002–16005.

120 Han, Y., Wang, Y.-G., Chen, W. et al. (2017). Hollow N-doped carbon spheres with isolated cobalt single atomic sites: superior electrocatalysts for oxygen reduction. *Journal of the American Chemical Society* 139 (48): 17269–17272.

121 Zitolo, A., Ranjbar-Sahraie, N., Mineva, T. et al. (2017). Identification of catalytic sites in cobalt-nitrogen-carbon materials for the oxygen reduction reaction. *Nature Communications* 8 (1): 957.

122 Xiao, M.L., Zhang, H., Chen, Y.T. et al. (2018). Identification of binuclear Co_2N_5 active sites for oxygen reduction reaction with more than one magnitude higher activity than single atom CoN_4 site. *Nano Energy* 46: 396–403.

123 Wu, J., Zhou, H., Li, Q. et al. (2019). Densely populated isolated single co-N site for efficient oxygen electrocatalysis. *Advanced Energy Materials* 9 (22): 1900149.

124 He, Y., Guo, H., Hwang, S. et al. (2020). Single cobalt sites dispersed in hierarchically porous nanofiber networks for durable and high-power PGM-free cathodes in fuel cells. *Advanced Materials* 32 (46): 2003577.

125 Xie, X., He, C., Li, B. et al. (2020). Performance enhancement and degradation mechanism identification of a single-atom Co-N-C catalyst for proton exchange membrane fuel cells. *Nature Catalysis* 3 (12): 1044–1054.

126 Zhang, Z., Zhao, X., Xi, S. et al. (2020). Atomically dispersed cobalt trifunctional electrocatalysts with tailored coordination environment for flexible rechargeable Zn-air battery and self-driven water splitting. *Advanced Energy Materials* 10 (48): 2002896.

127 Chen, Y., Gao, R., Ji, S. et al. (2021). Atomic-level modulation of electronic density at cobalt single-atom sites derived from metal–organic frameworks: enhanced oxygen reduction performance. *Angewandte Chemie International Edition* 60 (6): 3212–3221.

128 He, Y., Hwang, S., Cullen, D.A. et al. (2019). Highly active atomically dispersed CoN_4 fuel cell cathode catalysts derived from surfactant-assisted MOFs: carbon-shell confinement strategy. *Energy & Environmental Science* 12 (1): 250–260.

129 Miner, E.M., Fukushima, T., Sheberla, D. et al. (2016). Electrochemical oxygen reduction catalysed by Ni_3(Hexaiminotriphenylene)$_2$. *Nature Communications* 7 (1): 10942.

130 Li, J., Chen, S., Yang, N. et al. (2019). Ultrahigh-loading zinc single-atom catalyst for highly efficient oxygen reduction in both acidic and alkaline media. *Angewandte Chemie International Edition* 58 (21): 7035–7039.

131 Song, P., Luo, M., Liu, X. et al. (2017). Zn single atom catalyst for highly efficient oxygen reduction reaction. *Advanced Functional Materials* 27 (28): 1700802.

132 Wang, J., Li, H., Liu, S. et al. (2021). Turning on Zn 4s electrons in a N2-Zn-B2 configuration to stimulate remarkable ORR performance. *Angewandte Chemie International Edition* 60 (1): 181–185.

133 Li, J., Chen, M., Cullen, D.A. et al. (2018). Atomically dispersed manganese catalysts for oxygen reduction in proton-exchange membrane fuel cells. *Nature Catalysis* 1 (12): 935–945.

134 Liu, H., Jiang, L., Khan, J. et al. (2023). Decorating single-atomic Mn sites with FeMn clusters to boost oxygen reduction reaction. *Angewandte Chemie International Edition* 62 (3): e202214988.

135 Luo, E., Zhang, H., Wang, X. et al. (2019). Single-atom Cr–N4 sites designed for durable oxygen reduction catalysis in acid media. *Angewandte Chemie International Edition* 58 (36): 12469–12475.

136 Xiao, M.L., Zhu, J.B., Li, G.R. et al. (2019). A single-atom iridium heterogeneous catalyst in oxygen reduction reaction. *Angewandte Chemie-International Edition* 58 (28): 9640–9645.

137 Liu, S., Li, Z., Wang, C. et al. (2020). Turning main-group element magnesium into a highly active electrocatalyst for oxygen reduction reaction. *Nature Communications* 11 (1): 938.

138 Luo, F., Roy, A., Silvioli, L. et al. (2020). P-Block single-metal-site tin/nitrogen-doped carbon fuel cell cathode catalyst for oxygen reduction reaction. *Nature Materials* 19 (11): 1215–1223.

139 Wang, C., Wang, D., Liu, S. et al. (2020). Engineering the coordination environment enables molybdenum single-atom catalyst for efficient oxygen reduction reaction. *Journal of Catalysis* 389: 150–156.

140 Li, J.-C., Qin, X., Xiao, F. et al. (2021). Highly dispersive cerium atoms on carbon nanowires as oxygen reduction reaction electrocatalysts for Zn–air batteries. *Nano Letters* 21 (10): 4508–4515.

141 Zhu, M., Zhao, C., Liu, X. et al. (2021). Single atomic cerium sites with a high coordination number for efficient oxygen reduction in proton-exchange membrane fuel cells. *ACS Catalysis* 11 (7): 3923–3929.

142 Yin, L., Zhang, S., Sun, M. et al. (2023). Heteroatom-driven coordination fields altering single cerium atom sites for efficient oxygen reduction reaction. *Advanced Materials* 35 (28): 2302485.

143 Hu, H., Wang, J., Cui, B. et al. (2022). Atomically dispersed selenium sites on nitrogen-doped carbon for efficient electrocatalytic oxygen reduction. *Angewandte Chemie International Edition* 61 (3): e202114441.

144 Liu, J., Jiao, M., Lu, L. et al. (2017). High performance platinum single atom electrocatalyst for oxygen reduction reaction. *Nature Communications* 8: 15938.

145 Sui, R., Zhang, X., Wang, X. et al. (2022). Silver based single atom catalyst with heteroatom coordination environment as high performance oxygen reduction reaction catalyst. *Nano Research* 15 (9): 7968–7975.

146 Wang, X., An, Y., Liu, L. et al. (2022). Atomically dispersed pentacoordinated-zirconium catalyst with axial oxygen ligand for oxygen reduction reaction. *Angewandte Chemie International Edition* 61 (36): e202209746.

147 Ji, B., Gou, J., Zheng, Y. et al. (2023). Coordination chemistry of large-size yttrium single-atom catalysts for oxygen reduction reaction. *Advanced Materials* 35 (24): 2300381.

148 Wang, J., Huang, Z., Liu, W. et al. (2017). Design of N-coordinated dual-metal sites: a stable and active Pt-free catalyst for acidic oxygen reduction reaction. *Journal of the American Chemical Society* 139 (48): 17281–17284.

149 Wang, J., Liu, W., Luo, G. et al. (2018). Synergistic effect of well-defined dual sites boosting the oxygen reduction reaction. *Energy & Environmental Science* 11 (12): 3375–3379.

150 Lu, Z., Wang, B., Hu, Y. et al. (2019). An isolated zinc–cobalt atomic pair for highly active and durable oxygen reduction. *Angewandte Chemie International Edition* 58 (9): 2622–2626.

151 Zhou, W., Su, H., Cheng, W. et al. (2022). Regulating the scaling relationship for high catalytic kinetics and selectivity of the oxygen reduction reaction. *Nature Communications* 13 (1): 6414.

152 Zhu, P., Xiong, X., Wang, X. et al. (2022). Regulating the FeN_4 moiety by constructing Fe–Mo dual-metal atom sites for efficient electrochemical oxygen reduction. *Nano Letters* 22 (23): 9507–9515.

153 Xie, Y., Chen, X., Sun, K. et al. (2023). Direct oxygen-oxygen cleavage through optimizing interatomic distances in dual single-atom electrocatalysts for efficient oxygen reduction reaction. *Angewandte Chemie International Edition* 62 (17): e202301833.

154 Han, A., Wang, X., Tang, K. et al. (2021). An adjacent atomic platinum site enables single-atom iron with high oxygen reduction reaction performance. *Angewandte Chemie-International Edition* 60 (35): 19262–19271.

155 He, T., Chen, Y., Liu, Q. et al. (2022). Theory-guided regulation of FeN4 spin state by neighboring cu atoms for enhanced oxygen reduction electrocatalysis in flexible metal-air batteries. *Angewandte Chemie International Edition* 61 (27): e202201007.

156 Shao, Y., Dodelet, J.-P., Wu, G., and Zelenay, P. (2019). PGM-free cathode catalysts for PEM fuel cells: a mini-review on stability challenges. *Advanced Materials* 31 (31): 1807615.

157 Banham, D., Ye, S., Pei, K. et al. (2015). A review of the stability and durability of non-precious metal catalysts for the oxygen reduction reaction in proton exchange membrane fuel cells. *Journal of Power Sources* 285: 334–348.

158 Kumar, K., Dubau, L., Mermoux, M. et al. (2020). On the influence of oxygen on the degradation of Fe-N-C catalysts. *Angewandte Chemie-International Edition* 59 (8): 3235–3243.

159 Lefèvre, M. and Dodelet, J.-P. (2003). Fe-based catalysts for the reduction of oxygen in polymer electrolyte membrane fuel cell conditions: determination of the amount of peroxide released during electroreduction and its influence on the stability of the catalysts. *Electrochimica Acta* 48 (19): 2749–2760.

160 Trogadas, P., Parrondo, J., and Ramani, V. (2008). Degradation mitigation in polymer electrolyte membranes using cerium oxide as a regenerative free-radical scavenger. *Electrochemical and Solid-State Letters* 11 (7): B113.

161 Xie, H., Xie, X., Hu, G. et al. (2022). Ta-TiO_x nanoparticles as radical scavengers to improve the durability of Fe-N-C oxygen reduction catalysts. *Nature Energy* 7 (3): 281–289.

8

Developments and Prospects of Aqueous Batteries

Shuo Yang[1], *Shengmei Chen*[1], *and Chunyi Zhi*[1,2]

[1] *City University of Hong Kong, Department of Materials Science and Engineering, 83 Tat Chee Avenue, Kowloon, Hong Kong 999077, P. R. China*
[2] *City University of Hong Kong, Hong Kong Institute for Clean Energy, 83 Tat Chee Avenue, Kowloon, Hong Kong 999077, P. R. China*

8.1 Introduction

Excessive fossil energy consumption brings major concerns to environmental problems and energy crises, making it indispensable to search for alternative energy sources with environmental benignity and sustainability [1]. Electrochemical energy storage systems, especially batteries, have demonstrated their superiority among various technologies in energy density, flexibility, and scalability. Commercial Li-ion batteries (LIBs) occupy dominant positions in electric vehicles and rechargeable devices [2, 3]. However, the flammable nature of nonaqueous electrolytes has brought safety issues such as combustion and explosion. The scarcity of Li resources also leads to economic challenges for the application of LIBs, especially for grid-scale energy storage.

Aqueous batteries eliminating the use of flammable electrolytes stand out for their high safety, low cost, and easy operation [4]. The main obstacles for aqueous batteries between the prototype and practical applications are the narrow electrochemical stability window (ESW) of water and the unavoidable side reactions in aqueous media [5, 6]. The former confines the energy density of batteries, while the latter causes long-term stability problems. Extensive efforts have been devoted to addressing these issues and contributing to the prosperity of aqueous batteries. In this chapter, we introduce the recent developments of aqueous batteries based on different charge carriers (metal or nonmetal ions). The remaining challenges and corresponding solutions, including water decomposition, dendrite growth, and side reactions, are also involved in this part.

Electrochemical Energy Storage Devices: Non-Conventional Technologies and Materials,
First Edition. Edited by Yongbing Tang and Luojiang Zhang.

8.2 Aqueous Batteries Based on Monovalent Metal Ions

8.2.1 Aqueous LIBs

The earliest aqueous LIBs date from 1994s research, which used $LiMn_2O_4$ as the cathode and 5 M $LiNO_3$ aqueous solutions as the electrolyte [7]. The prepared $LiMn_2O_4||VO_2$ battery delivered an average voltage near 1.5 V and an energy density of 55 Wh kg^{-1}, which shows promise as a potential alternative for aqueous lead-acid and Ni-Cd batteries. Several materials, including metal oxides, polyanionic compounds, and Prussian blue analogs (PBAs), can act as cathodes for aqueous LIBs [8]. Among these materials, spinel $LiMn_2O_4$ is the most extensively investigated due to its availability in aqueous solutions [9]. Many efforts have been devoted to further developing the long-term stability of $LiMn_2O_4$, including electrolyte optimization, nanostructure design, and metal ion-doped strategies [10, 11]. For example, a remarkable capacity retention of 93% could be achieved after 10,000 cycles for nanoscale $LiMn_2O_4$ with porous morphology, while the capacity for bulky $LiMn_2O_4$ cathode decayed rapidly [11].

Unlike cathode materials, most anode materials used in nonaqueous LIBs are unsuitable for aqueous batteries as their low redox potentials are usually beyond the cathodic limit of aqueous electrolytes [4]. The hydrogen evolution reaction (HER) would occur on the surface of the anode and lead to the decomposition of the aqueous electrolyte. The vanadium-based compounds, including VO_2, LiV_3O_8, $H_2V_3O_8$, and $Na_{1+x}V_3O_8$, are first investigated as anode materials for early aqueous LIBs. Most of the vanadium-based anodes suffer from poor cycle life with rapid capacity deterioration during the cycling process, which is attributed to the dissolution of vanadium and the decomposition of water. Besides vanadium-based compounds, polyanionic materials, including $LiTi_2(PO_4)_3$, TiP_2O_7, and some organic materials, are also investigated as alternative anodes for aqueous batteries [12, 13].

In 2015, the concept of "water-in-salt" electrolyte (WISE) was proposed by adopting salt-concentrated electrolytes, making aqueous LIBs enter a new era [14]. In the WISE composed of highly concentrated lithium bis(trifluoromethane sulfonyl)imide (LiTFSI) electrolyte (21 M), the H_2O molecules were partly replaced by $TFSI^-$ anions in the Li^+ primary solvation sheath, contributing to the lower electrochemical activity of water (Figure 8.1a). Moreover, the electrochemical reduction of $TFSI^-$ anions could generate sufficient LiF and enable the formation of solid-electrolyte interphase (SEI), as proved by density functional theory (DFT) simulation, X-ray photoelectron spectroscopy (XPS) survey, and transmission electron microscopy (TEM) observation (Figure 8.1b,c).

Benefiting from the inactive water and SEI, the WISE could offer an ESW up to 3.0 V, much higher than pure H_2O (1.23 V). Consequently, the $LiMn_2O_4||Mo_6S_8$ battery demonstrated an open-circuit voltage (OPV) of 2.3 V and an energy density of 100 Wh kg^{-1}. After that, "water-in-bisalt" electrolyte (WIBE) containing 21 M LiTFSI and 7 M lithium trifluoromethane sulfonate (LiOTf) with wider ESW (3.13 V) allowed the use of TiO_2 anode and enabled high cell voltage with an average discharge voltage of 2.1 V [17]. To further expand the ESW, hydrate-melt

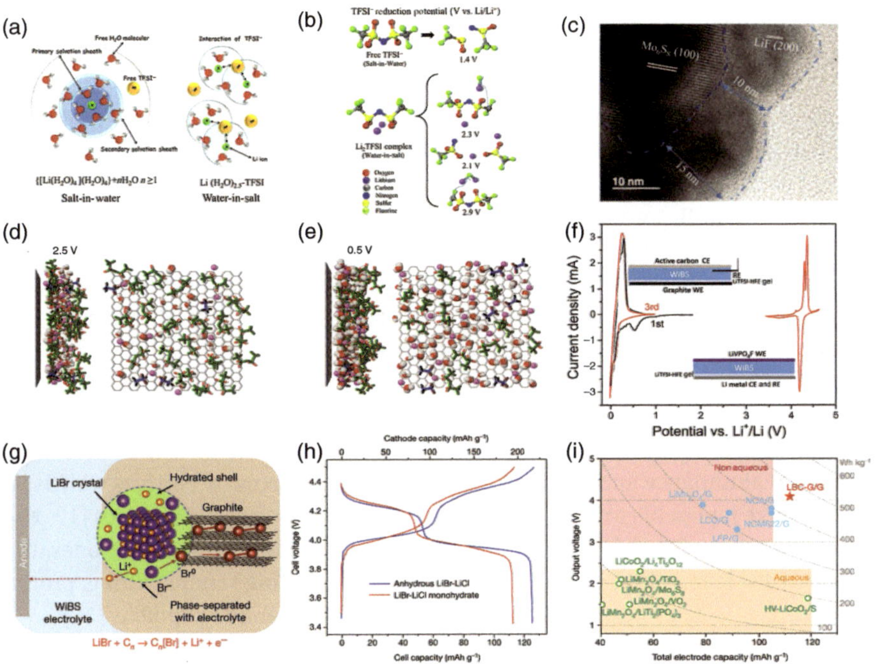

Figure 8.1 (a) Illustration of the evolution of the Li^+ primary solvation sheath in diluted and water-in-salt solutions. (b) Predicted reduction potentials based on the DFT calculations. (c) TEM observations of cycled Mo_6S_8 anode demonstrate the formation of LiF SEI. Source: Reproduced with permission of Suo et al. [14]. © 2015/American Association for the Advancement of Science - AAAS. Snapshots of inner-Helmholtz interfacial regions of the anode surface in WiBS (21 M LiTFSI + 7 M LiOTF) at (d) 2.5 V and (e) 0.5 V vs. Li, respectively. (f) 4.0 V stability window and stabilization of graphite lithiation in aqueous electrolyte. Source: Reproduced with permission [15]. Copyright 2017, Elsevier. (g) Schematic illustration of the halogen conversion–intercalation chemistry in graphite. (h) Typical galvanostatic charge–discharge (GCD) curves of two Li-ion full cells with conversion–intercalation chemistry. (i) Energy density of Li-ion full cells with conversion–intercalation chemistry compared with various state-of-the-art nonaqueous or aqueous LIBs. Source: Reproduced with permission [16]. Copyright 2019, Springer Nature.

electrolyte (HME) utilizing a eutectic system was prepared by dissolving LiTFSI and $LiN(SO_2C_2F_5)_2$ (LiBETI) in water [18]. Owing to the low water content (molar ratio of $H_2O : Li^+ = 2.0$) in $Li(TFSI)_{0.7}(BETI)_{0.3}{\cdot}2H_2O$ electrolyte and the decreased water activity, the ESW of electrolyte could be further expanded to 3.8 V.

Although increasing the salt concentration in water can expand the ESW of electrolytes, the cathodic stability, which is usually higher than 1.2 V vs. Li/Li^+, limits the use of anode materials, including silicon, graphite, and Li metal [5]. The "cathodic challenge" originates from the expulsion of anions by the anode surfaces with a negative charge and the preferential adsorption of water molecules when the electrode is polarized to low potential (Figure 8.1d,e) [15]. Therefore, most aqueous LIBs suffer lower discharge voltages and energy densities than nonaqueous LIBs. To solve this problem, a highly fluorinated ether (HFE) gel with strong hydrophobicity was adopted as an inhomogeneous additive to minimize competitive HER on the anode surface. Aqueous LIBs with discharge plateaus at or above 4.0 V were prepared using the graphite or Li metal anode (Figure 8.1f). Furthermore, a halogen conversion–intercalation chemistry in graphite was proposed to endow the aqueous LIBs with an energy density of 460 Wh kg^{-1}, even outperforming state-of-the-art nonaqueous LIBs (Figure 8.1g–i) [16]. In this work, a mixture of anhydrous LiBr and LiCl with graphite was used as the cathode material. During the charging process, halide anion Br^- was first oxidized to $Br^{0.05}$ and intercalated into the graphite interlayer, followed by the oxidization and intercalation of Cl^- (Figure 8.1g). Such a conversion–intercalation process with the reversible formation of graphite intercalation compound, i.e. $C_{3.5}[Br_{0.5}Cl_{0.5}]$, could yield a high capacity of 243 mAh g^{-1} and an attractive discharge voltage of 4.2 V (Figure 8.1h).

8.2.2 Aqueous Na-Ion Batteries (NIBs)

The natural abundance and even distribution of Na resources on earth as well as its low cost make it possible to mitigate the Li shortage by using Na in batteries [19, 20]. Although aqueous NIBs deliver a similar mechanism to the nonaqueous ones, the complicated side reactions in aqueous electrolytes, including water splitting, proton co-insertion, and electrode dissolution, hinder their further development [21].

Compared with LIBs, the larger radius of Na^+ (0.102 nm) limits the choices of electrode materials for aqueous NIBs. Many reported cathode materials suffer from limited specific capacity in aqueous media. For example, $Na_{0.44}MnO_2$, a widely explored cathode material for aqueous NIBs, showed a low specific capacity of 39 mAh g^{-1} at 0.2 C [22]. Partial substitution of Mn by Ti to form a novel Na-rich tunnel-type material ($Na_{0.66}[Mn_{0.66}Ti_{0.34}]O_2$) can provide a higher specific capacity of 76 mAh g^{-1} at 2 C and capacity retention of 89% after 400 cycles; however, it is still far from satisfactory [23].

The lack of high-performance anode materials is another problem associated with aqueous NIBs. Among various Ti-based anode materials used in nonaqueous NIBs, only NASICON-type $NaTi_2(PO_4)_3$ demonstrates its potential in aqueous NIBs [24]. The reversible insertion of Na^+ based on the Ti^{3+}/Ti^{4+} redox couple can provide a specific capacity of 133 mAh g^{-1} [25]. However, the unsatisfactory cycling stability originating from the side reactions between electrode material and water impedes its practical application. Following that, several inorganic materials have been developed as anode materials for aqueous NIBs, including vanadium-based materials ($V_2O_5{\cdot}0.6H_2O$, $Na_2V_6O_{16}{\cdot}nH_2O$, $Na_2VTi(PO_4)_3$, etc.), PBAs ($KMn[Cr(CN)_6]$), and metal oxides (MnO_3). [21]. Besides, organic anode materials have also attracted tremendous attention recently. For example, 1,4,5,8-naphthalenetetracarboxylic dianhydride (PNTCDA)-derived polyimide could provide a specific capacity of 140 mAh g^{-1} at a current density of 1 A g^{-1} [26]. Moreover, the electrode exhibited an extremely long life of 50,000 cycles with a capacity retention of 70% by coupling a liquid I^-/I_3^- cathode, outperforming most inorganic-based NIBs. Following this work, electrochemical performances of other polyimides were further investigated, like poly-(naphthalene-4-formyl ethylenediamine) (PNFE) and disodium naphthalenediimide (SNDI) [27, 28].

The WISEs, with the ability to expand the ESW, have also been adopted in aqueous NIBs. Owing to the intense interaction between Na^+ and trifluoromethane sulfonate anion (OTF^-), an electrolyte containing 9.26 M NaOTF could provide a wide ESW of 2.5 V by forming a Na^+-conducting SEI [29]. As a result, aqueous NIBs constructed with $NaTi_2(PO_4)_3$ anode and $Na_{0.66}[Mn_{0.66}Ti_{0.34}]O_2$ cathode delivered an energy density of 31 Wh kg^{-1} based on the total electrode mass. Using a 17 M $NaClO_4$ electrolyte with higher salt concentration can expand the ESW to 2.7 V, while the high solution of reduced products of $NaClO_4$ prevents the formation of the SEI layer, thus resulting in a high cathodic limit [30]. Therefore, a novel electrolyte containing 17 M $NaClO_4$ and 2 M NaOTF was developed to promote the formation of NaF-based SEI on the anode surface [31]. After NaOTF addition, the HER potential was broadened to 1.6 V vs. Na/Na^+, enabling $Na_3V_2(PO_4)_3$ with a redox potential of 1.7 V vs. Na/Na^+. As a result, a symmetric $Na_3V_2(PO_4)_3 || Na_3V_2(PO_4)_3$ was assembled with an energy density of 70 W h kg^{-1}. Several other types of electrolytes based on the concentrated solutions have been developed to expand the ESW of water, including 37 M sodium bis(fluorosulfonyl)imide (NaFSI), tetraethylammonium cation (TEA^+)-assisted WISE (9 M NaOTF + 22 M TEAOTF), and electrolyte with super-concentrated sugar solution [32]. However, the energy densities of reported aqueous NIBs are less than 100 Wh kg^{-1} owing to the lack of proper electrode materials with high specific capacity and output voltage.

8.2.3 Aqueous K-Ion Batteries (KIBs)

The large ionic radius of K^+ (0.138 nm), which may cause structural distortion of electrode materials, is one of the main obstacles to developing high-performance aqueous KIBs [33]. The state-of-the-art aqueous KIBs contain limited cathode materials, among which PBAs are demonstrated as the most promising candidates [32]. The large ion transport channels and rigid frameworks of PBAs may provide fast rate capability and long-term stability. For example, crystalline copper hexacyanoferrate (CuHCF) exhibited a high redox potential of 0.946 V vs. standard hydrogen electrode (SHE) and excellent rate performance with capacity retention of 67% when increasing the current density from 0.05 to 5 A g^{-1} [34]. However, the absence of K^+ in PBAs with one-electron transfer during the redox process led to their low specific capacity (<70 mAh g^{-1}). Therefore, K-rich PBA $K_2Fe^{II}[Fe^{II}(CN)_6]$ was developed to enable two one-electron transfer processes, providing a specific capacity up to 120 mAh g^{-1} at 0.2 A g^{-1} [35].

Similar to the cathode material, only a few works have been reported regarding anode materials for aqueous KIBs. MoO_3 was the first reported inorganic anode material, which exhibited poor cycling life (20 cycles) and low discharge capacity (59.5 mAh g^{-1}) [36]. The adoption of potassium acetate (KAc)-based WISE (30 M) enabled the use of $KTi_2(PO_4)_3$ anode with long cycle life (11,000 cycles with capacity retention of 68%). However, the specific capacity was still far from satisfactory (58 mAh g^{-1}) [37]. Organic materials are another type of promising anode material for aqueous KIBs. For example, the PNTCDA electrode based on a reversible enolization process delivered a stable specific capacity of 106.6 mAh g^{-1} after 300 cycles [38].

Organic 3,4,9,10-perylenetetracarboxylic diimide (PTCDI) with a stable specific capacity of 125 mAh g^{-1} was also demonstrated as a promising candidate for aqueous KIBs [39]. The full battery based on the PTCDI anode and KFeMnHCF cathode could output an energy density of 80 Wh kg^{-1}, even superior to most aqueous NIBs.

8.3 Aqueous Batteries Based on Multivalent Metal Ions

8.3.1 Aqueous Zn-Based Batteries (ZBs)

Among various energy storage systems, aqueous ZBs have drawn particular interest owing to the abundant resources, suitable redox potential (−0.763 V vs. SHE in the neutral electrolyte), and high theoretical specific capacity of Zn (820 mAh g^{-1} and 5854 mAh cm^{-3}) [40]. Some other metal anodes can also provide large theoretical volumetric capacities. For example, Al anode with the Al/Al^{3+} redox reaction exhibits volumetric capacities of 8046 mAh cm^{-3} [41]. However, the instability of these metal anodes (e.g. Mg, Ca, and Al) in aqueous solutions may lead to poor electrochemical performance [42]. In contrast, ZBs display higher reversibility in various electrolytes, which can be divided into alkaline ZBs and zinc-ion batteries (ZIBs) in neutral or acid electrolytes.

8.3.1.1 Alkaline ZBs

In aqueous alkaline electrolytes, the main electrochemical process of the anode is related to the transformation between Zn and $Zn(OH)_4{}^{2-}$ [43]:

$$Zn + 4OH^- \leftrightarrow Zn(OH)_4{}^{2-} + 2e^- \tag{8.1}$$

During the discharging process, Zn is dissolved in alkaline electrolytes and forms zincate ions. After that, ZnO will form on the surface of Zn anode attributed to the limited solubility of $Zn(OH)_4{}^{2-}$ [44]. In addition to surface passivation, the nonuniform deposition/dissolution process may lead to shape change and dendrite growth. Moreover, the low redox potential (−1.35 V vs. SHE) thermodynamically favors the HER. Many efforts have been made to solve these problems, including structural engineering, surface modifications, and additive introduction [43]. The structural-engineering strategy focuses on developing Zn anode with a large surface area and a porous structure. For example, Zn sponges with three-dimensional (3D) porous structures (Figure 8.2a) were demonstrated to be cycled for more than 100 cycles and 50,000 cycles at depths of discharge (DOD) of 40% and 1%, respectively [45]. Surface modifications with a protective layer on the anode materials can effectively suppress HER and morphology evolution. Several kinds of materials, such as TiO_2 and TiN_xO_y, have been demonstrated to enable Zn or ZnO anode with higher stability [48]. In terms of electrolyte additives, metal ions with lower reduction potential (Sn^{2+}, Bi^{3+}, etc.) can adjust the electrodeposition process of Zn^{2+} via an electrostatic shield mechanism [43]. Besides, some organic materials, like sodium dodecyl sulfate (SDS), may form a protective layer on the surface of the anode and prevent direct contact with the electrolyte, thus mitigating corrosion and passivation [49]. Moreover, the adsorbed organic molecules can enable uniform local current distribution, which is beneficial for suppressing dendrite growth. Although these strategies can contribute to addressing one or two issues for Zn anodes and thus enhance their electrochemical performance to some extent, it is still hard to develop alkaline Zn-based batteries with high capacity, preferable stability, and low cost simultaneously [50].

Unlike the Zn anode with a dissolution/deposition mechanism, the metal (Mn, Ni, Co) oxides/hydroxides used as the cathode materials are based on a protonation/deprotonation mechanism [51]. MnO_2 cathode can provide a high theoretical specific capacity of 617 mAh g^{-1} with a two-step reaction [52]. In the first-step reaction, the proton inserts into the lattice of MnO_2 and forms MnOOH. The MnOOH can be further converted into $Mn(OH)_2$ in the second-step reaction, accompanied by an additional electron transfer. However, achieving the theoretical capacity of Zn||MnO_2 batteries is difficult because of the formation of electrochemically inert byproducts of $ZnMn_2O_4$ and Mn_3O_4 [52]. Bi_2O_3 doping and using a LiOH/KOH mixture electrolyte can effectively alleviate the formation of $ZnMn_2O_4$ and Mn_3O_4, enabling Mn-Zn batteries with higher capacity and better rechargeability [53]. However, the formation of irreversible $LiMnO_4$ during the electrochemical process limited the reversible capacity (360 mAh g^{-1}). Therefore, a Cu-intercalated Bi-MnO_2 layered cathode was prepared to realize reversible two-electron reactions with a high capacity of 617 mAh g^{-1} and a long lifetime of 6000 cycles [46]. The presence of Cu^{2+} facilitated the charge transfer and stabilized Mn species during the electrochemical process, which allowed the complete transformation of MnO_2 from $Mn(OH)_2$ (Figure 8.2b).

Zn||$Ni(OH)_2$ batteries with Ni-based hydroxides (α-$Ni(OH)_2$) as cathode can provide a high OPV of ~1.75 V and an energy density of 372 Wh kg^{-1}, thus drawing tremendous attention since 1899 [5]. However, the formation of a thermodynamically stable β-$Ni(OH)_2$ phase during cycling may cause a reduced specific capacity [54]. Replacing Ni atoms with Co, Cu, Al, and Zn help stabilize α-$Ni(OH)_2$ by forming a layered double hydroxide (LDH) structure [52]. For example, the α-$Ni(OH)_2$ cathode doped by Al and Co atoms displayed a stabilized LDH structure with a large interlayer spacing of 0.71 nm [47]. After combing NiAlCo-LDH with carbon nanotubes, the battery with hybrid cathode showed a peak energy density of 324 Wh kg^{-1} and good stability over 2000 cycles (Figure 8.2c–e).

Alkaline ZBs have achieved great success in rate capability, cycling life, and energy density; however, most of them are based on relatively low areal/mass loading, which is unsatisfactory for practical application [55]. Moreover, the utilization, stability, and durability of Zn anode are still rather low. Therefore, developing high-performance alkaline ZBs with large area capacity, high-rate capability, long stability, and low cost remains challenging.

8.3.1.2 Neutral ZIBs

The use of mild neutral electrolytes in ZIBs alleviates the formation of passivated ZnO and dissolution of Zn compared with alkaline ZBs, which endows ZIBs with better reversibility [42]. The successful preparation of the rechargeable Zn||MnO_2

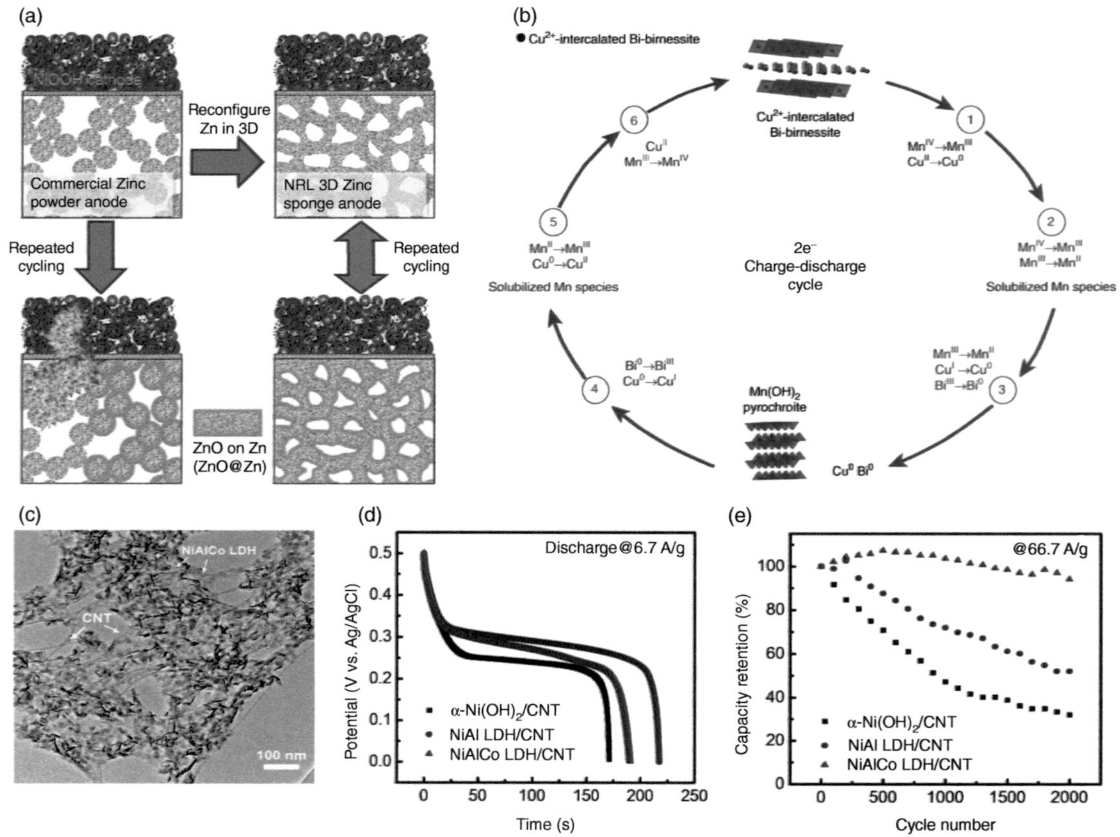

Figure 8.2 (a) Schematic of the effect of recharging Ni-Zn (conventional powder zinc anodes) vs. Ni-3D sponge Zn. Source: Reproduced with permission [45]. Copyright 2017, American Association for the Advancement of Science. (b) Electrochemical reactions for the regeneration cycle of Cu^{2+}-intercalated Bi-MnO_2. Source: Reproduced with permission [46]. Copyright 2017, Springer Nature. (c) TEM images of NiAlCo LDH/CNT hybrid. (d) GCD curves and (e) cycling performance of different electrode materials. Source: Reproduced with permission of Gong et al. [47]. © 2014/Royal Society of Chemistry.

battery in 1988 paved the way for ZIBs [56]. Several electrochemical mechanisms have been proposed for the Zn^{2+} storage in the MnO_2 cathode. Among various investigations, the Zn^{2+} intercalation–deintercalation mechanism and Zn^+/H^+ Co-insertion mechanism are the most widely accepted ones [57]. Owing to the high charge density and large ionic radius of Zn^{2+}, the repeated insertion/deintercalation of Zn^{2+} will lead to structural collapse of the MnO_2 lattice, resulting in unsatisfactory cycling performance. Designing a polyaniline (PANI)-intercalated MnO_2 with a polymer-reinforced layered structure can effectively improve the cycling stability of Zn||MnO_2 batteries [58]. Dissolution of Mn^{2+} is another issue in Zn||MnO_2 batteries, which may lead to the continuous loss of active materials and fast capacity decay [59]. The addition of $MnSO_4$ can shift the dissolution equilibrium of Mn^{2+} and mitigate capacity deterioration. As a result, a long-lifespan Zn||MnO_2 battery of 5000 cycles with a capacity retention of 93% could be achieved. Furthermore, an electrolytic Zn||MnO_2 battery based on an electrolysis/electrodeposition of MnO_2/Mn^{2+} mechanism with a high output voltage of 2 V was developed by tuning the pH of the electrolyte [60]. Therefore, it is possible to improve the energy densities of ZIBs by exploring novel electrochemistry.

V-based compounds, including bilayered vanadium pentoxide (V_2O_5) and metal vanadates (LiV_3O_8, $Na_{0.33}V_2O_5$, $Zn_{0.25}V_2O_5{\cdot}nH_2O$, $Ca_{0.25}V_2O_5{\cdot}nH_2O$, $Mg_xV_2O_5{\cdot}nH_2O$, etc.), have been frequently applied as cathode materials for ZIBs, and their studies have made great progress for ZIB development [61]. Many V-based materials exhibit high specific capacities (200–400 mAh g^{-1}) because of the multiple oxidation states of vanadium [51]. The existence of structure H_2O and interlayer metal ions in metal vanadates may lead to a longer lifespan and superior rate capability. Taking $Zn_{0.25}V_2O_5{\cdot}nH_2O$ as an example, the co-intercalated H_2O molecules buffered the charge density of Zn^{2+} and accelerated the charge transfer, thus enabling fast kinetics [62]. Moreover, the indigenous Zn^{2+} in the interlayer of $Zn_{0.25}V_2O_5$ ensured structural stability. However, the low operation voltages under 1.0 V limit the energy density of V-based materials, which has hindered their practical applications.

PBAs with 3D open frameworks and structural stability have attracted growing attention recently [63]. In 2015, zinc hexacyanoferrate (ZnHCF) was proposed as cathode material for ZIBs with a high discharge voltage of 1.73 V and a specific capacity of 60 mAh g^{-1} [64]. Later, the effects of particle morphology on the electrochemical performance were further explored, which revealed that the ZnHCF with a cubooctahedral shape delivered the highest (100) surface coverage and the best rate capability [65]. After the successful development of Zn||ZnHCF batteries, the Zn^{2+}-storage abilities in CuHCF, nickel hexacyanoferrate (NiHCF), and iron hexacyanoferrate (FeHCF) were further investigated [63]. To solve the low capacities in PBAs (typically lower than 80 mAh g^{-1}), a Na-Zn hybrid battery with $Na_2MnFe(CN)_6$ as the cathode and Zn as the anode was developed to enable a high specific capacity of 137 mAh g^{-1} [66].

In addition to the inorganic cathode, some organic materials, such as conductive polymers, carbonyl compounds, and organic radicals, are emerging as novel cathode materials for ZIBs [67]. Among these materials, quinone materials with sufficient active sites can provide high specific capacities. For example, calix[4]quinone (C4Q) could yield a high capacity of 335 mAh g^{-1} with an energy efficiency of 93% at 20 mA g^{-1}, exceeding many inorganic and organic materials [68]. However, the Zn||C4Q battery required an expensive ion-selective membrane to inhibit the dissolution of organic compounds. Besides, its output voltage was also less satisfying. The application of organic materials as cathodes for ZIBs is still in its infancy, which needs further investigations for understanding the electrochemical mechanism and searching for high-performing materials.

As for Zn anodes, the formation of zinc dendrite is an obstacle hindering the practical application of ZIBs. Building a proper artificial interface layer can prevent direct contact between the Zn anode and electrolyte, thus alleviating the side reactions like Zn corrosion and HER. Moreover, the interface layer with mechanical robustness can help to suppress dendrite growth [69]. It is noted that most artificial interface layers are built by an *ex situ* method, which may cause inadequate adhesion between the interface layer and the surface of Zn. Other strategies, such as electrolyte modification, current collector modification, and charge–discharge protocol design, are also proposed to achieve Zn anode with higher stability [70].

8.3.2 Other Aqueous Multivalent Metal-Ion Batteries

Compared with monovalent metal ions, multivalent metal ions (Zn^{2+}, Mg^{2+}, Ca^{2+}, Al^{3+}, etc.) transfer multiple charges per ion [71]. However, multivalent metal ions exhibit strong coulombic interactions with the electrode materials, which makes it challenging to develop proper electrode materials. The electrochemical plating/stripping potentials of multivalent metal ions (except for Zn^{2+}) are far from the ESW of aqueous electrolyte, thus hindering the direct use of metal anode.

Although water molecules were demonstrated to facilitate the insertion of Mg^{2+} into electrode materials several years ago, the aqueous Mg-ion battery (MIB) was not developed until 2017 (Figure 8.3a). In this work, $Na_{1.4}Ni_{1.3}Fe(CN)_6{\cdot}5H_2O$ with the ability to reversibly intercalate Mg^{2+} was chosen as the cathode material for aqueous MIBs [72]. Coupling with an aromatic polyimide anode, the battery exhibited an average voltage of 1 V (Figure 8.3b) and an overall energy density of 33 Wh kg^{-1}. After that, a concentrated electrolyte (4 M $MgTFSI_2$) with ESW of 2.0 V was proposed to enable aqueous MIBs with higher operation voltage (Figure 8.3c,d) [73]. The resultant MIBs based on $Li_3V_2(PO_4)_3$ cathode and poly pyromellitic dianhydride (PPMDA) yielded a high voltage of 1.9 V and an overall density of 62.4 Wh kg^{-1}. More recently, hydrated eutectic electrolyte coupling the $Mg(NO_3)_2{\cdot}6H_2O$ with acetamide further expanded the ESW to 2.7 V (Figure 8.3e–g) [74]. In the hydrated eutectic electrolyte, acetamide molecules dominated the primary solvation sheath of Mg^{2+} and suppressed the water activity. Moreover, the hydrogen bond network provided by water molecules may ensure fast Mg^{2+} transportation. However, the energy density of full batteries was still unsatisfactory (52.2 Wh kg^{-1}) owing to the limited specific capacity (38.1 mAh g^{-1}) of electrode materials.

The lack of proper electrode materials is one critical issue limiting the application of aqueous MIBs. Besides the organic anodes, only $FeVO_4$ is reported as inorganic

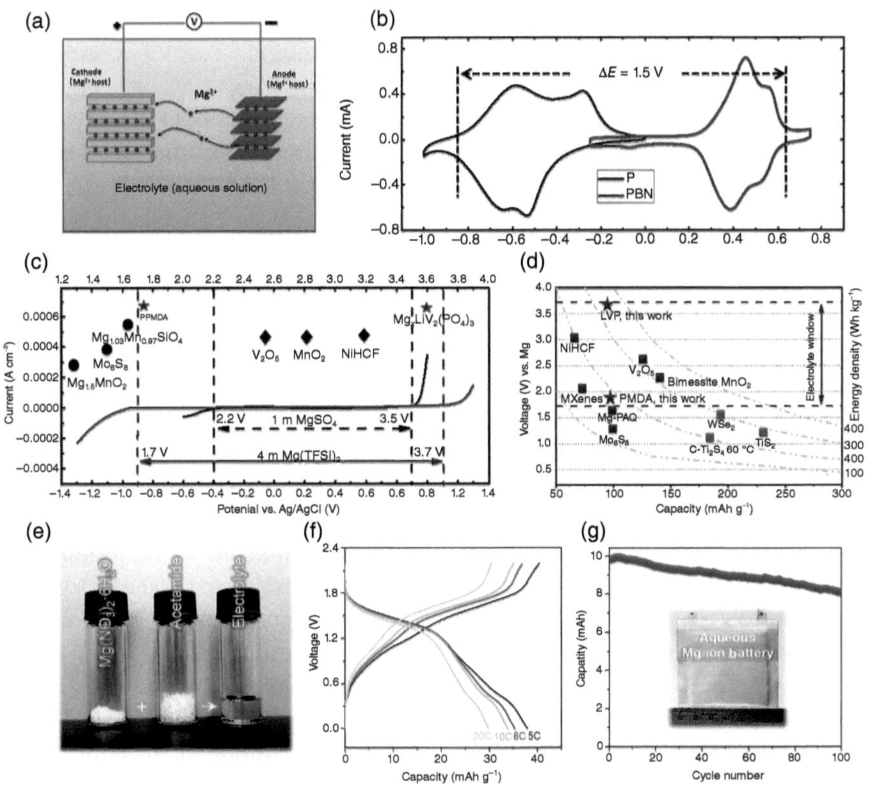

Figure 8.3 (a) Illustration of the first aqueous MIBs. (b) The CV curves of $Na_{1.4}Ni_{1.3}Fe(CN)_6 \cdot 5H_2O$ cathode and polyimide anode for aqueous MIBs. Source: Reproduced with permission [72]. Copyright 2017, American Chemical Society. (c) The ESW of 1 and 4 M $Mg(TFSI)_2$ aqueous electrolytes. (d) Performance comparison of different electrode materials for aqueous MIBs. Source: Reproduced with permission [73]. Copyright 2017, American Chemical Society. (e) Optical photo of hydrated eutectic electrolyte obtained by mixing $Mg(NO_3)_2 \cdot 6H_2O$ and acetamide. (f) GCD curves of the full MIB. (g) Cycling performance of the pouch-type MIB at 5 C. Source: Reproduced with permission of Zhu et al. [74]. © 2022/Royal Society of Chemistry.

anode material for aqueous MIBs [75]. The cathode materials for aqueous MIBs are relatively various, mainly including transition metal oxides (MnO_2, V_2O_5, MoO_3, etc.), PBAs, and some oxides with spinel structure ($MgMn_2O_4$). However, their electrochemical performance needs to be improved [76]. Therefore, future development of aqueous MIBs requires electrode materials with novel chemistry and high performance.

Aluminum anode with a volumetric capacity up to 8046 mAh cm^{-3} is one of the promising candidates for next-generation energy storage [41]. Room temperature ionic liquids (ILs) with wide ESW are the most common nonaqueous electrolytes for Al-ion batteries (AIBs) [77]. The early research on aqueous AIBs focuses on developing electrode materials that can accommodate/release Al^{3+} with long-term sustainability. Anatase TiO_2 with tunnels formed by stacked TiO_6 octahedrons was proposed for Al^{3+} storage in 2012 [78]. The TiO_2 anode delivered a redox peak for the reversible insertion/extraction of Al^{3+} at −1.26 and −0.84 V (vs. saturated calomel electrode, SCE), with a discharge capacity of 75 mAh g^{-1} at 4 mA cm^{-2} (Figure 8.4a,b). In 2018, the researchers found that metal Al could directly act as an anode for aqueous AIBs by replacing the passivating Al_2O_3 film on Al with a protective SEI layer [81]. The energy density of the Al||MnO_2 battery was estimated to be 235 Wh kg^{-1} based on the active and inactive materials. Following this work, $Al_xMnO_2 \cdot nH_2O$ prepared by introducing Al^{3+} into Mn_3O_4 was developed to address the lack of high-performance cathode materials for aqueous AIBs [82]. The adoption of aqueous electrolyte (5 M $Al(OTF)_3$) was demonstrated to facilitate ion transport and provide good cyclic durability. As a result, aqueous AIBs using this electrolyte exhibited a promising energy density of 481 Wh kg^{-1}. After that, several cathode materials based on metal Al anode were developed, including graphite, sulfur, and PBAs [41]. Even though great success has been achieved in the cathode materials, little research focuses on the sustainability of Al metal anode in aqueous electrolytes.

In 2020, the Zn-Al alloy was prepared to exhibit oxidation-resistant and dendrite-suppressed properties [83]. Following this work, the same group constructed an amorphous Al interlayer coating on the surface of Al (Al@a-Al), which could suppress HER and boost interfacial ions transfer kinetics (Figure 8.4c,d) [79]. Along with the long cycling life of the symmetric cell (800 hours), the Al@*a*-Al||Al_xMnO_2 full cell exhibited higher discharge potential and better sustainability.

In contrast to Mg and Al, the reversible electrodeposition of metallic Ca is rather difficult in nonaqueous electrolytes due to the formation of an ionically insulating passivation layer [84]. Moreover, the Ca^{2+} storage in electrodes with intercalation chemistry suffers from poor electrochemical performance because of the sluggish diffusion of Ca^{2+}. The sluggish diffusion of Ca^{2+} can be boosted by the shielding effect of water in aqueous electrolytes, thus lowering the diffusion barriers in inorganic crystals. The first aqueous Ca-ion full battery was assembled by coupling a CuHCF cathode with a polyimide (poly[*N*,*N*′-(ethane-1,2-diyl)-1,4,5,8-naphthalenetetracarboxiimide], PNDIE) anode (Figure 8.4e,f) [80]. During the discharging process, Ca^{2+} was inserted into the lattice of CuHCF to form $Ca_{0.3}$CuHCF, corresponding to a specific capacity of 65 mAh g^{-1}. Consequently, the aqueous PNDIE||CuHCF full battery provided an average operating voltage of 1.24 V with a specific capacity of 40 mAh g^{-1}. Up to date, only a handful of electrode materials have been developed for aqueous Ca-ion batteries with limited energy density (lower than 100 Wh kg^{-1}) [85].

8.4 Aqueous Batteries Based on Nonmetallic Ions

8.4.1 Aqueous Proton (H^+) Batteries

Nonmetallic ions based on abundant elements are emerging as novel charge carriers for aqueous batteries with unique cost and power density advantages, which are

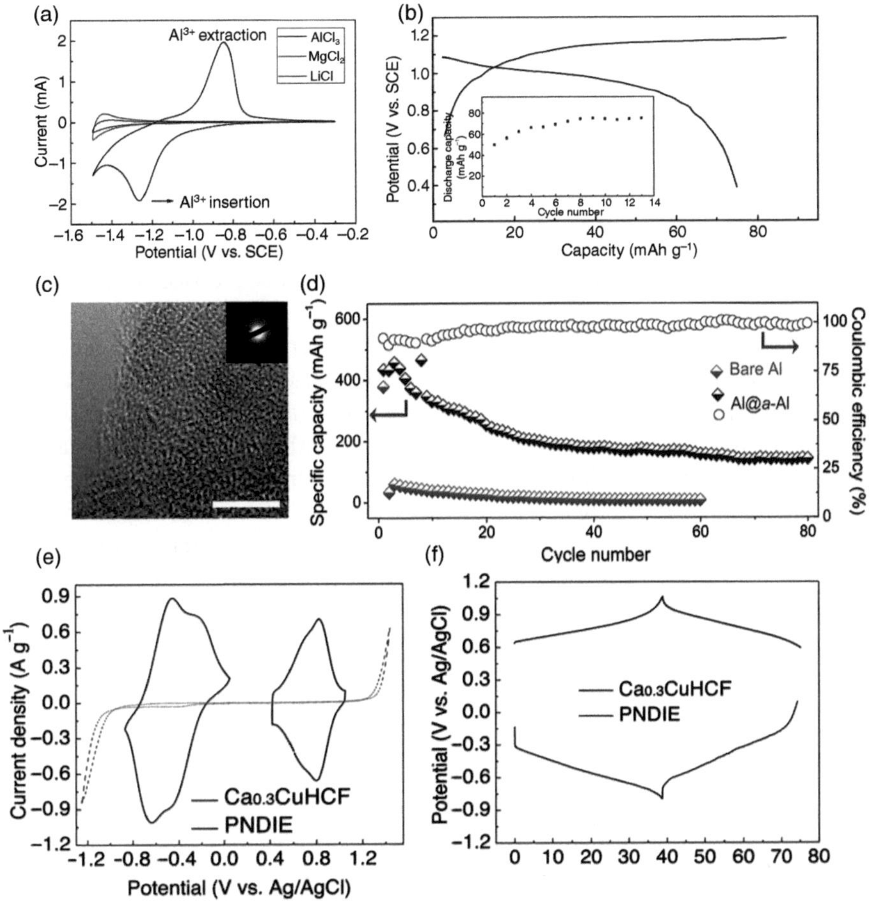

Figure 8.4 (a) CV curves of the anatase TiO_2 in 1 M $AlCl_3$, $MgCl_2$, and LiCl aqueous solutions. (b) GCD curve and cycling performance (the inserted figure) of the TiO_2 nanotube arrays. Source: Reproduced with permission [78]. Copyright 2012, The Royal Society of Chemistry. (c) High-resolution TEM image of Al@*a*-Al anode. (d) Cycling performance of Al@*a*-Al||Al_xMnO_2 and bare Al||Al_xMnO_2 cells. Source: Reproduced with permission of Yan et al. [79]. © 2022/American Chemical Society. (e) CV and (f) GCD curves of PNDIE anode and $Ca_{0.3}$CuHCF cathode in 2.5 M $Ca(NO_3)_2$ electrolyte. Source: Gheytani et al. [80]/John Wiley & Sons/CC BY 4.0.

especially suitable for large-scale energy storage [86]. Among various charge carriers for batteries, H^+ exhibits the lowest molar mass (1 g mol^{-1}) and smallest minimal ionic radius (0.89 fm). These natures of H^+ may endow proton batteries with high energy density and fast-charging performance. The adoption of H^+ charge carriers can date from lead-acid batteries in which H_2SO_4 reacted with PbO_2 to form $PbSO_4$ [87]. However, little attention is paid to proton insertion because of the great success of LIBs.

Organic materials resistant to acidic dissolution may act as promising electrode materials for aqueous H^+ batteries [88]. For example, quinone-based materials with C=O groups could reversibly host H^+ by forming C—OH groups. The first rocking-chair H^+ battery was assembled using a tetrachlorohydroquinone (TCHQ) and an anthraquinone (AQ) anode [89]. However, the energy density of TCHQ||AQ batteries was far from satisfactory due to the low output voltage of 0.6 V. After that, pyrene-4,5,9,10-tetraone (PTO) with the ability to store 4 H^+ per formula was demonstrated with a high specific capacity of 395 mAh g^{-1} [90]. In addition to small organic molecules, some organic polymers could act as H^+ hosts in acidic electrolytes [91]. The polymerization of organic monomers can help address the dissolution of organic molecules. However, it will introduce redox-inactive mass to lower the energy density of electrodes.

In contrast to organic materials with a chemical conversion mechanism, the metal oxides and PBAs usually undergo an H^+ insertion/extraction mechanism [92]. For example, MoO_3 with layer structure was proposed to enable reversible H^+ insertion [93]. During the discharging process, H^+ interacted with terminated oxygen atoms in MoO_3 and formed $H_{0.88}MoO_3$ with increased d-spacing. Although the reversible H^+ insertion was demonstrated to be a diffusion-limited process, MoO_3 exhibited a specific capacity of 88 mAh g^{-1} even at a high rate up to 100 C. After that, an aqueous H^+ battery with high MoO_3 mass loading (90.48 mg cm^{-2}) and high areal specific capacity (22.4 mAh cm^{-2}) was achieved by taking advantage of fast H^+ diffusion [94]. Some metal oxides, including MnO_2 and V_2O_5, could serve as host materials for H^+ even in mild electrolytes [86]. Ionized H^+ originating from H_2O was usually inserted with metal ions (e.g., Zn^{2+}) into the lattice of metal oxides.

The defective PBA ($Cu[Fe(CN)_6]_{0.63}\cdot 3.4H_2O$) with abundant lattice water molecules could form a hydrogen-bonding network, thus enabling diffusion-free Grotthuss topochemistry in aqueous H^+ batteries (Figure 8.5a) [95]. As a result, the electrode showed cyclic voltammetry (CV) curves with nearly identical redox pair (Figure 8.5b) and excellent rate capability with a specific capacity of 49 mAh g^{-1} at 4000 C. The negligible lattice evolution during the electrochemical performance endowed the electrode with a super-long lifetime of 0.73 million cycles at 500 C. After that, another hydrated PBA ($Ni[Fe(CN)_6]_{2/3}\cdot 4H_2O$) was developed to demonstrate its excellent performance at a low temperature of −40 °C [98].

Aqueous H_2SO_4 solutions with different concentrations are the most frequently used electrolytes in aqueous H^+ batteries, which possess the merits of high ionic conductivity, stable anions, and low cost. However, the H^+ in aqueous H_2SO_4 solutions may form H_3O^+ with a larger ionic radius [99]. Moreover, the corrosive nature of strong acid electrolytes may lead to the dissolution of electrode materials and the deterioration of electrochemical performance. Recently, a concentrated H_3PO_4 (9.5 M) electrolyte was exploited to improve the cycling performance of H^+ batteries [100]. Moreover, the melting point of −85 °C enabled the batteries with capacity retention of 55% at −78 °C compared to that at room temperature. Inspired by the wide ESW of organic solvent, the same group devised a novel nonaqueous electrolyte based on anhydrous H_3PO_4 and acetonitrile, enabling longer cycling stability of various proton electrodes [101].

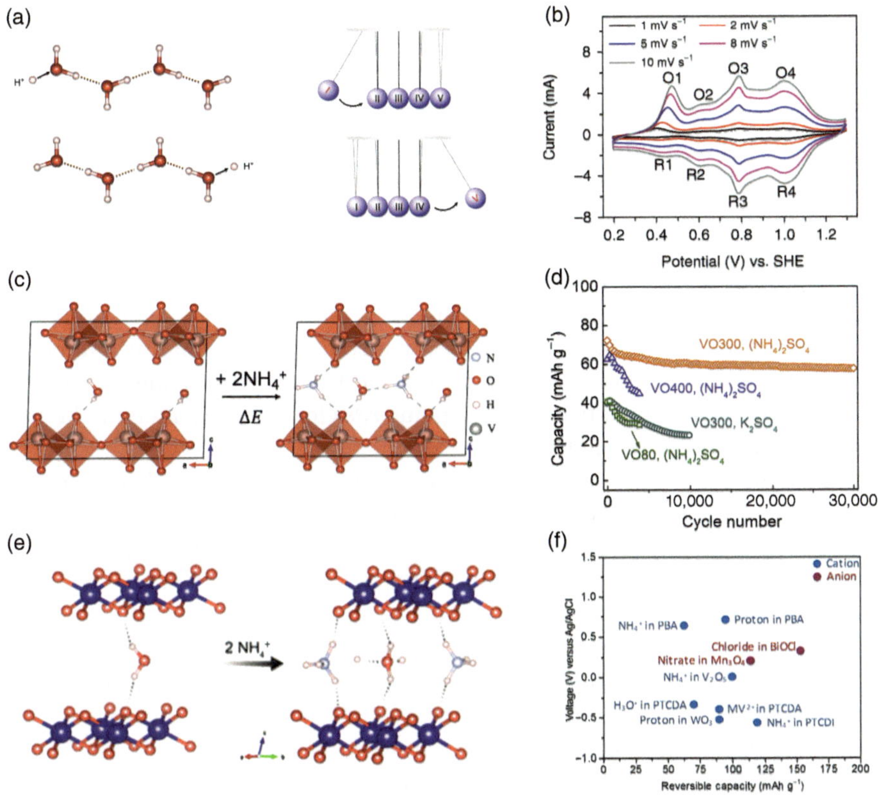

Figure 8.5 (a) Scheme of the Grotthuss proton transportation. (b) CV curves of $Cu[Fe(CN)_6]_{0.63}{\cdot}3.4H_2O$ at various scan rates. Source: Reproduced with permission [95]. Copyright 2019, Springer Nature. (c) Initial structure (left) and (right) lowest-energy configuration found for $V_2O_5{\cdot}0.5H_2O$ hosting ammonium. (d) Comparison of long-term stability of V_2O_5 electrodes in $(NH_4)_2SO_4$ and K_2SO_4 electrolytes at 5 A g^{-1}. Source: Reproduced with permission [96]. Copyright 2019, Elsevier. (e) The lowest energy configuration of hydrous MnO_2 hosting NH_4^+. Source: Reproduced with permission Song et al. [97] Copyright 2019, John Wiley & Sons. (f) Comparison of various electrode materials for nonmetal ion-based ABs. Source: Reproduced with permission [5]. Copyright 2020, American Association for the Advancement of Science.

8.4.2 Aqueous Ammonium-Ion (NH_4^+) Batteries

NH_4^+, with a relatively small hydrated ionic size (3.31 Å), less-corrosive nature, and high-dissociation property, is a promising alternative for H^+ as charge carriers [102]. In 2011, the investigation of NH_4^+ storage properties of NiHCF and CuHCF electrodes opened the avenue for developing aqueous NH_4^+ batteries [103]. Among various ions (Li^+, Na^+, K^+, and NH_4^+), the insertion of NH_4^+ in the electrode demonstrated the highest redox potential. After that, the first full cell based on NH_4^+ was assembled by coupling a Prussian white analog ($(NH_4)_{1.47}Ni[Fe(CN)_6]_{0.88}$) with a PTCDI anode, which could provide an average operation voltage of 1.0 V [104].

The bilayered V_2O_5 with V=O unit would form a strong hydrogen bond with the four terminated hydrogen atoms of NH_4^+ [96]. Therefore, the migration of NH_4^+ in V_2O_5 was accompanied by the breaking of original hydrogen bond and the formation of new hydrogen bond with another oxygen atom, which was different from the K^+ ions with ionic interaction with the host (Figure 8.5c). Attributed to the pronounced pseudocapacitive phenomenon, the V_2O_5 electrode exhibited faster rate capability and a longer lifetime of 30,000 cycles (Figure 8.5d). Several full cells based on other V-based materials were reported after that. For example, flexible fiber-shaped aqueous NH_4^+ consisting of $NH_4V_4O_{10}$ cathode and PANI anode was demonstrated to power the pressure sensor in the continuous test [105]. Besides V_2O_5, the formation of hydrogen bonds between NH_4^+ and host materials was also observed in other metal oxides [97]. In an electrodeposited MnO_x electrode, O atoms from MnO_2 and adjacent lattice water could both act as binding sites for NH_4^+, facilitating the fast electrode kinetics (Figure 8.5e).

In addition to inorganic materials, organic compounds with carbonyl groups could also serve as host materials for aqueous NH_4^+ batteries. For example, PTCDI, the reversible electrochemical enolization reaction of which could form an O-NH_4^+ structure and store two NH_4^+ per formula [104]. Furthermore, PNTCDA showed a similar NH_4^+ storage mechanism to PTCDI, which could provide a specific capacity of 107.7 mAh g^{-1} even at 10 A g^{-1} [106].

Several ammonium salts were adopted as the electrolytes for aqueous NH_4^+ batteries, including $(NH_4)_2SO_4$, NH_4Cl, NH_4Ac, and NH_4NO_3. Less attention has been paid to developing highly concentrated electrolytes and WISEs. In 2018, the WISE containing 25 M NH_4Ac was reported to expand the ESW to 2.95 V. However, the cycling performance of aqueous NH_4^+ batteries based on WISE and titanic acid electrodes needed further improvement [106]. More recently, a highly concentrated electrolyte based on 5.8 M aqueous $(NH_4)_2SO_4$ was demonstrated to enable batteries to provide better electrochemical performance [107].

8.4.3 Aqueous Anion-Based Batteries (ABs)

Some anions, such as hydroxide ions, halide anions, and polyatomic anions, are reported as charge carriers for aqueous batteries (Figure 8.5f) [108]. Rechargeable alkaline batteries typically contain metal oxide/hydroxide cathode, metal/metal oxide anode, and OH^- charge carriers. According to the reaction mechanism, the anode materials for alkaline can be categorized into dissolution–deposition-type anodes including Zn and ZnO and conversion-type anodes including Fe-based, Bi-based, and Co-based materials [52]. Furthermore, aqueous Zn-air batteries operated in alkaline electrolytes with extremely high theoretical energy density (1084 Wh kg^{-1}) are promising candidates for energy storage systems.

The storage of halide ions is usually based on a conversion mechanism. For example, BiOCl anode could be reversibly transformed into metal Bi and amorphous Bi_2O_3 during the charging process in a 1 M NaCl electrolyte [109]. At the same time, the Ag cathode would be oxidized to AgCl and release electrons.

However, the irreversibility of this reaction would lead to low coulombic efficiency at the initial cycle.

The insertion of polyatomic anions into the electrode is difficult in aqueous batteries, as the redox potential for insertion is usually beyond the anodic limit of aqueous solutions [86]. Recently, nitrate (NO_3^-) ions were found to be reversibly inserted into Mn_3O_4 by creating amorphous nano-regions [110]. Graphite, a commonly used electrode in dual-ion batteries, has been demonstrated to host various anions in WISE with broadened ESW, including $[ZnCl_x]^{2-x}$, $[MgCl_3]^-$, and $[LiCl_2]^-$ [111]. For example, the reversible insertion of Mg-Cl superhalides in graphite yielded a specific capacity of 150 mAh g^{-1} at 1.1 V vs. Ag/AgCl [112]. Some organic compounds, like coronene, can also store anions via an insertion-type mechanism [113]. In the WISE of 20 M LiTFSI, the $TFSI^-$ would form covalent bonding with the edges of coronene, thus minimizing structural contractions in the lattice after anion insertion. On the contrary, polytriphenylamine (PTPAn) stored $TFSI^-$ anions via adsorption on N^+ sites, which followed a donor–acceptor principle [114].

8.5 Challenges and Solutions of Aqueous Batteries

8.5.1 Water Decomposition

The thermodynamic potential of water decomposition comprising oxygen evolution reaction (OER) and HER ($2H_2O \rightarrow 2H_2 + O_2$) is calculated to be 1.23 V under 25 °C and normal atmospheric pressure according to the following equation:

$$\Delta G = -nF\Delta E_0 \tag{8.2}$$

where ΔG represents the change of Gibbs free energy, n is the transferred electron quantity, and F refers to the Faraday constant.

The low ESW of aqueous electrolytes confines the choice of electrode materials, suppresses the output voltages, and lowers the energy densities of aqueous batteries [5]. Moreover, the generation of hydrogen and oxygen on the surface of the electrode may lead to the low coulombic efficiency of batteries by partial consumption of electrons. Great efforts have been devoted to suppressing water decomposition and enlarging the ESW of aqueous electrolytes, including pH management, additive introduction, and novel electrolyte design. The pH of the electrolyte has an impact on HER and OER. A pH decrease in the electrolyte may negatively shift the potential of OER and HER, while an increase in pH will shift them positively [9]. Therefore, it is possible to manipulate the stability of aqueous batteries via a pH-management strategy. For example, the stability of olivine $FePO_4$ anode in LIBs could be improved by carefully adjusting the pH values of the electrolyte [115]. More recently, a decoupling Zn||MnO_2 battery was developed using the alkaline electrolyte for the anode and the acid electrolyte for the cathode [116]. A neutral electrolyte isolated these two kinds of electrolytes by bipolar membranes. The pH-decoupling design took advantage of the low redox potential of $Zn/Zn(OH)_4^{2-}$ and the high redox potential of Mn^{2+}/MnO_2 [60]. Based on this novel configuration, the aqueous Zn||MnO_2 battery provided a high OPV of 2.83 V, corresponding to an energy density of 90 Wh kg^{-1} (based on total mass). Besides Zn||MnO_2 battery, decoupling batteries based on Zn anode in alkaline electrolytes and Li-based cathode ($LiMn_2O_4$) in mild electrolytes also demonstrated a high energy density [117].

Introducing proper additives in the electrolyte can also expand the ESW of aqueous electrolytes. For example, introducing SDS into aqueous electrolyte could effectively suppress the HER and OER by forming an adsorbed layer on the surface of the electrode [66]. Other additives such as tris(trimethylsilyl) borate disodium (TMSB) and propane-1,3-disulfonate (PDSS) were also reported to have similar effects on expanding the ESW of electrolytes [5].

In terms of designing novel electrolytes, the development of WISE represents a new paradigm for high-performance aqueous batteries. The formation of SEI layer on the electrode and the unique solvation sheath with less free water contribute to the wide ESW up to 4.0 V [5]. The hybridization of aqueous electrolytes with other organic solvents like dimethyl carbonate (DMC) and propylene carbonate (PC) endows the electrolytes with both broad ESW and excellent safety [118]. Some ILs like 1-ethyl-3-methylimidazolium bis(trifluoromethane sulfonyl)imide (EMIM-TFSI) are emerging as attractive alternatives to organic solvents to form hybrid-solvent electrolytes. Introducing an aqueous solution can address the poor ionic conductivities in ILs while maintaining the wide ESW.

8.5.2 Dendrite Growth

The early research of dendrite growth focuses on Zn anode, as most other metal anode exhibits poor reversibility in aqueous media because of the negative electrochemical plating/stripping potentials or surface passivation [32]. The morphology and composition of Zn anode vary on the types of electrolyte. ZnO with ramification texture will form in the alkaline-based electrolytes, while Zn dendrites in mild electrolytes usually exhibit hexagonal morphology [70]. Moreover, the areal capacities and current densities impact the formation of dendrites by accumulation effect and surface polarization, respectively. The larger current densities with higher areal capacities will exacerbate the dendrite situation.

The strategies for solving Zn dendrite issues can be categorized into building an artificial interface layer, electrolyte modification, and electrode design [70]. The artificial interface layers are usually composed of inorganic and organic layers with robust mechanical properties or the ability to facilitate homogeneous current distribution. Generally, the artificial interface layers with good hydrophilicity, strong adhesion, mechanical robustness, and uniformly porous structures have been considered to address Zn dendrite problems well.

As for electrolyte modifications, using proper additives or developing novel electrolytes may help to control the electrodeposition process of Zn^{2+}. The inorganic additives enable a dendrite-free anode via an electrostatic shield mechanism, while the organic additives realize even Zn deposition by adsorbing on the anode surface to improve interface wettability or mediate the crystallographic orientation [70]. Novel electrolytes, including solid/quasisolid-state electrolytes, WISEs, or organic electrolytes, are becoming more important in preventing Zn dendrite growth [50]. For example, quasisolid gel electrolytes with stable electrolyte/electrode interface and minimum side reactions would help to improve the reversibility of Zn anode [119].

Since the successful application of 3D Zn sponges in alkaline ZBs, the electrode design with different Zn topography has drawn tremendous attention [45]. The Zn anode with a large surface area can provide more active sites, homogenize localized electric field, and facilitate uniform Zn^{2+} diffusion. Besides the Zn sponges, some current collectors, including carbon-based, copper-based, and metal-organic framework (MOF)-based current collectors, will enable Zn anode with 3D structures and a large surface area [50]

8.5.3 Side Reactions

Besides the inevitable thermodynamics-driven water splitting, the self-corrosion at the electrode/electrolyte interface may also cause the HER [120]. Therefore, self-corrosion is accompanied by electrolyte consumption and decreased coulombic efficiency. Some current collectors, such as Al and Cu, would also react with strong acidic and alkaline electrolytes. However, little work focuses on minimizing the current collector corrosion. Many strategies are devoted to restraining electrode corrosion, especially the Zn anode. Building an artificial interface layer on the surface of Zn is one easy way to manipulate the electrode/electrolyte interface. Some inorganic (e.g. $PbSO_4$) or organic (e.g. PEG-200) electrolyte additives are also considered corrosion inhibitors [121].

Another side reaction related to aqueous batteries is surface passivation, which is owing to the formation of insulating metal oxides on the surface of anode. Surface passivation is one of the main reasons for capacity decay as it will block the ion and charge transfer. Besides, the dissolution of electrode material may also deteriorate the electrochemical performance of aqueous batteries [9]. Surface protection via coating or surface stabilizing should be exploited to address such issues.

8.6 Conclusions and Future Perspectives

In summary, the developments of aqueous batteries based on different charge carriers are reviewed in this chapter. Aqueous batteries have achieved great success in high energy density, fast rate capability, and low-term stability. Among various energy-storage systems, aqueous metal-ion-based batteries have great potential to meet future energy-density requirements, while batteries based on nonmetal-ion charge carriers are more suitable for grid-scale energy storage.

Despite the significant advances, there are some potential challenges for practical applications. First, further understanding of the fundamental mechanisms is required for designing better aqueous batteries. Owing to the complicated side reactions in aqueous electrolytes, the electrochemical mechanisms in certain materials remain controversial. Combining *in situ* characterizations with theoretical simulations will help to address these issues. Second, the energy density should be further enhanced to make aqueous batteries competitive with commercial nonaqueous LIBs. Therefore, the ESW of aqueous electrolytes needs further expansion to ensure higher output voltages. Moreover, electrode materials with novel chemistry are in urgent need of storing more electrons per molar mass. Last, tremendous efforts should be devoted to converting experimental investigations into industrial production. For example, most of the investigations on the reported batteries are based on coin cells; however, many problems, such as Zn dendrite and self-discharge, will be exacerbated when the battery is scaled up.

References

1 Larcher, D. and Tarascon, J.-M. (2015). Towards greener and more sustainable batteries for electrical energy storage. *Nature Chemistry* 7 (1): 19–29.

2 Etacheri, V., Marom, R., Elazari, R. et al. (2011). Challenges in the development of advanced Li-ion batteries: a review. *Energy & Environmental Science* 4 (9): 3243–3262.

3 Li, H., Wang, Z., Chen, L., and Huang, X. (2009). Research on advanced materials for Li-ion batteries. *Advanced Materials* 21 (45): 4593–4607.

4 von Wald Cresce, A. and Xu, K. (2021). Aqueous lithium-ion batteries. *Carbon Energy* 3 (5): 721–751.

5 Chao, D., Zhou, W., Xie, F. et al. (2020). Roadmap for advanced aqueous batteries: from design of materials to applications. *Science Advances* 6 (21): eaba4098.

6 Liu, Z., Huang, Y., Huang, Y. et al. (2020). Voltage issue of aqueous rechargeable metal-ion batteries. *Chemical Society Reviews* 49 (1): 180–232.

7 Li, W., Dahn, J.R., and Wainwright, D.S. (1994). Rechargeable lithium batteries with aqueous electrolytes. *Science* 264 (5162): 1115–1118.

8 Bin, D., Wen, Y., Wang, Y., and Xia, Y. (2018). The development in aqueous lithium-ion batteries. *Journal of Energy Chemistry* 27 (6): 1521–1535.

9 Kim, H., Hong, J., Park, K.-Y. et al. (2014). Aqueous rechargeable Li and Na ion batteries. *Chemical Reviews* 114 (23): 11788–11827.

10 Tian, L. and Yuan, A. (2009). Electrochemical performance of nanostructured spinel $LiMn_2O_4$ in different aqueous electrolytes. *Journal of Power Sources* 192 (2): 693–697.

11 Qu, Q., Fu, L., Zhan, X. et al. (2011). Porous $LiMn_2O_4$ as cathode material with high power and excellent cycling for aqueous rechargeable lithium batteries. *Energy & Environmental Science* 4 (10): 3985–3990.

12 Wang, H., Huang, K., Zeng, Y. et al. (2007). Electrochemical properties of TiP_2O_7 and $LiTi_2(PO_4)_3$ as anode material for lithium ion battery with aqueous solution electrolyte. *Electrochimica Acta* 52 (9): 3280–3285.

13 Han, C., Zhu, J., Zhi, C., and Li, H. (2020). The rise of aqueous rechargeable batteries with organic electrode materials. *Journal of Materials Chemistry A* 8 (31): 15479–15512.

14 Suo, L., Borodin, O., Gao, T. et al. (2015). "Water-in-Salt" electrolyte enables high-voltage aqueous lithium-ion chemistries. *Science* 350 (6263): 938–943.

15 Yang, C., Chen, J., Qing, T. et al. (2017). 4.0 V Aqueous Li-ion batteries. *Joule* 1 (1): 122–132.

16 Yang, C., Chen, J., Ji, X. et al. (2019). Aqueous Li-ion battery enabled by halogen conversion–intercalation chemistry in graphite. *Nature* 569 (7755): 245–250.

17 Suo, L., Borodin, O., Sun, W. et al. (2016). Advanced high-voltage aqueous lithium-ion battery enabled by "Water-in-Bisalt" electrolyte. *Angewandte Chemie International Edition* 55 (25): 7136–7141.

18 Yamada, Y., Usui, K., Sodeyama, K. et al. (2016). Hydrate-melt electrolytes for high-energy-density aqueous batteries. *Nature Energy* 1 (10): 1–9.

19 Goikolea, E., Palomares, V., Wang, S. et al. (2020). Na-ion batteries-approaching old and new challenges. *Advanced Energy Materials* 10 (44): 2002055.

20 Palomares, V., Serras, P., Villaluenga, I. et al. (2012). Na-ion batteries, recent advances and present challenges to become low cost energy storage systems. *Energy & Environmental Science* 5 (3): 5884–5901.

21 Wang, Y., Liu, D., Sun, M., and Liu, J. (2021). Recent progress in electrode materials for aqueous sodium and potassium ion batteries. *Materials Chemistry Frontiers* 5: 7384–7402.

22 Kim, D.J., Ponraj, R., Kannan, A.G. et al. (2013). Diffusion behavior of sodium ions in $Na_{0.44}MnO_2$ in aqueous and non-aqueous electrolytes. *Journal of Power Sources* 244: 758–763.

23 Wang, Y., Mu, L., Liu, J. et al. (2015). A novel high capacity positive electrode material with tunnel-type structure for aqueous sodium-ion batteries. *Advanced Energy Materials* 5 (22): 1501005.

24 Bin, D., Wang, F., Tamirat, A.G. et al. (2018). Progress in aqueous rechargeable sodium-ion batteries. *Advanced Energy Materials* 8 (17): 1703008.

25 Park, S.I., Gocheva, I., Okada, S., and Yamaki, J.-I. (2011). Electrochemical properties of $NaTi_2(PO_4)_3$ anode for rechargeable aqueous sodium-ion batteries. *Journal of the Electrochemical Society* 158 (10): A1067.

26 Dong, X., Chen, L., Liu, J. et al. (2016). Environmentally-friendly aqueous Li (or Na)-ion battery with fast electrode kinetics and super-long life. *Science Advances* 2 (1): e1501038.

27 Deng, W., Shen, Y., Qian, J., and Yang, H. (2015). A polyimide anode with high capacity and superior cyclability for aqueous Na-ion batteries. *Chemical Communications* 51 (24): 5097–5099.

28 Kim, D.J., Jung, Y.H., Bharathi, K.K. et al. (2014). An aqueous sodium ion hybrid battery incorporating an organic compound and a Prussian blue derivative. *Advanced Energy Materials* 4 (12): 1400133.

29 Suo, L., Borodin, O., Wang, Y. et al. (2017). "Water-in-salt" electrolyte makes aqueous sodium-ion battery safe, green, and long-lasting. *Advanced Energy Materials* 7 (21): 1701189.

30 Lee, M.H., Kim, S.J., Chang, D. et al. (2019). Toward a low-cost high-voltage sodium aqueous rechargeable battery. *Materials Today* 29: 26–36.

31 Jin, T., Ji, X., Wang, P.F. et al. (2021). High-energy aqueous sodium-ion batteries. *Angewandte Chemie International Edition* 133 (21): 12050–12055.
32 Huang, J., Guo, Z., Ma, Y. et al. (2019). Recent progress of rechargeable batteries using mild aqueous electrolytes. *Small Methods* 3 (1): 1800272.
33 Hosaka, T., Kubota, K., Hameed, A.S., and Komaba, S. (2020). Research development on K-ion batteries. *Chemical Reviews* 120 (14): 6358–6466.
34 Wessells, C.D., Huggins, R.A., and Cui, Y. (2011). Copper hexacyanoferrate battery electrodes with long cycle life and high power. *Nature Communications* 2 (1): 550.
35 Su, D., McDonagh, A., Qiao, S.-Z., and Wang, G. (2017). High-capacity aqueous potassium-ion batteries for large-scale energy storage. *Advanced Materials* 29 (1): 1604007.
36 Schuppert, N.D., Mukherjee, S., Bates, A.M. et al. (2016). Ex-situ X-ray diffraction analysis of electrode strain at TiO_2 atomic layer deposition/α-MoO_3 interface in a novel aqueous potassium ion battery. *Journal of Power Sources* 316: 160–169.
37 Leonard, D.P., Wei, Z., Chen, G. et al. (2018). Water-in-salt electrolyte for potassium-ion batteries. *ACS Energy Letters* 3 (2): 373–374.
38 Wang, M., Wang, H., Zhang, H., and Li, X. (2020). Aqueous K-ion battery incorporating environment-friendly organic compound and Berlin green. *Journal of Energy Chemistry* 48: 14–20.
39 Jiang, L., Lu, Y., Zhao, C. et al. (2019). Building aqueous K-ion batteries for energy storage. *Nature Energy* 4 (6): 495–503.
40 Konarov, A., Voronina, N., Jo, J.H. et al. (2018). Present and future perspective on electrode materials for rechargeable zinc-ion batteries. *ACS Energy Letters* 3 (10): 2620–2640.
41 Yuan, D., Zhao, J., Manalastas, W. et al. (2020). Emerging rechargeable aqueous aluminum ion battery: status, challenges, and outlooks. *Nano Materials Science* 2 (3): 248–263.
42 Blanc, L.E., Kundu, D., and Nazar, L.F. (2020). Scientific challenges for the implementation of Zn-ion batteries. *Joule* 4 (4): 771–799.
43 Shang, W., Yu, W., Liu, Y. et al. (2020). Rechargeable alkaline zinc batteries: progress and challenges. *Energy Storage Materials* 31: 44–57.
44 Wang, N., Wan, H., Duan, J. et al. (2021). A review of zinc-based battery from alkaline to acid. *Materials Today Advances* 11: 100149.
45 Parker, J.F., Chervin, C.N., Pala, I.R. et al. (2017). Rechargeable nickel-3D zinc batteries: an energy-dense, safer alternative to lithium-ion. *Science* 356 (6336): 415–418.
46 Yadav, G.G., Gallaway, J.W., Turney, D.E. et al. (2017). Regenerable Cu-intercalated MnO_2 layered cathode for highly cyclable energy dense batteries. *Nature Communications* 8 (1): 14424.
47 Gong, M., Li, Y., Zhang, H. et al. (2014). Ultrafast high-capacity NiZn battery with NiAlCo-layered double hydroxide. *Energy & Environmental Science* 7 (6): 2025–2032.

48 Zhang, Y., Wu, Y., You, W. et al. (2020). Deeply rechargeable and hydrogen-evolution-suppressing zinc anode in alkaline aqueous electrolyte. *Nano Letters* 20 (6): 4700–4707.

49 Ghavami, R.K. and Rafiei, Z. (2006). Performance improvements of alkaline batteries by studying the effects of different kinds of surfactant and different derivatives of benzene on the electrochemical properties of electrolytic zinc. *Journal of Power Sources* 162 (2): 893–899.

50 Hao, J., Li, X., Zeng, X. et al. (2020). Deeply understanding the Zn anode behaviour and corresponding improvement strategies in different aqueous Zn-based batteries. *Energy & Environmental Science* 13 (11): 3917–3949.

51 Li, H., Ma, L., Han, C. et al. (2019). Advanced rechargeable zinc-based batteries: recent progress and future perspectives. *Nano Energy* 62: 550–587.

52 Huang, M., Li, M., Niu, C. et al. (2019). Recent advances in rational electrode designs for high-performance alkaline rechargeable batteries. *Advanced Functional Materials* 29 (11): 1807847.

53 Hertzberg, B.J., Huang, A., Hsieh, A. et al. (2016). Effect of multiple cation electrolyte mixtures on rechargeable Zn–MnO_2 alkaline battery. *Chemistry of Materials* 28 (13): 4536–4545.

54 Van der Ven, A., Morgan, D., Meng, Y.S., and Ceder, G. (2006). Phase stability of nickel hydroxides and oxyhydroxides. *Journal of the Electrochemical Society* 153 (2): A210.

55 Zhou, W., Zhu, D., He, J. et al. (2020). A scalable top-down strategy toward practical metrics of Ni–Zn aqueous batteries with total energy densities of 165 Wh kg^{-1} and 506 Wh L^{-1}. *Energy & Environmental Science* 13 (11): 4157–4167.

56 Shoji, T., Hishinuma, M., and Yamamoto, T. (1988). Zinc-manganese dioxide galvanic cell using zinc sulphate as electrolyte. Rechargeability of the cell. *Journal of Applied Electrochemistry* 18 (4): 521–526.

57 Ruan, P., Liang, S., Lu, B. et al. (2022). Design strategies for high-energy-density aqueous zinc batteries. *Angewandte Chemie International Edition* 61 (17): e202200598.

58 Huang, J., Wang, Z., Hou, M. et al. (2018). Polyaniline-intercalated manganese dioxide nanolayers as a high-performance cathode material for an aqueous zinc-ion battery. *Nature Communications* 9 (1): 2906.

59 Pan, H., Shao, Y., Yan, P. et al. (2016). Reversible aqueous zinc/manganese oxide energy storage from conversion reactions. *Nature Energy* 1 (5): 16039.

60 Chao, D., Zhou, W., Ye, C. et al. (2019). An electrolytic Zn–MnO_2 battery for high-voltage and scalable energy storage. *Angewandte Chemie International Edition* 58 (23): 7823–7828.

61 Wan, F. and Niu, Z. (2019). Design strategies for vanadium-based aqueous zinc-ion batteries. *Angewandte Chemie International Edition* 58 (46): 16358–16367.

62 Kundu, D., Adams, B.D., Duffort, V. et al. (2016). A high-capacity and long-life aqueous rechargeable zinc battery using a metal oxide intercalation cathode. *Nature Energy* 1 (10): 16119.

63 Song, M., Tan, H., Chao, D., and Fan, H.J. (2018). Recent advances in Zn-ion batteries. *Advanced Functional Materials* 28 (41): 1802564.

64 Zhang, L., Chen, L., Zhou, X., and Liu, Z. (2015). Towards high-voltage aqueous metal-ion batteries beyond 1.5 V: the zinc/zinc hexacyanoferrate system. *Advanced Energy Materials* 5 (2): 1400930.

65 Zhang, L., Chen, L., Zhou, X., and Liu, Z. (2015). Morphology-dependent electrochemical performance of zinc hexacyanoferrate cathode for zinc-ion battery. *Scientific Reports* 5 (1): 18263.

66 Hou, Z., Zhang, X., Li, X. et al. (2017). Surfactant widens the electrochemical window of an aqueous electrolyte for better rechargeable aqueous sodium/zinc battery. *Journal of Materials Chemistry A* 5 (2): 730–738.

67 Tie, Z. and Niu, Z. (2020). Design strategies for high-performance aqueous Zn/organic batteries. *Angewandte Chemie International Edition* 59 (48): 21293–21303.

68 Zhao, Q., Huang, W., Luo, Z. et al. (2018). High-capacity aqueous zinc batteries using sustainable quinone electrodes. *Science Advances* 4 (3): eaao1761.

69 Liu, F., Xu, R., Hu, Z. et al. (2019). Regulating lithium nucleation via CNTs modifying carbon cloth film for stable Li metal anode. *Small* 15 (5): 1803734.

70 Yang, Q., Li, Q., Liu, Z. et al. (2020). Dendrites in Zn-based batteries. *Advanced Materials* 32 (48): 2001854.

71 Ponrouch, A., Bitenc, J., Dominko, R. et al. (2019). Multivalent rechargeable batteries. *Energy Storage Materials* 20: 253–262.

72 Chen, L., Bao, J.L., Dong, X. et al. (2017). Aqueous Mg-ion battery based on polyimide anode and Prussian blue cathode. *ACS Energy Letters* 2 (5): 1115–1121.

73 Wang, F., Fan, X., Gao, T. et al. (2017). High-voltage aqueous magnesium ion batteries. *ACS Central Science* 3 (10): 1121–1128.

74 Zhu, Y., Guo, X., Lei, Y. et al. (2022). Hydrated eutectic electrolytes for high-performance Mg-ion batteries. *Energy & Environmental Science* 15 (3): 1282–1292.

75 Zhang, H., Ye, K., Zhu, K. et al. (2017). High-energy-density aqueous magnesium-ion battery based on a carbon-coated $FeVO_4$ anode and a Mg-OMS-1 cathode. *Chemistry – A European Journal* 23 (67): 17118–17126.

76 Liu, T., Cheng, X., Yu, H. et al. (2019). An overview and future perspectives of aqueous rechargeable polyvalent ion batteries. *Energy Storage Materials* 18: 68–91.

77 Li, C., Hou, C.-C., Chen, L. et al. (2021). Rechargeable Al-ion batteries. *EnergyChem* 3 (2): 100049.

78 Liu, S., Hu, J.J., Yan, N.F. et al. (2012). Aluminum storage behavior of anatase TiO_2 nanotube arrays in aqueous solution for aluminum ion batteries. *Energy & Environmental Science* 5 (12): 9743–9746.

79 Yan, C., Lv, C., Jia, B.-E. et al. (2022). Reversible Al metal anodes enabled by amorphization for aqueous aluminum batteries. *Journal of the American Chemical Society* 144 (25): 11444–11455.

80 Gheytani, S., Liang, Y., Wu, F. et al. (2017). An aqueous Ca-ion battery. *Advanced Science* 4 (12): 1700465.

81 Zhao, Q., Zachman, M.J., Al Sadat, W.I. et al. (2018). Solid electrolyte interphases for high-energy aqueous aluminum electrochemical cells. *Science Advances* 4 (11): aau8131.

82 Wu, C., Gu, S., Zhang, Q. et al. (2019). Electrochemically activated spinel manganese oxide for rechargeable aqueous aluminum battery. *Nature Communications* 10 (1): 73.

83 Yan, C., Lv, C., Wang, L. et al. (2020). Architecting a stable high-energy aqueous Al-ion battery. *Journal of the American Chemical Society* 142 (36): 15295–15304.

84 Ponrouch, A., Frontera, C., Bardé, F., and Palacín, M.R. (2016). Towards a calcium-based rechargeable battery. *Nature Materials* 15 (2): 169–172.

85 Han, C., Li, H., Li, Y. et al. (2021). Proton-assisted calcium-ion storage in aromatic organic molecular crystal with coplanar stacked structure. *Nature Communications* 12 (1): 2400.

86 Liang, G., Mo, F., Ji, X., and Zhi, C. (2021). Non-metallic charge carriers for aqueous batteries. *Nature Reviews Materials* 6 (2): 109–123.

87 Ruetschi, P. (1977). Review on the lead-acid battery science and technology. *Journal of Power Sources* 2 (1): 3–120.

88 Song, Z. and Zhou, H. (2013). Towards sustainable and versatile energy storage devices: an overview of organic electrode materials. *Energy & Environmental Science* 6 (8): 2280–2301.

89 Tomai, T., Mitani, S., Komatsu, D. et al. (2014). Metal-free aqueous redox capacitor via proton rocking-chair system in an organic-based couple. *Scientific Reports* 4 (1): 3591.

90 Liang, Y., Jing, Y., Gheytani, S. et al. (2017). Universal quinone electrodes for long cycle life aqueous rechargeable batteries. *Nature Materials* 16 (8): 841–848.

91 Emanuelsson, R., Sterby, M., Strømme, M., and Sjödin, M. (2017). An all-organic proton battery. *Journal of the American Chemical Society* 139 (13): 4828–4834.

92 Li, J., Yan, H., Xu, C. et al. (2021). Insights into host materials for aqueous proton batteries: structure, mechanism and prospect. *Nano Energy* 89: 106400.

93 Wang, X., Xie, Y., Tang, K. et al. (2018). Redox chemistry of molybdenum trioxide for ultrafast hydrogen-ion storage. *Angewandte Chemie International Edition* 57 (36): 11569–11573.

94 Su, Z., Ren, W., Guo, H. et al. (2020). Ultrahigh areal capacity hydrogen-ion batteries with MoO_3 loading over 90 mg cm^{-2}. *Advanced Functional Materials* 30 (46): 2005477.

95 Wu, X., Hong, J.J., Shin, W. et al. (2019). Diffusion-free Grotthuss topochemistry for high-rate and long-life proton batteries. *Nature Energy* 4 (2): 123–130.

96 Dong, S., Shin, W., Jiang, H. et al. (2019). Ultra-fast NH_4^+ storage: strong H bonding between NH_4^+ and bi-layered V_2O_5. *Chem* 5 (6): 1537–1551.

97 Song, Y., Pan, Q., Lv, H. et al. (2021). Ammonium-ion storage using electrodeposited manganese oxides. *Angewandte Chemie International Edition* 60 (11): 5718–5722.

98 Wu, X., Qiu, S., Xu, Y. et al. (2020). Hydrous nickel–iron Turnbull's blue as a high-rate and low-temperature proton electrode. *ACS Applied Materials & Interfaces* 12 (8): 9201–9208.

99 Donald, W.A., Leib, R.D., Demireva, M. et al. (2009). Directly relating reduction energies of gaseous $Eu(H_2O)_n{}^{3+}$, $n = 55–140$, to aqueous solution: the absolute SHE potential and real proton solvation energy. *Journal of the American Chemical Society* 131 (37): 13328–13337.

100 Jiang, H., Shin, W., Ma, L. et al. (2020). A high-rate aqueous proton battery delivering power below −78 °C via an unfrozen phosphoric acid. *Advanced Energy Materials* 10 (28): 2000968.

101 Xu, Y., Wu, X., Jiang, H. et al. (2020). A non-aqueous H_3PO_4 electrolyte enables stable cycling of proton electrodes. *Angewandte Chemie International Edition* 59 (49): 22007–22011.

102 Han, J., Varzi, A., and Passerini, S. (2022). The emergence of aqueous ammonium-ion batteries. *Angewandte Chemie International Edition* 61 (14): e202115046.

103 Wessells, C.D., Peddada, S.V., McDowell, M.T. et al. (2011). The effect of insertion species on nanostructured open framework hexacyanoferrate battery electrodes. *Journal of the Electrochemical Society* 159 (2): A98–A103.

104 Wu, X., Qi, Y., Hong, J.J. et al. (2017). Rocking-chair ammonium-ion battery: a highly reversible aqueous energy storage system. *Angewandte Chemie International Edition* 56 (42): 13026–13030.

105 Li, H., Yang, J., Cheng, J. et al. (2020). Flexible aqueous ammonium-ion full cell with high rate capability and long cycle life. *Nano Energy* 68: 104369.

106 Qiu, S., Xu, Y., Li, X. et al. (2021). Reinforced potassium and ammonium storage of the polyimide anode in acetate-based water-in-salt electrolytes. *Electrochemistry Communications* 122: 106880.

107 Han, J., Zarrabeitia, M., Mariani, A. et al. (2022). Concentrated electrolytes enabling stable aqueous ammonium-ion batteries. *Advanced Materials* 34 (32): 2201877.

108 Liu, Q., Wang, Y., Yang, X. et al. (2021). Rechargeable anion-shuttle batteries for low-cost energy storage. *Chem* 7 (8): 1993–2021.

109 Chen, F., Leong, Z.Y., and Yang, H.Y. (2017). An aqueous rechargeable chloride ion battery. *Energy Storage Materials* 7: 189–194.

110 Jiang, H., Wei, Z., Ma, L. et al. (2019). An aqueous dual-ion battery cathode of Mn_3O_4 via reversible insertion of nitrate. *Angewandte Chemie International Edition* 58: 5286–5291.

111 Kim, K.-I., Tang, L., Mirabedini, P. et al. (2022). $[LiCl_2]$ – superhalide: a new charge carrier for graphite cathode of dual-ion batteries. *Advanced Functional Materials* 32 (23): 2112709.

112 Kim, K.-I., Tang, L., Muratli, J.M. et al. (2022). A graphite||PTCDI aqueous dual-ion battery. *ChemSusChem* 15 (5): e202102394.

113 Rodríguez-Pérez, I.A., Zhang, L., Leonard, D.P., and Ji, X. (2019). Aqueous anion insertion into a hydrocarbon cathode via a water-in-salt electrolyte. *Electrochemistry Communications* 109: 106599.

114 Dong, X., Yu, H., Ma, Y. et al. (2017). All-organic rechargeable battery with reversibility supported by "Water-in-salt" electrolyte. *Chemistry – A European Journal* 23 (11): 2560–2565.

115 Wang, Y., Yang, S.-Z., You, Y. et al. (2018). High-capacity and long-cycle life aqueous rechargeable lithium-ion battery with the $FePO_4$ anode. *ACS Applied Materials & Interfaces* 10 (8): 7061–7068.

116 Zhong, C., Liu, B., Ding, J. et al. (2020). Decoupling electrolytes towards stable and high-energy rechargeable aqueous zinc–manganese dioxide batteries. *Nature Energy* 5 (6): 440–449.

117 Yuan, X., Wu, X., Zeng, X.-X. et al. (2020). A fully aqueous hybrid electrolyte rechargeable battery with high voltage and high energy density. *Advanced Energy Materials* 10 (40): 2001583.

118 Chen, S., Zhang, M., Zou, P. et al. (2022). Historical development and novel concepts on electrolytes for aqueous rechargeable batteries. *Energy & Environmental Science* 15: 1805–1839.

119 Tang, Y., Liu, C., Zhu, H. et al. (2020). Ion-confinement effect enabled by gel electrolyte for highly reversible dendrite-free zinc metal anode. *Energy Storage Materials* 27: 109–116.

120 Guan, K., Tao, L., Yang, R. et al. (2022). Anti-corrosion for reversible zinc anode via a hydrophobic interface in aqueous zinc batteries. *Advanced Energy Materials* 12 (9): 2103557.

121 Guo, S., Qin, L., Zhang, T. et al. (2021). Fundamentals and perspectives of electrolyte additives for aqueous zinc-ion batteries. *Energy Storage Materials* 34: 545–562.

9

Progress and Perspectives of Flow Batteries: Material Design and Engineering

Mengqi Zhang[1,2], Changkun Zhang[1], Xianfeng Li[1], and Guihua Yu[3]

[1] *Dalian Institute of Chemical Physics, Division of Energy Storage, Chinese Academy of Sciences, Dalian National Laboratory for Clean Energy, 457 Zhongshan Road, Dalian 116023, China*
[2] *University of Chinese Academy of Sciences, 19A Yuquan Road, Shijingshan District, Beijing 100049, China*
[3] *The University of Texas at Austin, Walker Department of Mechanical Engineering, Materials Science and Engineering Program, Austin, TX 78712, USA*

9.1 Introduction

The increasing energy demand and environmental pollution necessitate the development of renewable energy, such as solar energy and wind energy [1]. However, the inherent intermittent and fluctuating nature of renewable energy will endanger the electric grid if directly integrated [2]. Therefore, it is of vital importance to develop large-scale electrical energy storage (EES) systems to smoothen the power output of renewable electricity.

Flow battery (FB), one of the most promising large-scale EES technologies, has aroused wide attention due to its decoupled output energy (the volume of electrolyte) and power (the size of stacks), long lifetime, and high safety [3, 4]. In a typical FB, the positive and negative electrolytes are pumped into the cell and redox reactions occur on the electrodes to achieve the conversion between chemical energy and electrical energy. The anode and cathode are separated by an ion-exchange or porous membrane (Figure 9.1).

The first modern FB system was proposed by Thaller of the National Aeronautics and Space Administration (NASA) of the United States in 1974. NASA mainly investigated the Fe/Cr system. At the same time in Japan, the National Institute of Advanced Industrial Science and Technology was conducting basic research and developing the Fe/Cr system. The Zn/Br system was first demonstrated in 1977 [5]. Subsequently in 1985, Professor Skyllas-Kazacos from the University of New South Wales University (UNSW) conducted considerable research on the vanadium flow battery (VFB), developed the current VFB in sulfuric acid chemistry, and promoted the solubility and stability of VO^{2+} largely [6]. Currently, VFBs have been demonstrated at kW and multi-MW levels for different applications. Many VFB companies have been established including the SEI, Australian Vanadium Ltd., Rongke Power Corp., CellCube, VRB Energy, Prudent Energy, Invinity Energy

Electrochemical Energy Storage Devices: Non-Conventional Technologies and Materials,
First Edition. Edited by Yongbing Tang and Luojiang Zhang.

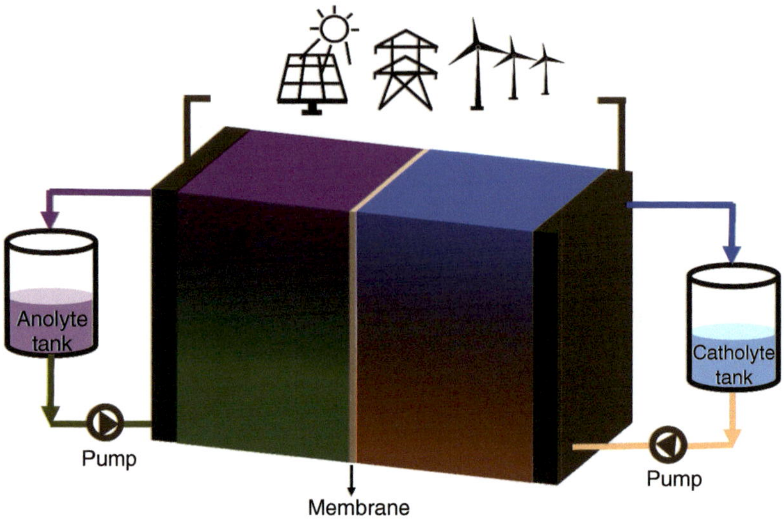

Figure 9.1 Schematic illustration of a typical FB.

Systems, Singapore's VFlowTech, and Largo Clean Energy. Moreover, different types of FBs including Zn/Fe, all-iron, organic, and new systems have been proposed and rapid achievement has been achieved for FB technologies over the past several decades.

Generally, the properties of FB can be evaluated by energy density, power density, coulombic efficiency (CE), voltage efficiency (VE), and energy efficiency (EE). The energy density can be calculated as below:

$$\text{Energy density} = \frac{nCFV}{\mu_V} \tag{9.1}$$

where n represents the number of electrons transferred in the reaction, C represents the lower concentration of two electrolytes, F is Faraday's constant, V is the working voltage of FB, and μ_V is the volume factor ($\mu_V = 1 +$ [lower electrolyte concentration]/[higher electrolyte concentration]). Therefore, to increase the energy density, the solubility of redox-active species (RAS) should be high. The redox potentials should be high for positive active materials and low for negative active materials to get a high operating voltage.

The power density mainly depends on the cell voltage, kinetics of active materials, and conductivity of membranes and electrolytes. CE is the ratio between the discharge capacity and charge capacity. Generally, the CE is below 100% due to the crossover of redox species and side reactions, such as hydrogen evolution reaction. VE is the ratio between discharge voltage and charge voltage at constant current, which is relative to the polarization including activation polarization, ohmic polarization, and concentration polarization. EE equals CE times VE, which reflects the energy utilization. In large-scale EES systems, EE is a key parameter to measure the applicability and economy of energy storage technology.

In this chapter, we will summarize the recent progress on the key components of FBs, including electrolytes, membranes, electrodes, and bipolar plates (BPs). Following this, we will introduce some novel FBs. Finally, we will present the main challenges and perspectives on FBs.

9.2 Research Progress on the Electrolyte

9.2.1 VFB

The VFB is regarded as the most promising large-scale EES system, which has been extensively studied [7]. V^{2+}/V^{3+} and VO^{2+}/VO_2^+ are utilized as negative and positive redox couples, respectively. The standard open-circuit voltage (OCV) of VFB is 1.26 V [8]. The cross contamination of electrolytes can be significantly diminished, because both electrolytes use the same active element. The electrolytes can be easily regenerated by remixing the anolyte and catholyte. However, the main challenges of VFB are the low energy density and high cost. The concentration of electrolyte is typically below 2 M, and the energy density of VFB is ~25 Wh l^{-1}.

The solubility and stability of vanadium ions are significantly influenced by temperature and supporting electrolyte. Sulfuric acid is commonly used as supporting electrolyte. When the vanadium and H_2SO_4 concentration are 2.0 and 5.0 M, respectively, V^{2+}, V^{3+}, and VO^{2+} tend to participate at temperatures below 10 °C, while V^{5+} is more stable at low temperatures but participates at temperatures above 40 °C [9]. The concentration of H_2SO_4 also influences the solubility; with the concentration increase of sulfuric acid, the solubility of V^{2+}, V^{3+}, and VO^{2+} decreases due to common ion effect, while the solubility of VO_2^+ increases [10].

Aiming to improve the solubility and stability of vanadium electrolytes, some studies focus on the optimization of supporting electrolytes. It is found that the solubility and stability of vanadium ions are significantly improved in mixed supporting electrolytes (CH_3SO_3H and H_2SO_4) [11]. However, the high viscosity of CH_3SO_3H and different complexations may lead to high solution resistance and slow electrochemical kinetics of the VO^{2+}/VO_2^+ couple. In addition, the mixed acids (HCl and H_2SO_4) can increase the vanadium ion concentration to 2.5 M and broaden the operating temperature range from 10–40 °C to −5–50 °C [12]. VO_2^+ shows enhanced stability in the mixed electrolytes at elevated temperatures due to the formation of stable $VO_2Cl(H_2O)_2$. Several companies have adopted this strategy to increase the energy density of VFBs [13]. Some stabilizing additives are also introduced to improve the solubility and stability of vanadium electrolytes [14, 15]. The additives containing K^+, phosphate, and polyphosphate ions are inappropriate as they can react with VO_2^+, while polyacrylic acid and its mixture with CH_3SO_3H can stabilize all four vanadium ions [16].

9.2.2 Zinc–Bromine FB

Zinc–bromine flow battery (ZBFB) is a promising technology for distributed energy storage due to its high theoretical energy density (440 Wh kg^{-1}) and low cost. In a typical ZBFB, during the charge process, the Zn^{2+} ions are reduced to Zn metal and deposited on the surface of the negative electrode, while Br^- ions are oxidized to Br_2. The reactions reverse during the discharge process.

The formation of Zn dendrites and the high corrosion and diffusivity of bromine limit its industrialization and commercialization. Considerable efforts have been devoted to addressing these problems and improving the overall performance of ZBFB.

9.2.2.1 Suppress the Formation and Growth of Zn Dendrites

During the charging process, the severe concentration polarization of Zn^{2+} leads to uneven zinc deposition. The inhomogeneity of electrode surface is prone to form Zn dendrites, thus causing a short circuit of battery.

The Zn^{2+} concentration prominently influences the zinc stripping/plating processes. When the concentration was above 0.4 M, the dense planar Zn was formed due to instantaneous nucleation, while below 0.3 M, the mossy Zn was obtained resulting from progressive nucleation [17].

The organic and inorganic additives can also be introduced into electrolytes to regulate the deposition morphology of zinc. Organic molecules, such as polyethylenimine (PEI), ammonium compounds, and glucose, can absorb on zinc nucleus and regulate the current distribution or form an electrostatic shield, resulting in more uniform Zn deposition [18–20]. In addition, some organic additives, such as ethylene glycol (EG) and ethanol, can coordinate with Zn^{2+} and influence the solvation structure of Zn^{2+} [21, 22]. Inorganic additives, such as Cr^{3+} and Pd^{2+}, influence the deposition kinetics and growth mechanism of Zn [23, 24]. The surfactant can suppress the zinc dendrites by preventing uneven dissolution of Zn during discharge process [25].

9.2.2.2 Suppress the Self-Discharge

The generated bromine during charging can easily permeate into the negative side and react with Zn, leading to serious self-discharge. Therefore, the bromine sequestering/complexing agents have been introduced into the electrolyte to suppress bromine crossover [26].

Bromine complexing agents, such as *N*-ethyl-*N*-methyl pyrrolidinium bromide (MEP) and *N*-methyl-*N*-ethyl morpholinium bromide (MEM), tend to combine with polybromide anions to form complex (Br_nQ). The complex can not only decrease the saturated vapor pressure of bromine but also form the bromine-rich oil phase with higher density, which is easily separated from the aqueous solution, thus inhibiting the bromine diffusion [27, 28]. However, the inhomogeneous distribution of catholyte compromises the battery performance. Li et al. reported a new complexing agent (3-chloro-2-hydroxypropyltrimethyl ammonium chloride, CHA) [29]. CHA complexed with polybromide anions (Br_3^-, Br_5^-) and formed homogeneous catholyte, due to the hydroxyl group in CHA, which effectively prolonged the battery lifetime and improved CE (Figure 9.2). Some other functional additives can also be introduced to improve the properties of electrolytes. For

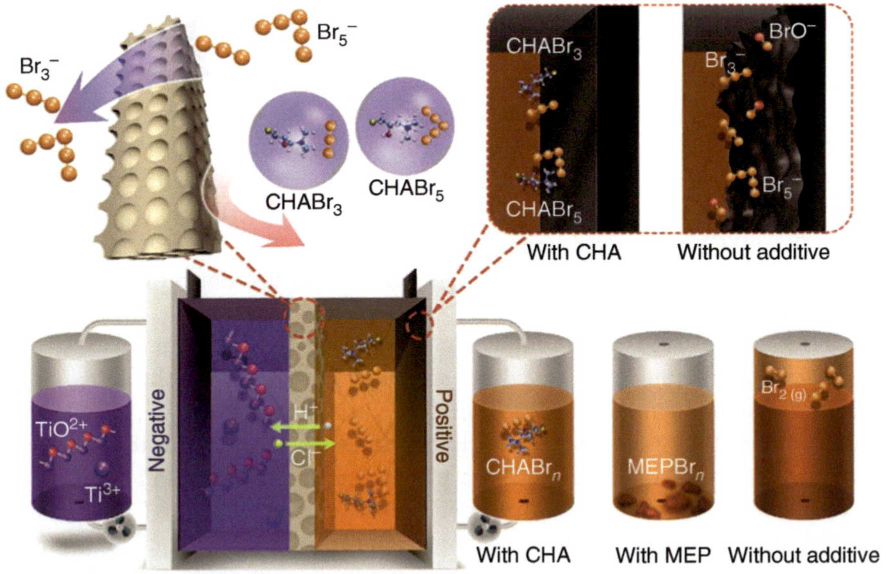

Figure 9.2 The working principle of the titanium-bromine FB with nearly homogeneous phase positive electrolyte (with CHA as an additive, lower-right corner) and low corrosivity (top-right corner). Source: Reproduced with permission Li et al. [29]. Copyright 2020, Wiley-VCH.

example, methanesulfonic acid can increase the electrolyte conductivity, inhibit zinc dendrite, and improve the kinetics of Zn^{2+}/Zn and Br_2/Br^- [30].

9.2.3 Zinc–Iron FB

Zinc–iron FB is also a promising technology for distributed electrochemical energy storage due to its low cost, high voltage, and environmental friendliness. In alkaline zinc–iron FB, $Zn(OH)_4{}^{2-}/Zn$ and $Fe(CN)_6{}^{4-}/Fe(CN)_6{}^{3-}$ are the negative and positive active redox couples, respectively. However, the zinc dendrites and low solubility of $Fe(CN)_6{}^{4-}/Fe(CN)_6{}^{3-}$ ($0.4\ mol\ l^{-1}$) restrict its practical application.

The solubility of $K_4Fe(CN)_6$ in KOH solution is limited due to the "common ion effect." When employing $Na_4Fe(CN)_6$ and KOH as positive RAS and supporting electrolyte, respectively, the solubility of positive electrolyte can reach 1 M in 3 M KOH [31]. Recently, the "diverse ion effect" was proposed to improve solubility by mixing $K_4Fe(CN)_6$ with NaOH, and the maximum solubility of $Fe(CN)_6{}^{4-}$ can reach 1.46 M in 0.5 M NaOH [32]. Density functional theory (DFT) revealed that the absorption energy of $K_4Fe(CN)_6$ in NaOH solution was higher than that of $K_4Fe(CN)_6$ in KOH solution. The strong interaction between $K_4Fe(CN)_6$ and NaOH reduced the lattice energy of $K_4Fe(CN)_6$, resulting in high solubility. What is more, the diverse ion effect also improved the redox potential by 14.16 mV. The FB with 1.46 M $Fe(CN)_6{}^{4-}$ delivered a high-capacity retention of 98.51% at $100\ mA\ cm^{-2}$. The FB displayed high energy density ($73.64\ Wh\ l^{-1}$) and power density ($656.81\ mW\ cm^{-2}$).

In addition, replacing the cation ions (Na^+, K^+) with NH_4^+, the $(NH_4)_3[Fe(CN)_6]$/$(NH_4)_4[Fe(CN)_6]$, exhibited high solubility of 1.91 M/1.60 M in water, mainly due to the strong hydrogen bonding between the hydrophilic NH_4^+ and water molecules [33]. A symmetry-breaking strategy was recently proposed by replacing parts of cyanide ligands with water-soluble bipyridine ligands 2,2′-bipyridine-4,4′-dicarboxylic acid (H_2Dcbpy) [34]. The asymmetric complexes $Na_4[Fe^{II}(Dcbpy)_2(CN)_2]$, $Na_4[Fe^{II}(Dcbpy)(CN)_4]$, $K_4[Fe^{II}(Dcbpy)_2(CN)_2]$, and $K_4[Fe^{II}(Dcbpy)(CN)_4]$ showed enhanced solubility of 1.09, 1.02, 1.22, and 1.12 M, which can be attributed to the enhanced polarizability and stronger intermolecular interactions between water.

However, there are some conflicting perspectives on the chemical stability of ferri-/ferrocyanide in alkaline solutions in recent publications. The ferri-/ferrocyanide symmetric FB exhibited fast capacity decay in alkaline solution (pH = 14). Some research attributed the capacity decay to the chemical decomposition of ferri-/ferrocyanide via CN^- ligand dissociation and subsequent irreversible hydroxylation [35]. It has been also observed that $K_3[Fe(CN)_6]$ underwent CN^-/OH^- exchange reaction, leading to the $Fe(OH)_3$ precipitate [36]. While other researchers reported that ferri-/ferrocyanide did not suffer from chemical decomposition [37].

Ventosa and coworkers evaluated the stability of ferri-/ferrocyanide in alkaline solution by the symmetric FB and found that the capacity fading resulted from electrochemical oxygen evolution reaction (OER) rather than the chemical decomposition of ferri-/ferrocyanide [38]. However, recent report revealed that ferri-/ferrocyanide solutions were stable at pH = 14 in the dark or indoor light, the capacity fading was due to the electroless OER which enhanced the electroless reduction of ferricyanide to ferrocyanide [39].

Neutral zinc–iron FB mainly employs Zn/Zn^{2+} and Fe^{2+}/Fe^{3+} as negative and positive redox couples, respectively. Compared with alkaline zinc–iron FB, the neutral system is less corrosive and more environmentally friendly. The concentration of positive electrolyte can reach 2.0 M, endowing a high theoretical energy density. But Fe^{2+}/Fe^{3+} is prone to hydrolyze, which is the main challenge for the neutral zinc–iron FB. The glycine additive can coordinate with iron ions, forming $Fe^{3+}(Gly)_2$ and $Fe^{2+}(Gly)_2$ [40]. Cluster models revealed that the bond energy and radius of iron-glycine were much larger than that of iron-malic acid and iron-malonic acid (Figure 9.3a,b). Therefore, glycine can effectively suppress hydrolysis and crossover of Fe^{2+}/Fe^{3+}. The neutral zinc–iron FB achieved a high energy density of 56.3 Wh l^{-1}.

9.2.4 All-Iron FBs

All-iron FBs feature low cost and environmental friendliness. The first all-iron FB was reported by L. W. Hruska and R. Savinell in 1981, by employing Fe^{2+}/Fe and Fe^{3+}/Fe^{2+} redox couples as negative and positive electrolytes, respectively [42]. Nevertheless, the all-iron hybrid FB suffers from the formation of iron dendrite and hydrogen evolution reaction. To address this problem, numerous studies have focused on the all-liquid iron-based FBs with soluble iron complexes redox couples

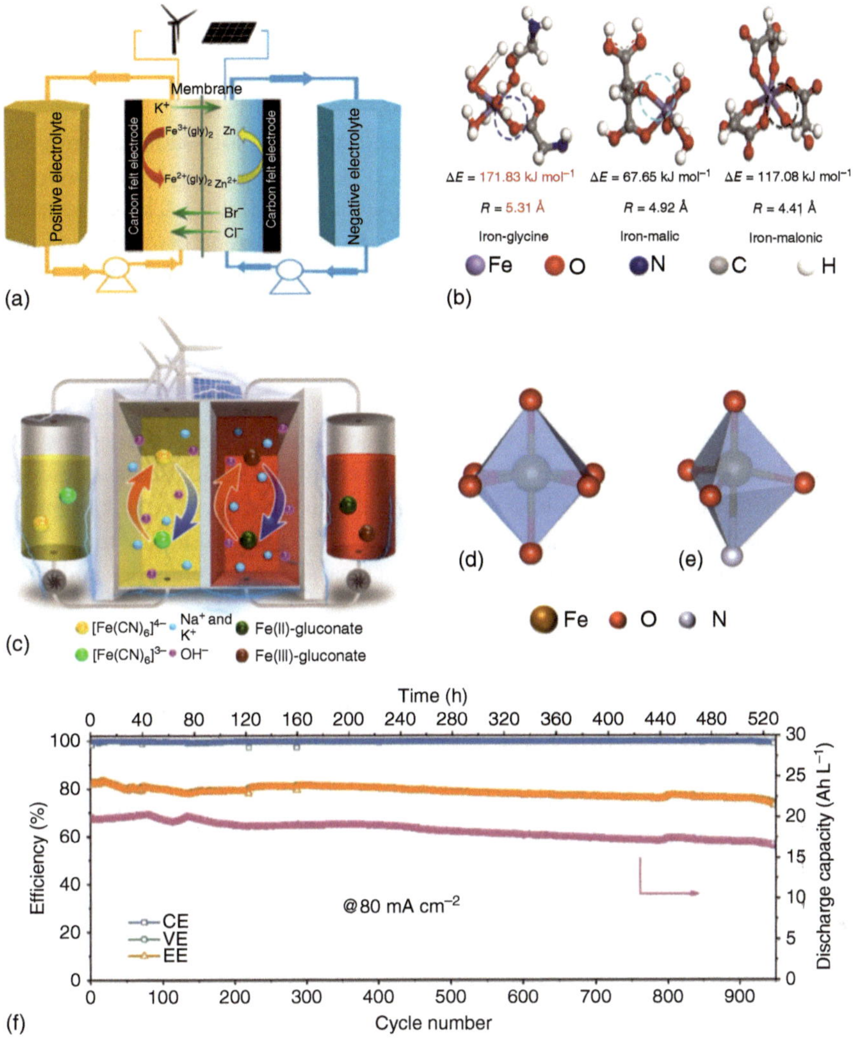

Figure 9.3 (a) Representation of neutral zinc–iron FB. (b) Theoretical calculations of bond energies and radii of iron-glycine, iron-malic, and iron-malonic. Source: Reproduced with permission Xie et al. [40]. Copyright 2017, Wiley-VCH. (c) Schematic diagram of the alkaline all-iron FB. (d) The iron core of FeOOH. (e) The iron core of iron-triethanolamine complex. (f) Cycling performance of alkaline all-iron FB at the current density of 80 mA cm^{-2} at 25 °C. Source: Reproduced with permission Liu et al. [41]. Copyright 2022, Elsevier.

as electrolytes. Among many Fe organometallic complexes, Fe-triethanolamine (Fe[TEA]) has high solubility and fast kinetics. Yan and coworkers reported an all-soluble all-iron FB assembled with Fe(TEA) redox pair and iron-cyanide redox pair in alkaline condition [43]. The FB displayed a cell voltage of 1.34 V and 110 cycles at 40 mA cm^{-2} with a CE of 93% and an EE of 73%. However, the bonds of Fe(TEA) complex are prone to dissociate. Fe irons are reduced to solid form, thus inducing dendrites and hydrogen evolution reaction. To improve the stability of iron complex, 3-(*N*,*N*-bis [2-hydroxyethyl] amino)-2-hydroxypropanesulfonic acid (DIPSO) and 2,2-bis (hydroxymethyl)-2,2′,2′-nitrilotriethanol (BIS-TRIS) were also employed as the iron ion ligands [44, 45]. The FBs exhibited better performance due to the stronger bonding between these ligands and iron ions. However, the lifespan of all-iron FBs is still below 250 cycles. Recently, a more stable all-iron FB for long-duration energy storage was achieved based on $[Fe(CN)_6]^{3-}/[Fe(CN)_6]^{4-}$ and the ferric/ferrous-gluconate ($Gluc^-$) complexes as RASs [41]. The complexes exhibited high stability because of their strong six-coordination between iron ions and gluconate (Figure 9.3c–f). The FB demonstrated a long lifespan (more than 520 hours) with a CE of ~99% and an EE of 83% at 80 mA cm^{-2}.

9.2.5 Aqueous Organic FBs

Aqueous organic flow battery (AOFB) depends on the redox reactions of water-soluble organic RASs, which are composed of earth-abundant elements (such as C, H, O, and N). More importantly, their structures are diverse and properties (such as solubility, redox potential, and stability) can be tuned by molecular engineering or electrolyte optimization. AOFB is regarded as a promising candidate for large-scale EES.

9.2.5.1 Ferrocene Derivatives

Ferrocene (Fc) is a typical sandwich metallocene that contains two aromatic cyclopentadienyl ligands (Cp^-) and an iron center. Ferrocene can undergo a reversible one-electron transfer reaction at 0.4 V (vs. standard hydrogen electrode [SHE]). The low solubility of Fc hinders its application in AOFB. The water-soluble quaternary ammonium and anionic sulfonated groups were used to modify the ferrocene derivatives [46–48]. The hydrophilic and electron-withdrawing ammonium groups on the (ferrocenylmethyl)-trimethylammonium chloride (FcNCl) and N^1-ferrocenyl-methyl-N^1, N^1, N^2, N^2, N^2-pentamethylpropane-1,2-diaminium dibromide (FcN_2Br_2) exhibited high solubility and high redox potential (0.61 V vs. normal hydrogen electrode, [NHE]) [46]. The capacity retention of FcNCl/*N*, *N*′-dimethyl-4,4′-bipyridinium dichloride (MV) AOFB decreased with increasing electrolyte concentration, which can be attributed to the collision of two ferrocenium species in a second-order process. While for the bis((3-trimethylammonio)propyl) ferrocene dichloride (BTMAP-Fc), the two positive quaternary ammonium groups suppressed the bimolecular decomposition by coulombic repulsion and inhibited crossover through charge and size exclusion [47].

9.2.5.2 TEMPO Derivatives

TEMPO is a typical nitroxide radical undergoing a reversible one-electron redox reaction at 0.8–1 V (vs. SHE). The delocalization of unpaired electron over N, O and steric-hindering effect of adjacent four methyl groups endow TEMPO with high stability. A water-soluble TEMPO derivative, 4-hydroxy-2,2,6,6-tetramethylpiperidin-1-oxyl (4-HO-TEMPO), was used as the cathode RAS in AOFB [49]. 4-HO-TEMPO exhibited a high solubility (2.1 M in water) and high potential. When paired with

MV, the AOFB delivered an OCV of 1.25 V. But the capacity retention decreased with electrolyte concentration, which may be caused by the crossover of stability of charge-neutral TEMPO and self-catalyzed oxidation of hydroxyl groups. Aiming to enhance potential voltage as well as molecular stability, many works have been dedicated to the substituent engineering (e.g. $-N(CH_3)_3$, $-OCH_2CH_2CH_2N(CH_3)_3$) at 4-position of six-membered ring TEMPO nitroxide radicals. When introducing a hydrophilic and election-withdrawing substituent ($-N(CH_3)_3^+$) to the 4-position, *N*,*N*,*N*,2,2,6,6-heptamethylpiperidinyloxy-4-ammonium chloride (TEMPTMA) displayed a high solubility (3.2 M in 0.3 M NaCl; 2.3 M in 1.5 M NaCl) and a high potential of 0.79 V (vs. Ag/AgCl) [50].

Aiming to improve the stability of TEMPO derivatives, Xu and coworkers decorated 4-HO-TEMPO with ammonium alkyl group, yielding 4-[3-(trimethylammonio) propoxy]-2,2,6,6-tetramethylpiperidine-1-oxyl (TMAP-TEMPO) chloride [51]. The coulombic repulsion of positively charged groups suppressed biomolecular interactions. N^1,N^1,N^1,N^3,N^3,2,2,6,6-nonamethyl-N^3-(piperidinyloxy)propane-1,3-bis (ammonium)dichloride (N_2-TEMPO) was a new TEMPO derivative [52]. The strong electrostatic repulsion between positively charged ammonium of N_2-TEMPO and AMV anion-exchange membrane suppressed the molecule crossover. In addition, the strong electrostatic repulsion of two N_2-TEMPO improved electrochemical stability during cycling. Some organic cations frequently utilized in ionic liquids are also incorporated with TEMPO to improve solubility. For the methylimidazole-functionalized 4-acetylamino-2,2,6,6-tetramethylpiperidine-*N*-oxyl derivative (MIAcNH-TEMPO), the "π–π"-conjugated imidazolium and "*p*–π"-conjugated acetylamino delocalized the electron density of N–O head, thus preventing the radical and oxoammonium forms from electrophilic/nucleophilic attack [53].

To explore the ring structure effect on physicochemical properties, two five-membered ring nitroxide radicals with 160 mV higher potential than six-membered ring nitroxide radicals were proposed [54]. The enhanced redox potential is largely contributed by lowered standard free energy of reduction reaction and the charge population sum of N–O radical head. However, the solubility of this type of RAS is not desirable (less than 0.1 M).

9.2.5.3 Viologen Derivatives

Viologens (Vi^{2+}), namely *N*,*N'*-disubstituted 4,4'-bipyridines, undergo two-step single-electron reductions at −0.45 and −0.76 V (vs. SHE) in neutral-pH solutions. The first step (from Vi^{2+} to Vi^+) is highly reversible, while the second step (from Vi^+ to Vi^0) is irreversible, due to the insolubility of charge neutral Vi^0. Viologens are highly potential anolyte materials owing to their high solubility, low cost, and fast electrochemical kinetics. The simplest viologen derivative, MV, has a high solubility of 2.5 M in water and 2.4 M in 1.5 M NaCl [49, 50]. The radical cation $[MV]^{+\cdot}$ is easy to dimerize followed by a disproportionation reaction to $[MV]^{2+}$ and $[MV]^0$, thus causing capacity loss. The 1,1'-bis[3-(trimethylammonio)propyl]-4,4'-bipyridinium tetrachloride ($[NPr]_2V$) by grafting two positively charged ammonium pendant arms was synthesized to improve stability of viologen derivatives [55]. The electrostatic repulsion between the ammonium groups inhibited dimerization side reaction. Negatively charged sulfonate groups are also applied to enhance the charge repulsion, yielding 1,1′-bis(3-sulfonatopropyl)-4,4′-bipyridinium, $(SPr)_2V$ [33, 56]. While functionalized by four methyl groups on the 2,2′,6,6′-position of $(SPr)_2V$, the steric hindrance of methyl groups reduced the electrostatic attraction between SO_3^- and quaternary ammonium groups, leading to the low membrane permeation [57].

To make use of the second redox reaction to increase the energy density, Liu and coworkers reported two viologen derivatives, 1-methyl-1′-[3-(trimethylammonio) propyl]-4,4′-bipyridinium trichloride, $[(Me)(NPr)V]Cl_3$, 1,1′-bis[3-(trimethylammonio)propyl]-4,4′-bipyridinium tetrabromide, $[(NPr)_2V]Br_4$ [58]. Due to the hydrophilic ammonium groups, both $[(Me)(NPr)V]Cl_3$ and $[(NPr)_2V]Br_4$ exhibited two reversible redox reactions. Li and coworkers reported a neutral (1,1′)-di(2-ethanol)-4,4′-bipyridinium dibromide $(2HO\text{-}V)Br_2$ [59]. The hydrophilic hydroxyl groups increased the solubility of charge-zero state. Furthermore, a π-conjugation extended viologen, 4,4′-(thiazolo[5,4-d]thiazole-2,5-diyl)bis(1-(3-(trimethylammonio)propyl)pyridin-1-ium) tetrachloride, $[(NPr)_2TTz]Cl_4$ was synthesized by introducing a planar thiazolo[5,4-d]thiazole (TTz) between two pyridiniums to reduce the potential gap between the two-step redox reactions [60]. The enhanced resonance and electron delocalization stabilized the reduced state molecules. $[(NPr)_2TTz]Cl_4$ displayed two near redox potentials. Similar findings have been also reported in the extended bipyridinium derivatives (exBPs, exBP, exBP-Me, and exBP-DMe) [61]. All exBPs displayed a one-step two-electron redox reaction at −0.75 to −0.91 V (vs. SHE), due to the extended conjugation reduced energy difference between lowest unoccupied molecular orbital (LUMO) and β-highest occupied molecular orbital (β-HOMO) of the singly reduced state.

9.2.5.4 Quinone Derivatives

Quinones, including benzoquinones (BQs), naphthoquinones (NQs), and anthraquinones (AQs), are typically low-cost, fast-kinetics active materials. They can be utilized as positive or negative active materials in acid, alkaline, or neutral-pH solutions. BQs are extensively applied in AOFB due to low cost and fast kinetics. Narayanan and coworkers coupled 4,5-dihydroxybenzene-1,3-disulfonic acid (BQDS) catholyte with anthraquinone-2-sulfonic acid (AQS) and anthraquinone-2, 6-disulfonic acid (AQDS) anolyte to assemble all-quinone AOFB [62, 63]. BQDS was susceptible to the Michael reaction with water and eventually converted to fully substituted 1,2,4,6-tetrahydroxybenzene-3,5-disulfonic acid. Functionalizing 1,4-hydroquinone with four (dimethylamino)methyl groups (2,3,5,6-tetrakis((dimethylamino)methyl)hydro quinone (FQH_2)) was an effective way to inhibit the Michael reaction, as all positions were replaced with substituents [64]. The Zn/FQH_2 AOFB with a three-electrolyte, two-membrane configuration displayed a high OCV of 2.0 V and a capacity of 22 Ah l^{-1}.

AQs with extended π-conjugated structures exhibit more negative potential and higher electrochemical stability than BQs and NQs. A low-cost and stable AQ derivative, 9,10-anthraquinone-2,7-disulfonic acid (AQDS), was initially reported for the metal-free FBs [65]. However, as the anode RAS, its high potential of 0.21 V (vs. SHE) limited the cell voltage. AQs can also be applied in alkaline solution. 2,6-Dihydroxyanthraquinone (2,6-DHAQ) exhibited a low potential of −0.68 V (vs. SHE) and a solubility of >0.6 M in 1 M KOH [66].

The AOFB consisting of 0.5 M DHAQ and 0.4 M $K_4Fe(CN)_6$ showed a capacity fade rate of 0.1% per cycle at 100 mA cm^{-2}. The carboxylate terminal groups functionalized deviates 4,4′-((9,10-anthraquinone-2,6-diyl)dioxy)dibutyrate (2,6-DBEAQ) exhibited high electrochemical stability, while the oxidized state of 2,6-DBEAQ was prone to γ-hydroxybutyrate cleavage induced by water or hydroxide at high temperature [67]. To address this problem, carboxylate groups were replaced by phosphonate groups to achieve better stability at pH of 9 and 12 [68]. The relationship between the properties of AQs and water-solubilizing groups and the way bound to the redox core was systematically investigated [69]. It was found that the terminal electron-withdrawing groups increased the redox potential in the order of $SO_3^- > CO_2^- > PO_3^{2-}$. Besides, the unsaturated C=C bonds can extend π-conjugation and enhance the resonance effect of electron-withdrawing groups, thus influencing the redox potential. AQ derivatives can also be reported in pH-neutral solutions [70, 71].

A potentially green and economical *in situ* electrosynthetic method for AQ-based electrolytes without the use of hazardous oxidants or precious metal catalysts was demonstrated. The as-generated electrolytes, which are extremely stable, can be immediately used in a redox FB without separation or purification [72].

9.2.5.5 Heterocyclic Aromatics

Phenothiazine derivatives are promising positive active materials for AOFB. A highly stable methylene blue (MB) was proposed as a positive electrolyte with a high solubility (≈1.8 M in H_2O + acetic acid [AA] + H_2SO_4 mixture) [73]. At a high concentration of 1.5 M, the V/MB AOFB displayed a high capacity of 71 Ah l^{-1} and a low-capacity fade rate of 0.074% per cycle. Furthermore, a kW-scale MB-based AOFB stack was assembled for the first time and demonstrated ultrastable cycling performance, proving the feasibility of practical application of AOFB [74].

The alloxazine can undergo a two-electron redox reaction in alkaline solution. The carboxylic acid group functionalized alloxazine isomeric molecule, alloxazine 7/8-carboxylic acid (ACA) can be used as the anode RAS in the alkaline solution. When coupled with ferri-/ferrocyanide as catholyte, the ACA-based AOFB delivered a capacity retention of 91% over 400 cycles at a concentration of 0.5 M [75]. The solubility of another alloxazine derivative, sodium salt of flavin mononucleotide (FMN-Na), was promoted up to 1.5 M with the presence of nicotinamide molecule in the alkaline solution [76]. However, FMN-Na is susceptible to hydrolysis reaction at high temperatures and dimerization via π-stacking.

Phenazine derivatives are also promising anode RASs for AOFB. 7,8-Dihydroxyphenazine-2-sulfonic acid (DHPS) exhibited a high solubility of 1.45 M in 1 M NaOH [77]. The AOFB delivered a high capacity of 67 Ah l^{-1}. In addition, a series of amino acid-substituted phenazine derivatives were investigated and found that the substituted positions prominently influenced the redox potential, solubility, and stability. 1,6-Substituted phenazine derivatives exhibited high electrochemical stability [78], while the 2,7- and 1,8-substituted derivatives were prone to tautomerization from the reduced forms. The same group also synthesized three propionic-acid-functionalized phenazine derivatives at 1,6-, 1,8-, and 2,7-positions. All derivatives exhibited high solubility and excellent stability for both reduced and oxidized states at evaluated temperature (50 °C) [79].

9.2.6 Nonaqueous FBs

In nonaqueous FBs, the aprotic electrolytes can provide a wide electrochemical window, and various organic RASs such as the metal complexes, polycyclic aromatic hydrocarbons, pyridinium-based, benzophenone, azobenzene, *N*-methyl phthalimide, phenazine, and 9-fluorenone have been studied since Matsuda et al. designed the first nonaqueous FB based on the ruthenium tris(2,21-bipyridine) ($[Ru(bpy)_3]^{2+}$) metal compound in 1988 [3, 80–82].

As shown in Figure 9.4, nonaqueous FBs with cell voltages of >4.0 V can be achieved based on the various nonaqueous chemistries. There are several methods to enhance the solubility of RASs in a nonaqueous solution, including modifying with alkyl or ether chains to lower the molecular interaction between the RASs [83], eutectic electrolytes [84–88], and liquid or low-melting RASs [89]. Although highly soluble RASs can be achieved, the low ionic conductivity and high viscosity of highly concentrated electrolytes limit the development of nonaqueous FBs. Another critical issue is the lack of the appropriate membrane in nonaqueous FBs, and the currently used separators in nonaqueous system have low ionic conductivity (Nafion, FAPQ, LATP, Al_2O_3).

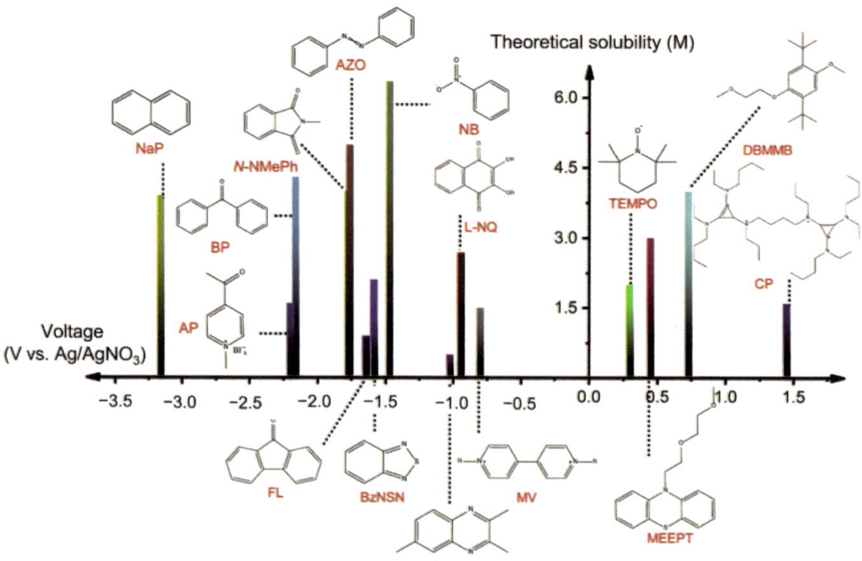

Figure 9.4 The solubility, redox potential, and molecular structure of organic RASs for nonaqueous FBs.

9.3 Research Progress on the Membrane

The membrane is one of the key components of an FB, playing a significant role in separating the positive and negative electrolytes and transporting charge-carrier ions to complete the internal electrical circuit. An ideal membrane should possess high ion conductivity, ion selectivity, good chemical and mechanical stability, and low cost [90]. According to the morphology and proton conductive mechanism, the membranes can be classified into two types: ion-exchange membranes (IEMs) and porous membranes.

9.3.1 Ion-Exchange Membranes

IEMs contain ion-exchange groups, which can transfer ions. IEMs can be divided into perfluorinated, partially fluorinated, and nonfluorinated membranes.

Perfluorinated membranes, like Nafion, are commonly used in FB [91]. The hydrophobic Teflon backbone endows the membrane with excellent chemical and mechanical stability, while the hydrophilic terminal sulfonic acid groups provide ion conductivity. Several models were put forward to interpret the morphology of Nafion, such as cluster-network model [92] and water channel model [93]. Nafion membrane features high proton conductivity and good stability; however, the low ionic selectivity limits its application in traditional inorganic FBs. While, in AOFBs, Nafion membranes can inhibit the crossover of redox-active molecules with bigger molecular size than inorganic ions and have been used in different organic systems.

Aiming to improve the ion selectivity of Nafion, considerable research have focused on the modification of Nafion, such as doping inorganic particles (SiO_2, TiO_2, zirconium phosphates, etc.) [94–96], introducing top layer (PEI, polypyrrole) by coating or interfacial polymerization [97–99]. Qiu and coworkers prepared the Nafion/SiO_2 membrane by *in situ* sol–gel method [95]. The hybrid membrane can effectively reduce the permeation of vanadium ions, because the polar clusters were filled with SiO_2 nanoparticles. The VFB with Nafion/SiO_2 membrane displayed higher CE and EE than that with Nafion.

Partially fluorinated membranes are mainly based on fluorocarbon polymers with good chemical stability, such as poly(ethylene-*co*-tetrafluoroethylene) (ETFE) and poly(vinylidene difluoride) (PVDF) [100, 101]. Some ion exchange groups are introduced to the membrane by radiation grafting.

Nonfluorinated membranes have received wide attention due to their low cost, high ion selectivity, and good thermal and mechanical stability. The membranes are mainly based on nonfluorinated aromatic polymers, such as polybenzimidazole (PBI), sulfonated poly(arylene thioether ketone ketone), sulfonated poly(ether ether ketone) (SPEEK), quaternized poly(phthalazinone ether ketone) (QAPPEK), and poly[(fluorenealkylene)-*co*-(biphenyl alkylene)] (PFBA-QA) [102–105]. These membranes can be divided into cation- and anion-exchange membranes. Many nonfluorinated membranes display lower vanadium permeability than Nafion. However, the stability under the strong acid and oxidative operating conditions is poor. The degradation mechanism of SPEEK and polysulfone IEMs was investigated by *ex situ* and *in situ* tests [106, 107]. It was suggested that the ion exchange groups induced strong electrophilic carbon centers, which were susceptible to be attacked by VO_2^+. The introduction of ion exchange groups weakens the stability of nonfluorinated membranes.

Recently, an amphoteric ion-rectification PBI membrane was reported [108]. The membrane can be used in both acidic and alkaline conditions due to its unique structures. To alleviate the Donnan effect, acid swelling is conducted to promote the transport of H^+ and OH^- via the Grotthuss mechanism. More importantly, the acid-swelled PBI membrane exhibited excellent performance in pilot-scale battery stacks. The 3 kW VFB stack assembled with H_3PO_4-PBI membrane displayed an EE of 80% at 200 mA cm^{-2}. The 1 kW alkaline Zn-Fe stack exhibited an EE of 88% at 80 mA cm^{-2} for 300 cycles. The H_3PO_4-PBI membrane can be used in a wide pH range.

Notably, a pilot-scale synthesis and roll-to-roll manufacturing of SPEEK membranes for FBs with high stability was achieved [109]. It is verified that the membrane with cation-exchange groups mainly transports OH^- through the Grotthuss mechanism in alkaline electrolytes owing to the hydrogen-bond network formed in SPEEK membrane. The membrane displayed excellent performance from lab-scale to kW-scale stacks of alkaline zinc–iron FBs. The 4000 W stack stably ran over 800 hours with an EE of 85.46% at 80 mA cm^{-2}.

9.3.2 Porous Membranes

To overcome the high cost of perfluorinated membranes and the poor stability of nonfluorinated membranes, Zhang et al. first proposed the concept of pore size exclusion [110]. The porous membranes can separate vanadium ions from protons by their difference in Stokes radius. To confirm this concept, polyacrylonitrile (PAN) nanofiltration (NF) membranes were prepared via phase inversion method. The membrane had a finger-like support layer and a thin selective layer (Figure 9.5a). The vanadium/proton selectivity increased with the decreasing pore size distribution. The performance of VFB with the NF membrane was comparable to Nafion.

However, there is a trade-off between selectivity and ion conductivity. The membranes with larger pores possess higher ion conductivity while decreasing selectivity [113]. A myriad of studies have focused on improving ion selectivity while still keeping the proton conductivity.

Regulating the morphology of porous membranes is an effective strategy to improve the performance. A highly symmetric, sponge-like porous membrane with positively charged groups was designed (Figure 9.5b) [111]. Micron-sized cells can be filled with sulfuric acid, promoting the protons transfer, while positively charged groups improved the selectivity via the Donnan exclusion mechanism. These sponge-like pore structures diminished the trade-off between selectivity and ion conductivity.

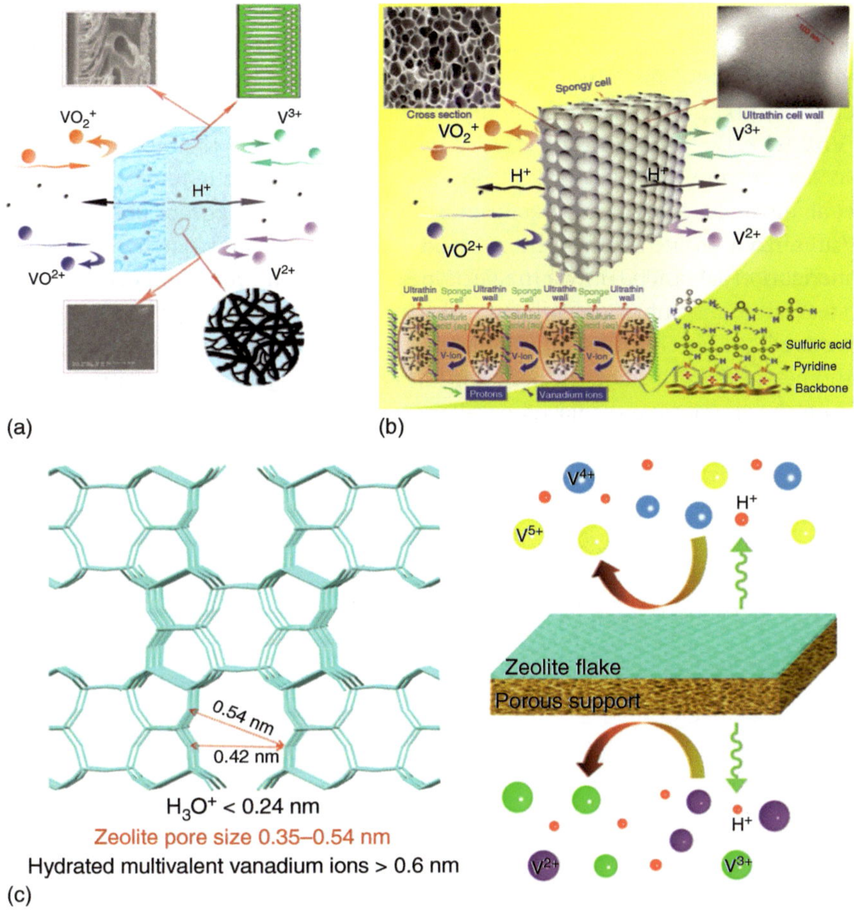

Figure 9.5 (a) Schematic principle of NF membranes in VFB. Source: Reproduced with permission of Zhang et al. [110]. © 2011/Royal Society of Chemistry. (b) Principle of designed membranes with symmetric spongy structures in a VFB. Source: Reproduced with permission of Zhang et al. [111]. © 2013/Royal Society of Chemistry. (c) Design principles of a VFB with a porous membrane bearing a zeolite flake layer. Source: Reproduced with permission Yuan et al. [112]. Copyright 2016, Wiley-VCH.

Introducing inorganic nanoparticles into porous membranes is also an effective way. Zhang et al. introduced SiO_2 into the pores of PAN NF membranes via *in situ* hydrolysis of tetraethylorthosilicate (TEOS) [114]. The decreased pore size increased the selectivity. Meanwhile, the silica nanoparticles can improve the hydrophilicity of membranes, thus facilitating the proton transport. The VFB assembled with the membrane displayed a CE of 98% and an EE of 79% at 80 mA cm^{-2}.

In addition, composite membranes have aroused wide attention due to their unique structures. A composite membrane consists of a porous support layer, which provides high mechanical stability as well as high ion conductivity, and a dense selective layer, which enhances the selectivity. More importantly, the support layer and selective layer can be tuned separately [113]. Introducing an ultrathin

ZSM-35 zeolite flake onto a poly(ether sulfone)-based porous membrane, the pore size of ZSM-35 was between the hydrated Stokes radius of protons (<0.24 nm) and vanadium ions (>0.6 nm), thus separating vanadium ions and protons effectively (Figure 9.5c) [112]. While the porous supporting layer ensured the ion conductivity. The VFB with the composite membrane demonstrated a CE of >99% and an EE of >81% at 200 mA cm^{-2}. To further improve the ion conductivity of membrane, Dai et al. introduced an ultrathin polyamide selective layer on porous polyethersulfone/sulfonated polyetheretherketone blend (PES/SPEEK) substrate by interfacial polymerization [115]. The ultrathin selective layer with sub 1 nm pores can separate hydrated vanadium ions from protons effectively, while the protons can transport in selective layer via Grotthuss mechanism and Vehicle mechanism. The composite membranes broke the trade-off between ion conductivity and selectivity. A VFB based on the membrane displayed excellent performance with an EE of >80% at 260 mA cm^{-2}.

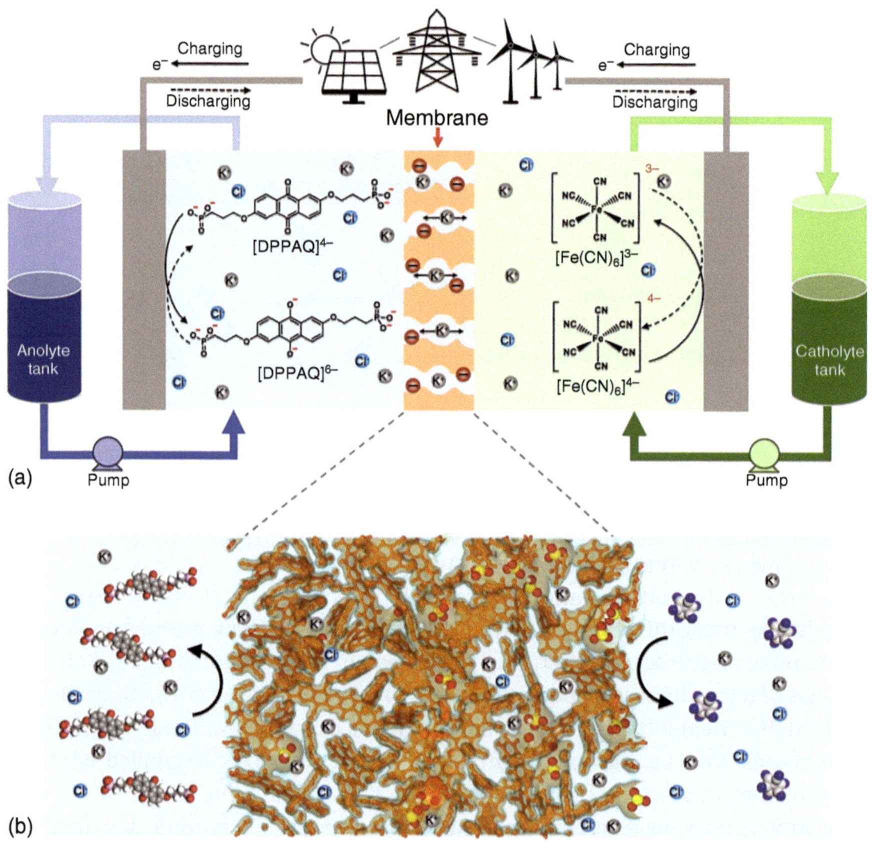

Figure 9.6 (a) Schematics of an aqueous organic redox flow battery for grid-scale energy storage. (b) Schematic showing the polymer ion-sieving membrane with subnanometer-sized pores that enable fast transport of charge-carrying ions while limiting the crossover of redox-active species. Source: Reproduced with permission Ye et al. [117]. Copyright 2022, Springer.

More recently, polymers of intrinsic microporosity (PIMs) have been used as membranes for AOFBs to break the trade-off between ion conductivity and selectivity. The PIMs possessed interconnected micropores less than 2 nm to achieve size sieving, and ion-exchange groups were introduced to PIMs to improve ion conductivity. An amidoxime-functionalized PIM (AO-PIMs) was used in the AOFB as the size-selective ion-exchange membrane [116]. The 2,6-DPPAQ/$K_4Fe(CN)_6$ AOFBs assembled with PIM membranes exhibited low-capacity fade rate. However, due to the weak dissociation of amidoxime groups, the ion conductivity was low. To enhance ion conductivity, the same group introduced negatively charged sulfonate groups to spirobifluorene-based PIM polymer (PIM-SBF), obtaining sulfonated PIMs (sPIM-SBFs) [117]. The membrane with interconnected subnanometer pores can inhibit the crossover of redox-active molecules and transport salt ions, while the sulfonate groups can promote the transport of K^+ by electrostatic attraction (Figure 9.6a, b). The 2,6-DPPAQ/$K_4Fe(CN)_6$ AOFB assembled with the membrane displayed a low-capacity decay rate.

9.4 Electrodes and Bipolar Plates

9.4.1 Electrodes

Deemed as a core component of FB, the electrode plays an important role in providing electrochemical active sites for the reaction of redox couples. The properties of the electrode significantly influence the EE and operating current density of an FB system [118]. An ideal electrode should possess high electrical conductivity, high stability, and large surface area. Carbon felt (CF) and graphite felt (GF) are the most commonly used electrode materials for VFB due to their low cost, high conductivity, and high stability in strong acid conditions. However, the low electrochemical activity, poor kinetic reversibility, and few surface defects hinder their application [119]. Intensive efforts have been devoted to modifying the carbon materials, including surface treatments (such as electrochemical oxidation, acid treatment, and thermal activation) and surface modification (such as coating or depositing electrocatalysts on the surface).

Introducing functional groups, such as carboxylic and hydroxyl, can effectively improve the hydrophilicity and electrocatalytic activity of electrode. Sun and Skallas-Kazacos improved the electrochemical activity of GF by thermal treatment and acid treatment, respectively [120, 121]. The concentration of functional groups (–OH and –COOH) increased dramatically after thermal treatment. These functional groups served as the active sites for VO_2^+/VO^{2+} redox reactions. In addition, the content of –OH and –COOH on GF can also be increased via electrochemical activation [122].

Coating or depositing electrocatalysts is also an effective way to increase the electrochemical activity of electrode. The electrocatalysts can be classified into carbon- and metal-based nanomaterials. Carbon-based electrocatalysts mainly include carbon nanoparticles, carbon nanotubes, graphene oxide, and nitrogen-doped carbon, which possess high specific surface area and excellent electrical conductivity

[123–129]. Yan and coworkers investigated various carbon nanotube-based electrocatalysts such as multiwalled carbon nanotubes (MWCNTs) and single-walled carbon nanotubes (SWCNTs) [125, 126]. They found that the MWCNTs functionalized with hydroxyl groups and carboxyl groups can facilitate the VO_2^+/VO^{2+} redox reactions. In addition, the carboxyl functional groups exhibited the best catalytic performance among oxygen-containing functional groups. Metal-based electrocatalysts have been introduced onto electrodes as well. Sun and Skyllas-Kazakos prepared the metalized graphite fiber electrode by ion exchange with the solutions containing Pt^{4+}, Pd^{2+}, Au^{4+}, Mn^{2+}, Te^{4+}, In^{3+}, and Ti^{3+} and investigated their electrochemical behaviors [130]. The electrode modified with Ir exhibited the best electrocatalytic activity for vanadium redox species, while Pt-, Pd-, and Au-modified electrodes facilitated hydrogen evolution reaction.

In addition, Bi is one of the promising metal-based electrocatalysts due to its low cost, nontoxicity, and high conductivity. Bi-modified GF was first prepared by immersing GF in Bi_2O_3 solution followed by thermal reduction at 450 °C in air [131]. Bi mainly presented at the edge of the holes on the fibers and acted as active sites for VO_2^+/ VO^{2+} redox reactions. The same group also decorated a GF with Bi nanoparticles by electrodeposition [132]. The Bi nanoparticles can improve the reversibility of V^{3+}/V^{2+} redox reaction and inhibit hydrogen evolution. Some transition metal oxides also possess high catalytic activity; for example, PdO_2 has catalytic effects on VO_2^+/VO^{2+} redox reactions [133]. TiO_2 can accelerate V^{2+}/V^{3+} redox reactions [134], while Nb_2O_5 and Mn_3O_4 have catalytic effects on both reactions [135, 136].

9.4.2 Bipolar Plates

BPs are one of the essential components of FB, connecting each cell electrically, separating adjacent cells, and guiding the electrolyte flow. An ideal BP should possess excellent conductivity, high mechanical strength, high corrosion resistance, and high compactness to prevent electrolytes leakage [137]. The materials of BP can be classified into three types: metallic, graphitic, and carbon-polymer composites.

The metallic BPs possess excellent electrical conductivity. But metallic BPs without a protective layer are prone to suffer from surface corrosion [138]. The dissolved metal ions would contaminate the electrolytes. Therefore, metallic materials are not appropriate electrodes. The graphitic BPs with high chemical stability and high conductivity are commonly used in VFB. However, the low mechanical strength and high manufacturing cost limit further application.

Carbon-polymer-based BPs have obtained wide attention for their low cost and better mechanical strength. Carbon-polymer composites are commonly fabricated by mixing polymer and conducting fillers through compression and injection molding [139, 140]. The polymers contain thermoplastic and thermoset polymers, such as polypropylene, polyethylene, and PVDF, which endow the BPs with high mechanical strength [141–143]. Various conducting carbonaceous materials are introduced to enhance the electrical conductivity, such as carbon fiber, carbon black, and graphite powder [144–146]. The conductivity usually depends on the species, amounts of conductive particles, and their connectivity. Carbon black and carbon fiber were added into polypropylene elastomer [146]. The conductivity increased with the amounts of conductive filler, and carbon black exhibited better conductivity than carbon fiber at the same content. In addition, when carbon black and carbon fiber were simultaneously introduced, the conductivity prominently enhanced, because the carbon fiber can bridge carbon black particles to form new conductive paths. A low-carbon-content BP with both high electrical conductivity and flexural strength was designed by exploiting graphite powder and graphene as conductive filler [147]. The graphite powders were prone to aggregate in the polymer, resulting in high electrical resistance, while graphene sheets can connect the aggregated islands to form a conductive network.

9.5 Other Novel FBs

9.5.1 The Semisolid FBs (SSFBs)

The SSFB mainly exploits flowable suspension of solid particles as electrolytes, which consist of active materials, conductive additives, supporting salts, and solvents. The SSFB can achieve high energy density, which breaks through the bottleneck of solubility [89]. The first proof-of-concept for the SSFB was conducted by Chiang and coworkers [148] The SSFB consisting of $LiCoO_2$ (20 vol% [10.2 M] and 1.5% Ketjen black) as the catholyte and $Li_4Ti_5O_{12}$ (10 vol% [2.3 M] and 2% Ketjen black) as the anolyte reached the capacity of ~170 mAh g^{-1} at 1/8 C. The author calculated that the SSFB based on $LiCoO_2/Li_4Ti_5O_{12}$ can achieve 397 Wh l^{-1}. Some other electrode materials for Li-ion batteries have also been applied in SSFB, such as S, Si, $LiNi_{1/3}Co_{1/3}Mn_{1/3}O_2$, and organic molecules [149–152]. In addition, this concept can extend to aqueous electrolytes [153, 154].

However, the main challenge is the high viscosity of high-concentration suspension, resulting in high flow pressure and low conductivity. What is worse, the solid particles may aggregate and block flow channel. A new electrode architecture was designed by using nanoscale conductor particles to form flowable electrically percolating networks [155]. Different from conventional stationary current collectors, the conductive fluid acted as current collector, which prominently increased the electroactive zone. The lithium-polysulfide battery with the new electrode architecture displayed a capacity that is sixfold higher than that in conventional architecture.

9.5.2 The Redox-Targeting-Based FB

The redox-targeting-based FB is based on the chemical reactions between the redox mediators and battery materials. The redox mediators dissolved in the electrolyte generate power via redox reaction, while energy is stored in solid materials via the redox-mediated chemical reactions [156]. In principle, any reversible redox species with standard potential close to the Fermi level of the battery material is potentially suitable for the redox targeting reaction. Wang et al. first proposed the concept of redox targeting battery materials to eliminate the need for conductive additives of insulating battery materials [157]. The concept has extended to nonaqueous [158, 159] and aqueous FB [160, 161].

In 2013, Wang and coworkers reported the first redox flow lithium-ion battery, employing $LiFePO_4$ (3.45 V vs. Li^+/Li) as the active Li^+-storage material and 1,10-dibromoferrocene ($FcBr_2$, 3.55 V vs. Li^+/Li) and ferrocene (Fc, 3.25 V vs. Li^+/Li) as the redox mediators [162]. The redox flow lithium battery assembled with $LiFePO_4$ and $Li_4Ti_5O_{12}$ of 50% porosity as the cathodic and anodic storage materials delivered 6–12 times higher energy density than VFB. Recently, an all-organic redox-targeting FB was reported [163]. By tuning the steric and electronic properties of organic redox mediators and viologen-based insoluble polymer, the battery exhibited a high state of charge (SOC) (>85%) and high VE (>75%).

The redox-targeting FB approach can also be applied in aqueous system to improve the effective concentration of redox species. To increase the cell energy density, $LiFePO_4$ (0.21 V vs. Ag/AgCl) was introduced as solid energy storage material to Zn/Fc-SO_3Na FB. Fc-SO_3Na (0.13 V vs. Ag/AgCl) and BrFc-SO_3Na (0.41 V vs. Ag/AgCl) were used as redox mediators. The catholyte volumetric capacity reached up to 293.5 Ah l^{-1} [48].

9.6 FB Systems and Applications

In recent years, the VFBs have been widely used in various application scenarios, such as renewable energy source generation, smart grids, and off-grid power supplies. Skyllas-Kazacos and coworkers first fabricated a 1 kW VFB stack in 1984 at UNSW [8]. Currently, the VFB has been widely investigated by SEI (Japan), Pacific Northwest National Laboratory (USA), Fraunhofer UMSICHT (Germany), and so on.

In Japan, SEI has been studying FB technologies since 1980s and has installed a 15 MW/60 MWh VFB energy storage power station in 2015. SEI's 2 MW/8 MWh VFB project in California was inaugurated in 2017 and the biggest project of its type in the United States to date has been used to trial the use of VFBs for microgrid applications. In 2022, the 17 MWh/51 MWh VFB system was successfully integrated into the grid by SEI on the Japanese island of Hokkaido.

In China, the Dalian Institute of Chemical Physics, Chinese Academy of Sciences (DICP, CAS) cooperating with Dalian Rongke Power Technology Development Co. Ltd. (RKP) has achieved great progress in FB materials, including membranes, electrolytes, electrodes, and bipolar plates. The DICP-RKP cooperation group fabricated a 5 MW/10 MWh VFB energy storage system on a 50 MW wind farm in Liaoning Province belonging to Longyuan Electric Power Company in 2012, which achieved smooth output. Currently, the 200 MW/800 MWh national VFB demonstration project is being built by DICP-RKP group and the first phase 100 MW/400 MWh has been connected to the grid.

In addition, the ZBFB has reached the industrial stage. Redflow Ltd. (Australia), EnSync Energy Systems (USA), Lotte Chemical Corp. (Korea), Primus Power (USA), and Zbest Power Co. Ltd. (China) have been devoted to the research and commercialization of ZBFB. In 2018, Redflow installed a 1 kWh ZBFB module and a 600 kWh ZBFB system for smart grids. In 2021, Redflow completed the installation of a 2 MWh ZBFB system in California. DICP developed the first international 5 kW Zn-Br single FB system in 2017, which has been applied in an optical storage power supply system for Shanxi Huayin Technology Co. Ltd. (Ankang, Shanxi Province). Recently, DICP successfully developed a 30 kWh ZBFB system, which can be applied to distributed energy and household energy storage.

9.7 Conclusions and Remaining Challenges

In conclusion, we have summarized recent developments of the crucial components of FB, including electrolytes, membranes, electrodes, and bipolar plates. Despite enormous progress achieved, more efforts should be made to acquire an ideal FB with high energy density, low cost, and high stability.

The properties of electrolytes have a vital influence on the performance of a FB. The main challenges of VFB are the low energy density and high cost. The Zn-based FBs feature high energy density and low cost, but the problems caused by zinc dendrites should be further solved. Organic RASs are regarded as promising candidates to supplement conventional metal-based RASs. However, the critical limitations are the stability of the organic RASs. Molecular engineering and electrolytes optimization should be taken to further improve solubility and stability. More importantly, the decomposition mechanisms need to be clarified to further guide molecular engineering. Membrane is a key component of an FB. An ideal membrane should possess high ion conductivity and selectivity. Therefore, further investigation should be focused on exploring more appropriate membrane materials and designing the structures of membranes. The electrodes and bipolar plates with high electrical conductivity and high stability are preferred. More attention should be paid to increase electrical conductivity while keeping mechanical stability.

Last but not least, building new *in situ* methodologies together with computational simulations is a must to quest high-performance and low-cost FB, which can help better understand the phenomena inside the new battery chemistry at relevant timescales and interfaces.

References

1 Kammen, D.M. and Sunter, D.A. (2016). City-integrated renewable energy for urban sustainability. *Science* 352 (6288): 922–928.

2 Yang, Z., Zhang, J., Kintner-Meyer, M.C. et al. (2011). Electrochemical energy storage for green grid. *Chemical Reviews* 111 (5): 3577–3613.

3 Zhang, C. and Li, X. (2021). Perspective on organic flow batteries for large-scale energy storage. *Current Opinion in Electrochemistry* 30: 100836.

4 Zhao, Z., Liu, X., Zhang, M. et al. (2023). Development of flow battery technologies using the principles of sustainable chemistry. *Chemical Society Reviews* 52 (17): 6031–6074.

5 Lim, H., Lackner, A., and Knechtli, R. (1977). Zinc-bromine secondary battery. *Journal of the Electrochemical Society* 124 (8): 1154.

6 Kazacos, M., Cheng, M., and Skyllas-Kazacos, M. (1990). Vanadium redox cell electrolyte optimization studies. *Journal of Applied Electrochemistry* 20 (3): 463–467.

7 Wang, W., Luo, Q., Li, B. et al. (2013). Recent progress in redox flow battery research and development. *Advanced Functional Materials* 23 (8): 970–986.

8 Ulaganathan, M., Aravindan, V., Yan, Q. et al. (2016). Recent advancements in all-vanadium redox flow batteries. *Advanced Materials Interfaces* 3 (1): 1500309.

9 Choi, C., Kim, S., Kim, R. et al. (2017). A review of vanadium electrolytes for vanadium redox flow batteries. *Renewable and Sustainable Energy Reviews* 69: 263–274.

10 Cheng, M. (1991). *Electrolyte Optimization and Studies for the Vanadium Redox Flow Battery*. Australia: University of New South Wales.

11 Chu, Y., Liu, C., Ren, H. et al. (2016). Electrochemical performance of VO_2^+/VO^{2+} redox couple in the H_2SO_4-CH_3SO_3H solutions. *International Journal of Electrochemical Science* 11: 1987–1996.

12 Li, L., Kim, S., Wang, W. et al. (2011). A stable vanadium redox-flow battery with high energy density for large-scale energy storage. *Advanced Energy Materials* 1 (3): 394–400.

13 Wei, X., Nie, Z., Luo, Q. et al. (2013). Nanoporous polytetrafluoroethylene/silica composite separator as a high-performance all-vanadium redox flow battery membrane. *Advanced Energy Materials* 3 (9): 1215–1220.

14 Wu, X., Liu, J., Xiang, X. et al. (2014). Electrolytes for vanadium redox flow batteries. *Pure and Applied Chemistry* 86 (5): 661–669.

15 Chang, F., Hu, C., Liu, X. et al. (2012). Coulter dispersant as positive electrolyte additive for the vanadium redox flow battery. *Electrochimica Acta* 60: 334–338.

16 Zhang, J., Li, L., Nie, Z. et al. (2011). Effects of additives on the stability of electrolytes for all-vanadium redox flow batteries. *Journal of Applied Electrochemistry* 41 (10): 1215–1221.

17 Wang, S., Wang, Z., Yin, Y. et al. (2021). A highly reversible zinc deposition for flow batteries regulated by critical concentration induced nucleation. *Energy and Environmental Science* 14 (7): 4077–4084.

18 Banik, S.J. and Akolkar, R. (2015). Suppressing dendritic growth during alkaline zinc electrodeposition using polyethylenimine additive. *Electrochimica Acta* 179: 475–481.

19 Kim, M., Yun, D., and Jeon, J. (2019). Effect of a bromine complex agent on electrochemical performances of zinc electrodeposition and electrodissolution in zinc–bromide flow battery. *Journal of Power Sources* 438: 227020.

20 Sun, P., Ma, L., Zhou, W. et al. (2021). Simultaneous regulation on solvation shell and electrode interface for dendrite-free Zn ion batteries achieved by a low-cost glucose additive. *Angewandte Chemie International Edition* 60 (33): 18247–18255.

21 Li, B., Nie, Z., Vijayakumar, M. et al. (2015). Ambipolar zinc-polyiodide electrolyte for a high-energy density aqueous redox flow battery. *Nature Communications* 6: 6303.

22 Chang, N., Li, T., Li, R. et al. (2020). An aqueous hybrid electrolyte for low-temperature zinc-based energy storage devices. *Energy and Environmental Science* 13 (10): 3527–3535.

23 Bae, S., Lee, J., and Kim, D.S. (2019). The effect of Cr^{3+}-functionalized additive in zinc-bromine flow battery. *Journal of Power Sources* 413: 167–173.

24 Chang, G., Liu, S., Fu, Y. et al. (2019). Inhibition role of trace metal ion additives on zinc dendrites during plating and striping processes. *Advanced Materials Interfaces* 6 (23): 1901358.

25 Yang, J.H., Yang, H.S., Ra, H.W. et al. (2015). Effect of a surface active agent on performance of zinc/bromine redox flow batteries: improvement in current efficiency and system stability. *Journal of Power Sources* 275: 294–297.

26 Yin, Y., Yuan, Z., and Li, X. (2021). Rechargeable aqueous zinc-bromine batteries: an overview and future perspectives. *Physical Chemistry Chemical Physics* 23 (46): 26070–26084.

27 Wu, Y., Huang, P.W., Howe, J.D. et al. (2019). In operando visualization of the electrochemical formation of liquid polybromide microdroplets. *Angewandte Chemie International Edition* 58 (43): 15228–15234.

28 Kautek, W., Conradi, A., Fabjan, C., and Bauer, G. (2001). In situ FTIR spectroscopy of the Zn–Br battery bromine storage complex at glassy carbon electrodes. *Electrochimica Acta* 47 (5): 815–823.

29 Li, X., Xie, C., Li, T. et al. (2020). Low-cost titanium-bromine flow battery with ultrahigh cycle stability for grid-scale energy storage. *Advanced Materials* 32 (49): e2005036.

30 Wu, M., Zhao, T., Wei, L. et al. (2018). Improved electrolyte for zinc-bromine flow batteries. *Journal of Power Sources* 384: 232–239.

31 Yuan, Z., Duan, Y., Liu, T. et al. (2018). Toward a low-cost alkaline zinc-iron flow battery with a polybenzimidazole custom membrane for stationary energy storage. *iScience* 3: 40–49.

32 Wang, G., Zou, H., Xu, Z. et al. (2022). Unlocking the solubility limit of ferrocyanide for high energy density redox flow batteries. *Materials Today Energy* 28: 101061.

33 Luo, J., Hu, B., Debruler, C. et al. (2019). Unprecedented capacity and stability of ammonium ferrocyanide catholyte in pH neutral aqueous redox flow batteries. *Joule* 3 (1): 149–163.

34 Li, X., Gao, P., Lai, Y.-Y. et al. (2021). Symmetry-breaking design of an organic iron complex catholyte for a long cyclability aqueous organic redox flow battery. *Nature Energy* 6 (9): 873–881.

35 Luo, J., Sam, A., Hu, B. et al. (2017). Unraveling pH dependent cycling stability of ferricyanide/ferrocyanide in redox flow batteries. *Nano Energy* 42: 215–221.

36 Hu, M., Wang, A., Luo, J. et al. (2023). Cycling performance and mechanistic insights of ferricyanide electrolytes in alkaline redox flow batteries. *Advanced Energy Materials* 13: 2203762.

37 Goulet, M.-A. and Aziz, M.J. (2018). Flow battery molecular reactant stability determined by symmetric cell cycling methods. *Journal of the Electrochemical Society* 165 (7): A1466–A1477.

38 Páez, T., Martínez-Cuezva, A., Palma, J., and Ventosa, E. (2020). Revisiting the cycling stability of ferrocyanide in alkaline media for redox flow batteries. *Journal of Power Sources* 471: 228453.

39 Fell, E., De Porcellinis, D., Jing, Y. et al. (2022). Long-term stability of ferri-/ferrocyanide as an electroactive component for redox flow battery applications: on the origin of apparent capacity fade. *Journal of the Electrochemical Society* 170 (7): 070525.

40 Xie, C., Duan, Y., Xu, W. et al. (2017). A low-cost neutral zinc-iron flow battery with high energy density for stationary energy storage. *Angewandte Chemie International Edition* 56 (47): 14953–14957.

41 Liu, X., Li, T., Yuan, Z., and Li, X. (2022). Low-cost all-iron flow battery with high performance towards long-duration energy storage. *Journal of Energy Chemistry* 73: 445–451.

42 Hruska, L. and Savinell, R. (1981). Investigation of factors affecting performance of the iron-redox battery. *Journal of the Electrochemical Society* 128 (1): 18.

43 Gong, K., Xu, F., Grunewald, J.B. et al. (2016). All-soluble all-iron aqueous redox-flow battery. *ACS Energy Letters* 1 (1): 89–93.

44 Shin, M., Noh, C., Chung, Y., and Kwon, Y. (2020). All iron aqueous redox flow batteries using organometallic complexes consisting of iron and 3-[Bis (2-hydroxyethyl) amino]-2-hydroxypropanesulfonic acid ligand and ferrocyanide as redox couple. *Chemical Engineering Journal* 398: 125631.

45 Shin, M., Oh, S., Jeong, H. et al. (2022). Aqueous redox flow battery using iron 2,2-bis(hydroxymethyl)-2,2′,2′-nitrilotriethanol complex and ferrocyanide as newly developed redox couple. *International Journal of Energy Research* 46 (6): 8175–8185.

46 Hu, B., DeBruler, C., Rhodes, Z., and Liu, T.L. (2017). Long-cycling aqueous organic redox flow battery (AORFB) toward sustainable and safe energy storage. *Journal of the American Chemical Society* 139 (3): 1207–1214.

47 Beh, E.S., De Porcellinis, D., Gracia, R.L. et al. (2017). A neutral pH aqueous organic–organometallic redox flow battery with extremely high capacity retention. *ACS Energy Letters* 2 (3): 639–644.

48 Yu, J., Salla, M., Zhang, H. et al. (2020). A robust anionic sulfonated ferrocene derivative for pH-neutral aqueous flow battery. *Energy Storage Materials* 29: 216–222.

49 Liu, T., Wei, X., Nie, Z. et al. (2016). A total organic aqueous redox flow battery employing a low cost and sustainable methyl viologen anolyte and 4-HO-TEMPO catholyte. *Advanced Energy Materials* 6 (3): 1501449.

50 Janoschka, T., Martin, N., Hager, M.D., and Schubert, U.S. (2016). An aqueous redox-flow battery with high capacity and power: the TEMPTMA/MV system. *Angewandte Chemie International Edition* 55 (46): 14427–14430.

51 Liu, Y., Goulet, M.-A., Tong, L. et al. (2019). A long-lifetime all-organic aqueous flow battery utilizing TMAP-TEMPO radical. *Chem* 5 (7): 1861–1870.

52 Hu, B., Hu, M., Luo, J., and Liu, T.L. (2022). A stable, low permeable TEMPO catholyte for aqueous total organic redox flow batteries. *Advanced Energy Materials* 12 (8): 2102577.

53 Fan, H., Hu, B., Li, H. et al. (2022). Conjugate-driven electron density delocalization of piperidine nitroxyl radical for stable aqueous zinc hybrid flow batteries. *Angewandte Chemie International Edition* 61 (17): e202115908.

54 Hu, B., Fan, H., Li, H. et al. (2021). Five-membered ring nitroxide radical: a new class of high-potential, stable catholytes for neutral aqueous organic redox flow batteries. *Advanced Functional Materials* 31 (35): 2102734.

55 Hu, B., Tang, Y., Luo, J. et al. (2018). Improved radical stability of viologen anolytes in aqueous organic redox flow batteries. *Chemical Communications* 54 (50): 6871–6874.

56 DeBruler, C., Hu, B., Moss, J. et al. (2018). A sulfonate-functionalized viologen enabling neutral cation exchange, aqueous organic redox flow batteries toward renewable energy storage. *ACS Energy Letters* 3 (3): 663–668.

57 Li, H., Fan, H., Hu, B. et al. (2021). Spatial structure regulation: a rod-shaped viologen enables long lifetime in aqueous redox flow batteries. *Angewandte Chemie International Edition* 60 (52): 26971–26977.

58 DeBruler, C., Hu, B., Moss, J. et al. (2017). Designer two-electron storage viologen anolyte materials for neutral aqueous organic redox flow batteries. *Chem* 3 (6): 961–978.

59 Liu, W., Liu, Y., Zhang, H. et al. (2019). A highly stable neutral viologen/bromine aqueous flow battery with high energy and power density. *Chemical Communications* 55 (33): 4801–4804.

60 Luo, J., Hu, B., Debruler, C., and Liu, T.L. (2018). A pi-conjugation extended viologen as a two-electron storage anolyte for total organic aqueous redox flow batteries. *Angewandte Chemie International Edition* 57 (1): 231–235.

61 Tang, G., Liu, Y., Li, Y. et al. (2022). Designing robust two-electron storage extended bipyridinium anolytes for pH-neutral aqueous organic redox flow batteries. *JACS Au* 2 (5): 1214–1222.

62 Yang, B., Hoober-Burkhardt, L., Wang, F. et al. (2014). An inexpensive aqueous flow battery for large-scale electrical energy storage based on water-soluble organic redox couples. *Journal of the Electrochemical Society* 161 (9): A1371–A1380.

63 Yang, B., Hoober-Burkhardt, L., Krishnamoorthy, S. et al. (2016). High-performance aqueous organic flow battery with quinone-based redox couples at both electrodes. *Journal of the Electrochemical Society* 163 (7): A1442–A1449.

64 Park, M., Beh, E.S., Fell, E.M. et al. (2019). A high voltage aqueous zinc–organic hybrid flow battery. *Advanced Energy Materials* 9 (25): 1900694.

65 Huskinson, B., Marshak, M.P., Suh, C. et al. (2014). A metal-free organic–inorganic aqueous flow battery. *Nature* 505 (7482): 195–198.

66 Lin, K., Chen, Q., Gerhardt, M.R. et al. (2015). Alkaline quinone flow battery. *Science* 349 (6255): 1529–1532.

67 Kwabi, D.G., Lin, K., Ji, Y. et al. (2018). Alkaline quinone flow battery with long lifetime at pH 12. *Joule* 2 (9): 1894–1906.

68 Ji, Y., Goulet, M.A., Pollack, D.A. et al. (2019). A phosphonate-functionalized quinone redox flow battery at near-neutral pH with record capacity retention rate. *Advanced Energy Materials* 9 (12): 1900039.

69 Jing, Y., Fell, E.M., Wu, M. et al. (2021). Anthraquinone flow battery reactants with nonhydrolyzable water-solubilizing chains introduced via a generic cross-coupling method. *ACS Energy Letters* 7 (1): 226–235.

70 Hu, B., Luo, J., Hu, M. et al. (2019). A pH-neutral, metal-free aqueous organic redox flow battery employing an ammonium anthraquinone anolyte. *Angewandte Chemie International Edition* 58 (46): 16629–16636.

71 Zhu, Y., Li, Y., Qian, Y. et al. (2021). Anthraquinone-based anode material for aqueous redox flow batteries operating in nondemanding atmosphere. *Journal of Power Sources* 501: 229984.

72 Jing, Y., Wu, M., Wong, A.A. et al. (2020). In situ electrosynthesis of anthraquinone electrolytes in aqueous flow batteries. *Green Chemistry* 22 (18): 6084–6092.

73 Zhang, C., Niu, Z., Peng, S. et al. (2019). Phenothiazine-based organic catholyte for high-capacity and long-life aqueous redox flow batteries. *Advanced Materials* 31 (24): 1901052.

74 Zhang, Y., Li, F., Li, T. et al. (2023). Insights into an air-stable methylene blue catholyte towards kW-scale practical aqueous organic flow batteries. *Energy and Environmental Science* 16 (1): 231–240.

75 Lin, K., Gómez-Bombarelli, R., Beh, E.S. et al. (2016). A redox-flow battery with an alloxazine-based organic electrolyte. *Nature Energy* 1 (9): 16102.

76 Orita, A., Verde, M.G., Sakai, M., and Meng, Y.S. (2016). A biomimetic redox flow battery based on flavin mononucleotide. *Nature Communications* 7: 13230.

77 Hollas, A., Wei, X., Murugesan, V. et al. (2018). A biomimetic high-capacity phenazine-based anolyte for aqueous organic redox flow batteries. *Nature Energy* 3 (6): 508–514.

78 Pang, S., Wang, X., Wang, P., and Ji, Y. (2021). Biomimetic amino acid functionalized phenazine flow batteries with long lifetime at near-neutral pH. *Angewandte Chemie International Edition* 60 (10): 5289–5298.

79 Xu, J., Pang, S., Wang, X. et al. (2021). Ultrastable aqueous phenazine flow batteries with high capacity operated at elevated temperatures. *Joule* 5 (9): 2437–2449.

80 Ding, Y., Zhang, C., Zhang, L. et al. (2018). Molecular engineering of organic electroactive materials for redox flow batteries. *Chemical Society Reviews* 47 (1): 69–103.

81 Zhao, Z., Zhang, C., and Li, X. (2022). Opportunities and challenges of organic flow battery for electrochemical energy storage technology. *Journal of Energy Chemistry* 67: 621–639.

82 Matsuda, Y., Tanaka, K., Okada, M. et al. (1988). A rechargeable redox battery utilizing ruthenium complexes with non-aqueous organic electrolyte. *J. Appl. Electrochem.* 18: 909–914. https://doi.org/10.1007/BF01016050.

83 Suttil, J.A., Kucharyson, J.F., Escalante-Garcia, I.L. et al. (2015). Metal acetylacetonate complexes for high energy density non-aqueous redox flow batteries. *Journal of Materials Chemistry A* 3 (15): 7929–7938.

84 Zhang, C., Zhang, L., and Yu, G. (2020). Eutectic electrolytes as a promising platform for next-generation electrochemical energy storage. *Accounts of Chemical Research* 53 (8): 1648–1659.

85 Zhang, C., Zhang, L., Ding, Y. et al. (2018). Eutectic electrolytes for high-energy-density redox flow batteries. *ACS Energy Letters* 3 (12): 2875–2883.

86 Zhang, C., Chen, H., Qian, Y. et al. (2021). General design methodology for organic eutectic electrolytes toward high-energy-density redox flow batteries. *Advanced Materials* 33 (15): 2008560.

87 Zhang, C., Qian, Y., Ding, Y. et al. (2019). Biredox eutectic electrolytes derived from organic redox-active molecules: high-energy storage systems. *Angewandte Chemie International Edition* 58 (21): 7045–7050.

88 Zhang, C., Niu, Z., Ding, Y. et al. (2018). Highly concentrated phthalimide-based anolytes for organic redox flow batteries with enhanced reversibility. *Chem* 4 (12): 2814–2825.

89 Ding, Y., Zhang, C., Zhang, L. et al. (2019). Pathways to widespread applications: development of redox flow batteries based on new chemistries. *Chem* 5 (8): 1964–1987.

90 Zuo, P., Ye, C., Jiao, Z. et al. (2023). Near-frictionless ion transport within triazine framework membranes. *Nature* 617 (7960): 299–305.

91 Li, X., Zhang, H., Mai, Z. et al. (2011). Ion exchange membranes for vanadium redox flow battery (VRB) applications. *Energy and Environmental Science* 4 (4): 1147–1160.

92 Hsu, W.Y. and Gierke, T.D. (1983). Ion transport and clustering in Nafion perfluorinated membranes. *Journal of Membrane Science* 13 (3): 307–326.

93 Schmidt-Rohr, K. and Chen, Q. (2008). Parallel cylindrical water nanochannels in Nafion fuel-cell membranes. *Nature Materials* 7 (1): 75–83.

94 Teng, X., Zhao, Y., Xi, J. et al. (2009). Nafion/organic silica modified TiO_2 composite membrane for vanadium redox flow battery via in situ sol–gel reactions. *Journal of Membrane Science* 341 (1–2): 149–154.

95 Xi, J., Wu, Z., Qiu, X., and Chen, L. (2007). Nafion/SiO_2 hybrid membrane for vanadium redox flow battery. *Journal of Power Sources* 166 (2): 531–536.

96 Sang, S., Wu, Q., and Huang, K. (2007). Preparation of zirconium phosphate (ZrP)/Nafion1135 composite membrane and H^+/VO^{2+} transfer property investigation. *Journal of Membrane Science* 305 (1–2): 118–124.

97 Luo, Q., Zhang, H., Chen, J. et al. (2008). Modification of Nafion membrane using interfacial polymerization for vanadium redox flow battery applications. *Journal of Membrane Science* 311 (1–2): 98–103.

98 Xi, J., Wu, Z., Teng, X. et al. (2008). Self-assembled polyelectrolyte multilayer modified Nafion membrane with suppressed vanadium ion crossover for vanadium redox flow batteries. *Journal of Materials Chemistry* 18 (11): 1232–1238.

99 Zeng, J., Jiang, C., Wang, Y. et al. (2008). Studies on polypyrrole modified Nafion membrane for vanadium redox flow battery. *Electrochemistry Communications* 10 (3): 372–375.

100 Qiu, J., Zhai, M., Chen, J. et al. (2009). Performance of vanadium redox flow battery with a novel amphoteric ion exchange membrane synthesized by two-step grafting method. *Journal of Membrane Science* 342 (1–2): 215–220.

101 Qiu, J., Zhao, L., Zhai, M. et al. (2008). Pre-irradiation grafting of styrene and maleic anhydride onto PVDF membrane and subsequent sulfonation for application in vanadium redox batteries. *Journal of Power Sources* 177 (2): 617–623.

102 Liu, S., Wang, L., Ding, Y. et al. (2014). Novel sulfonated poly (ether ether ketone)/polyetherimide acid-base blend membranes for vanadium redox flow battery applications. *Electrochimica Acta* 130: 90–96.

103 Mai, Z., Zhang, H., Li, X. et al. (2011). Sulfonated poly(tetramethydiphenyl ether ether ketone) membranes for vanadium redox flow battery application. *Journal of Power Sources* 196 (1): 482–487.

104 Zhang, S., Yin, C., Xing, D. et al. (2010). Preparation of chloromethylated/quaternized poly(phthalazinone ether ketone) anion exchange membrane materials for vanadium redox flow battery applications. *Journal of Membrane Science* 363 (1–2): 243–249.

105 Zhao, Z., Dai, Q., Li, X. et al. (2022). A highly stable membrane for vanadium flow batteries (VFBs) enabled by the selective degradation of ionic side chains. *Journal of Materials Chemistry A* 10 (2): 762–771.

106 Yuan, Z., Li, X., Hu, J. et al. (2014). Degradation mechanism of sulfonated poly (ether ether ketone)(SPEEK) ion exchange membranes under vanadium flow battery medium. *Physical Chemistry Chemical Physics* 16 (37): 19841–19847.

107 Yuan, Z., Li, X., Zhao, Y., and Zhang, H. (2015). Mechanism of polysulfone-based anion exchange membranes degradation in vanadium flow battery. *ACS Applied Materials & Interfaces* 7 (34): 19446–19454.

108 Dai, Q., Xing, F., Liu, X. et al. (2022). High-performance PBI membranes for flow batteries: from the transport mechanism to the pilot plant. *Energy and Environmental Science* 15 (4): 1594–1600.

109 Yuan, Z., Liang, L., Dai, Q. et al. (2022). Low-cost hydrocarbon membrane enables commercial-scale flow batteries for long-duration energy storage. *Joule* 6 (4): 884–905.

110 Zhang, H., Zhang, H., Li, X. et al. (2011). Nanofiltration (NF) membranes: the next generation separators for all vanadium redox flow batteries (VRBs)? *Energy and Environmental Science* 4 (5): 1676–1679.

111 Zhang, H., Zhang, H., Zhang, F. et al. (2013). Advanced charged membranes with highly symmetric spongy structures for vanadium flow battery application. *Energy and Environmental Science* 6 (3): 776–781.

112 Yuan, Z., Zhu, X., Li, M. et al. (2016). A highly ion-selective zeolite flake layer on porous membranes for flow battery applications. *Angewandte Chemie International Edition* 55 (9): 3058–3062.

113 Wu, J., Dai, Q., Zhang, H., and Li, X. (2020). Recent development in composite membranes for flow batteries. *ChemSusChem* 13 (15): 3805–3819.

114 Zhang, H., Zhang, H., Li, X. et al. (2012). Silica modified nanofiltration membranes with improved selectivity for redox flow battery application. *Energy and Environmental Science* 5 (4): 6299–6303.

115 Dai, Q., Liu, Z., Huang, L. et al. (2020). Thin-film composite membrane breaking the trade-off between conductivity and selectivity for a flow battery. *Nature Communications* 11: 13.

116 Ye, C.C., Tan, R., Wang, A. et al. (2022). Long-life aqueous organic redox flow batteries enabled by amidoxime-functionalized ion-selective polymer membranes. *Angewandte Chemie International Edition* 61 (38): e202207580.

117 Ye, C., Wang, A., Breakwell, C. et al. (2022). Development of efficient aqueous organic redox flow batteries using ion-sieving sulfonated polymer membranes. *Nature Communications* 13: 3184.

118 Kim, K.J., Park, M.-S., Kim, Y.-J. et al. (2015). A technology review of electrodes and reaction mechanisms in vanadium redox flow batteries. *Journal of Materials Chemistry A* 3 (33): 16913–16933.

119 Park, M., Ryu, J., and Cho, J. (2015). Nanostructured electrocatalysts for all-vanadium redox flow batteries. *Chemistry – An Asian Journal* 10 (10): 2096–2110.

120 Sun, B. and Skyllas-Kazacos, M. (1992). Modification of graphite electrode materials for vanadium redox flow battery application—I. Thermal treatment. *Electrochimica Acta* 37 (7): 1253–1260.

121 Sun, B. and Skyllas-Kazacos, M. (1992). Chemical modification of graphite electrode materials for vanadium redox flow battery application—part II. Acid treatments. *Electrochimica Acta* 37 (13): 2459–2465.

122 Zhang, W., Xi, J., Li, Z. et al. (2013). Electrochemical activation of graphite felt electrode for VO_2^+/VO^{2+} redox couple application. *Electrochimica Acta* 89: 429–435.

123 Shao, Y., Wang, X., Engelhard, M. et al. (2010). Nitrogen-doped mesoporous carbon for energy storage in vanadium redox flow batteries. *Journal of Power Sources* 195 (13): 4375–4379.

124 Han, P., Wang, H., Liu, Z. et al. (2011). Graphene oxide nanoplatelets as excellent electrochemical active materials for VO^{2+}/and V^{2+}/V^{3+} redox couples for a vanadium redox flow battery. *Carbon* 49 (2): 693–700.

125 Li, W., Liu, J., and Yan, C. (2011). Multi-walled carbon nanotubes used as an electrode reaction catalyst for VO_2^+/VO^{2+} for a vanadium redox flow battery. *Carbon* 49 (11): 3463–3470.

126 Li, W., Liu, J., and Yan, C. (2012). The electrochemical catalytic activity of single-walled carbon nanotubes towards VO_2^+/VO^{2+} and V^{3+}/V^{2+} redox pairs for an all vanadium redox flow battery. *Electrochimica Acta* 79: 102–108.

127 Wei, G., Jia, C., Liu, J., and Yan, C. (2012). Carbon felt supported carbon nanotubes catalysts composite electrode for vanadium redox flow battery application. *Journal of Power Sources* 220: 185–192.

128 Li, W., Liu, J., and Yan, C. (2013). Modified multiwalled carbon nanotubes as an electrode reaction catalyst for an all vanadium redox flow battery. *Journal of Solid State Electrochemistry* 17 (5): 1369–1376.

129 Ulaganathan, M., Jain, A., Aravindan, V. et al. (2015). Bio-mass derived mesoporous carbon as superior electrode in all vanadium redox flow battery with multicouple reactions. *Journal of Power Sources* 274: 846–850.

130 Sun, B. and Skyllas-Kazakos, M. (1991). Chemical modification and electrochemical behaviour of graphite fibre in acidic vanadium solution. *Electrochimica Acta* 36 (3–4): 513–517.

131 González, Z., Sánchez, A., Blanco, C. et al. (2011). Enhanced performance of a bi-modified graphite felt as the positive electrode of a vanadium redox flow battery. *Electrochemistry Communications* 13 (12): 1379–1382.

132 Suárez, D.J., González, Z., Blanco, C. et al. (2014). Graphite felt modified with bismuth nanoparticles as negative electrode in a vanadium redox flow battery. *ChemSusChem* 7 (3): 914–918.

133 Wu, X., Xu, H., Lu, L. et al. (2014). PbO_2-modified graphite felt as the positive electrode for an all-vanadium redox flow battery. *Journal of Power Sources* 250: 274–278.

134 Vazquez-Galvan, J., Flox, C., Fabrega, C. et al. (2017). Hydrogen-treated rutile TiO_2 shell in graphite-core structure as a negative electrode for high-performance vanadium redox flow batteries. *ChemSusChem* 10 (9): 2089–2098.

135 Li, B., Gu, M., Nie, Z. et al. (2014). Nanorod niobium oxide as powerful catalysts for an all vanadium redox flow battery. *Nano Letters* 14 (1): 158–165.

136 Kim, K.J., Park, M.-S., Kim, J.-H. et al. (2012). Novel catalytic effects of Mn_3O_4 for all vanadium redox flow batteries. *Chemical Communications* 48 (44): 5455–5457.

137 Duan, Z., Qu, Z., Ren, Q., and Zhang, J. (2021). Review of bipolar plate in redox flow batteries: materials, structures, and manufacturing. *Electrochemical Energy Reviews* 4 (4): 718–756.

138 Satola, B. (2021). Bipolar plates for the vanadium redox flow battery. *Journal of the Electrochemical Society* 168 (6): 060503.

139 Wulfsberg, J., Herrmann, A., Ziegmann, G. et al. (2014). Combination of carbon fibre sheet moulding compound and prepreg compression moulding in aerospace industry. *Procedia Engineering* 81: 1601–1607.

140 Tiusanen, J., Vlasveld, D., and Vuorinen, J. (2012). Review on the effects of injection moulding parameters on the electrical resistivity of carbon nanotube filled polymer parts. *Composites Science and Technology* 72 (14): 1741–1752.

141 Satola, B., Komsiyska, L., and Wittstock, G. (2018). Corrosion of graphite-polypropylene current collectors during overcharging in negative and positive vanadium redox flow battery half-cell electrolytes. *Journal of the Electrochemical Society* 165 (5): A963–A969.

142 Liao, W., Jiang, F., Zhang, Y. et al. (2020). Highly-conductive composite bipolar plate based on ternary carbon materials and its performance in redox flow batteries. *Renewable Energy* 152: 1310–1316.

143 Jiang, F., Liao, W., Ayukawa, T. et al. (2021). Enhanced performance and durability of composite bipolar plate with surface modification of cactus-like carbon nanofibers. *Journal of Power Sources* 482: 228903.

144 Hwang, I.U., Yu, H.N., Kim, S.S. et al. (2008). Bipolar plate made of carbon fiber epoxy composite for polymer electrolyte membrane fuel cells. *Journal of Power Sources* 184 (1): 90–94.

145 Chen, H., Xia, X.-H., Yang, L. et al. (2016). Preparation and characterization of graphite/resin composite bipolar plates for polymer electrolyte membrane fuel cells. *Science and Engineering of Composite Materials* 23 (1): 21–28.

146 Zhang, J., Zhou, T., Xia, L. et al. (2015). Polypropylene elastomer composite for the all-vanadium redox flow battery: current collector materials. *Journal of Materials Chemistry A* 3 (5): 2387–2398.

147 Liao, W., Zhang, Y., Zhou, X. et al. (2019). Low-carbon-content composite bipolar plates: a novel design and its performance in vanadium redox flow batteries. *ChemistrySelect* 4 (8): 2421–2427.

148 Duduta, M., Ho, B., Wood, V.C. et al. (2011). Semi-solid lithium rechargeable flow battery. *Advanced Energy Materials* 1 (4): 511–516.

149 Hamelet, S., Larcher, D., Dupont, L., and Tarascon, J.M. (2013). Silicon-based non aqueous anolyte for Li redox-flow batteries. *Journal of the Electrochemical Society* 160 (3): A516–A520.

150 Biendicho, J.J., Flox, C., Sanz, L., and Morante, J.R. (2016). Static and dynamic studies on $LiNi_{1/3}Co_{1/3}Mn_{1/3}O_2$-based suspensions for semi-solid flow batteries. *ChemSusChem* 9 (15): 1938–1944.

151 Chen, H., Zou, Q., Liang, Z. et al. (2015). Sulphur-impregnated flow cathode to enable high-energy-density lithium flow batteries. *Nature Communications* 6: 5877.

152 Chen, H., Zhou, Y., and Lu, Y.-C. (2018). Lithium–organic nanocomposite suspension for high-energy-density redox flow batteries. *ACS Energy Letters* 3 (8): 1991–1997.

153 Li, Z., Smith, K.C., Dong, Y. et al. (2013). Aqueous semi-solid flow cell: demonstration and analysis. *Physical Chemistry Chemical Physics* 15 (38): 15833–15839.

154 Ventosa, E., Amedu, O., and Schuhmann, W. (2018). Aqueous mixed-cation semi-solid hybrid-flow batteries. *ACS Applied Energy Materials* 1 (10): 5158–5162.

155 Fan, F.Y., Woodford, W.H., Li, Z. et al. (2014). Polysulfide flow batteries enabled by percolating nanoscale conductor networks. *Nano Letters* 14 (4): 2210–2218.

156 Yan, R. and Wang, Q. (2018). Redox-targeting-based flow batteries for large-scale energy storage. *Advanced Materials* 30 (47): e1802406.

157 Wang, Q., Zakeeruddin, S.M., Wang, D. et al. (2006). Redox targeting of insulating electrode materials: a new approach to high-energy-density batteries. *Angewandte Chemie International Edition* 45 (48): 8197–8200.

158 Jia, C., Pan, F., Zhu, Y.G. et al. (2015). High–energy density nonaqueous all redox flow lithium battery enabled with a polymeric membrane. *Science Advances* 1 (10): e1500886.

159 Zhou, M., Chen, Y., Zhang, Q. et al. (2019). $Na_3V_2(PO_4)_3$ as the sole solid energy storage material for redox flow sodium-ion battery. *Advanced Energy Materials* 9 (30): 1901188.

160 Cheng, Y., Wang, X., Huang, S. et al. (2019). Redox targeting-based vanadium redox-flow battery. *ACS Energy Letters* 4 (12): 3028–3035.

161 Yu, J., Fan, L., Yan, R. et al. (2018). Redox targeting-based aqueous redox flow lithium battery. *ACS Energy Letters* 3 (10): 2314–2320.

162 Huang, Q., Li, H., Grätzel, M., and Wang, Q. (2013). Reversible chemical delithiation/lithiation of $LiFePO_4$: towards a redox flow lithium-ion battery. *Physical Chemistry Chemical Physics* 15 (6): 1793–1797.

163 Wong, C.M. and Sevov, C.S. (2021). All-organic storage solids and redox shuttles for redox-targeting flow batteries. *ACS Energy Letters* 6 (4): 1271–1279.

10

Hybrid Capacitor

Lin Liu[1], Tianyi Wang[2], Hong Gao[3], Chengyin Wang[2], and Guoxiu Wang[3]

[1] *Suzhou University, School of Chemistry and Chemical Engineering, 1769 Xuefu Avenue, Education Zone, Suzhou, Anhui 234000, China*
[2] *Yangzhou University, College of Chemistry and Chemical Engineering, 180 Siwangting Road, Yangzhou 225002, China*
[3] *University of Technology, School of Mathematical and Physical Sciences, Centre for Clean Energy Technology, 15 Broadway, Sydney, NSW 2007, Australia*

10.1 Introduction

In recent decades, the imperious demands for renewable and sustainable energies have attracted worldwide attention. The demand can be fulfilled by the development of environmentally friendly energy conversion and energy storage devices like capacitors, batteries, and other energy storage devices. Among these energy storage devices, the capacitor has drawn a great deal of attention due to its high power density, long cycle life, and wide operating temperature range [1]. With the advancement of energy storage technology, HC has been developed, which has combined energy storage of electronic double-layer capacitor (EDLC) and pseudocapacitor. Hence, HC demonstrates excellent electrochemical performance including high specific capacitance and energy storage capability.

Generally, a hybrid capacitor at least consists of an anode, cathode, electrolyte, and separator. According to the composition, the electrode used in HC can be divided into composite electrodes, redox-asymmetric electrodes, and battery-type electrodes [2]. The composite electrodes are mainly comprised of carbon-based materials and metal oxides/conducting polymers, which have integrated EDL capacitive materials with pseudocapacitive materials. Composite electrode-type HC has dual electrodes, one composite electrode as the anode and the other composite electrode as the cathode. HC containing the redox-asymmetric electrode can be fabricated by using pseudocapacitive materials and their composites as the anode and carbon-based materials as the cathode. As for battery-type HC, it can be composed of the battery-type electrode as the anode and the capacitor-type electrode as the cathode. Metal-ion HCs have been widely fabricated, such as Li-ion, Na-ion, K-ion, Ca-ion, Al-ion, and Zn-ion HCs [3–5].

HC is being researched to improve energy density, cycle life, safety, and kinetics. The performance of HCs is mainly dependent on the electrode materials. Therefore,

Electrochemical Energy Storage Devices: Non-Conventional Technologies and Materials,
First Edition. Edited by Yongbing Tang and Luojiang Zhang.

tremendous studies have been done for designing and synthesizing excellent electrode materials to get high power density and long cycling life. Various types of materials have been investigated as electrode, such as carbon-based materials, metal oxides, conducting polymers, MXenes, graphitic carbon nitride, and metal-organic frameworks (MOFs) [3, 6, 7]. It is particularly noteworthy that the potential gap between anode materials and cathode materials can increase the cell's working voltage. Therefore, it is also required to improve operative voltage range and specific capacitance to achieve an overall boosted performance of HC. In addition, the performance of HC is closely related to the balance of anode and cathode reaction kinetics.

The key to improving the performance of HCs is to master the energy storage mechanism and properly control the influencing factors of HCs. The energy storage principle of HCs is based on charging and discharging processes in the anode/cathode and electrolyte interface [7–9], which is the same as that of conventional capacitors. The total capacitance is contributed by the EDL capacitance and pseudocapacitance. The EDLC is generated by reversible electrostatic adsorption/desorption of the electrolyte ions in the charge/discharge process, while the pseudocapacitance originates in repeatable redox reactions at particular potentials. The EDLC charge storage is a surface process, which takes place directly across the interface by means of surface dissociation, adsorption of ions from solution, and defects in the crystal lattice of electrode materials. Therefore, the surface characteristics of the electrode materials are crucial in charge storage and capacitance. There exists a wide range of EDLC materials, especially carbon-based materials. The key point of designing these EDLC materials is to attain high surface area and porous structure, which enables the acceleration of the overall ion mobility. The pseudocapacitance is induced by diverse reactions of pseudocapacitive materials (metal oxides and conducting polymers) on the electrode surface. The collaboration of EDLC and pseudocapacitance facilitates high energy density, high power density, and long cycling performance.

In addition, the electrolyte volume and concentration also affect the performance of HC, since the electrolyte provides ions during charge and discharge. There have been two main types of electrolytes, including aqueous electrolytes and nonaqueous electrolytes. The aqueous electrolytes facilitate high ion migration owing to their low viscosity, while the nonaqueous electrolytes can achieve high potential windows and stable performance of HC. The nonaqueous electrolytes, especially gel electrolytes, have opened up new ways for fabricating flexible solid-state HC.

Current research on HC mainly focuses on electrode materials and physical design and fabrication. As mentioned earlier, different types of materials have been investigated as electrodes for HC. Throughout these advances in electrode materials, it is of great significance to prepare electrode materials with high surface area, tunable 3D structures, and uniform morphology [10]. Moreover, several studies have been devoted to the fabrication and design of HC. The most common design of HC is the cell design, where all components of the HC can be assembled. In addition to cell design, cylindrical cell and pouch-type cell designs have also been employed in HC [11, 12]. Recently, printable and flexible solid-state HCs have been developed, which has provoked researchers' interest [13–15]. The flexible HC can be integrated with a charging unit or other function units, and thus can further broaden its applications in the microelectronics and promote commercialization of HC [14, 15].

This chapter introduces the basic composition, energy storage mechanism, performance assessment, and recent advances in HC. Considering the great variety of HC, it has been divided into three types: composite electrode type, redox-asymmetric type, and battery electrode type. The energy storage mechanism, electrode materials, and recent advances in each type of HC have been discussed, enabling a clear understanding of HC fabrication and design. Moreover, the composition and performance evaluation of HC have been explicitly addressed, allowing further understanding of the relationship between the composition and properties of HC. Recent research on HC has been discussed and interpreted with a view to stimulating more works on the design, fabrication, and application of high-performance energy storage devices.

10.2 The Formation, Energy Storage Mechanism, and Performance Evaluation of Hybrid Capacitor

10.2.1 The Compositions of Hybrid Capacitor

10.2.1.1 Anode and Cathode

As mentioned earlier, hybrid capacitor has combined EDL capacitive materials with pseudocapacitive materials. According to capacitive characteristic of electrode materials, they can be classified into three types: composite electrode, redox-asymmetric, and battery-type electrodes. The classification of the hybrid electrodes is based on the dominant capacitance behavior. The composite electrodes are mainly comprised of carbon and conducting polymer/metal oxide composites. Carbon materials with high surface area play a role in EDL capacitance, while conducting polymers or metal oxides have been served as pseudocapacitance. The HC using redox-asymmetric electrode has been fabricated by using metal oxides as the anode and active carbon as the cathode. The electrochemical performance of asymmetric HCs depends on the whole capacitor cell, which exhibits better performance than the symmetric ones.

The battery-type HC consists of battery-type electrode and supercapacitor electrode. Metal-ion hybrid capacitor (MIHC) belongs to the battery-type HC. There have been many electrode materials ranging from carbon materials to pseudocapacitive materials such as metal oxide, conducting polymers, MOFs, and MXenes. The battery-type electrode serves as the anode and the capacitor-type electrode as the cathode.

10.2.1.2 Electrolytes

The overall performance of HC is closely associated with electrolyte material. The electrolytes affect the performance of HCs in the following aspects: [16] (i) electrical conductivity, (ii) anion or cation adsorption, and (iii) dielectric features.

Generally, there are two main types of electrolytes: aqueous and nonaqueous electrolytes. Aqueous electrolytes, such as H_2SO_4, KOH, and KCl, have been commonly used in the capacitor. The selection of aqueous electrolyte should take account of the composition and proper concentrations of HC. The use of aqueous electrolytes can enhance specific capacitance of HCs because of their excellent ionic conductivity. However, the main limitation of aqueous electrolytes lies in the restricted operative voltage window.

The nonaqueous electrolytes mainly include organic electrolytes and ionic liquids, enabling HC to work in wide operating voltage windows. However, organic electrolytes such as acetonitrile have low ionic conductivity, toxicity, and flammability, which restrict their applications in HC. In addition, ion-conducting polymer electrolytes have been used for capacitor fabrication in both solid and gel form due to their good electrical, mechanical, and thermodynamic properties. Proton-conducting polymer electrolytes are commonly used in fabricating solid-state capacitors, which can be processed into the thin film with good mechanical stability and good conductivity. Verma et al. synthesized a silver ion-conducting solid polymer electrolyte based on activated carbon (AC) that can improve the ionic conductivity and then used in the solid-state capacitor [17]. Poly(2-acrylamido-2-methyl-1-propanesulfonic acid) (PAMPS) is also used as an electrolyte for fabricating flexible symmetric capacitor, and the ionic conductivity and capacitance can be improved by using inorganic additives [18]. Peng et al. designed a self-healing hydrogel electrolyte by using diol-borate ester bonding cross-linked poly(vinyl alcohol) (B-PVA) combined with graphite oxide (GO) [19]. The as-prepared B-PVA/KCl/GO hydrogel electrolyte behaved good ionic conductivity and self-healing capability and had great potential in the application of flexible solid-state capacitors. With the growing number of microcapacitors and wearable devices, hydrogel electrolytes would get increasing attention in the fabrication of flexible and wearable capacitors.

10.2.1.3 Separator

The quality of separator would affect the overall performance of HCs as well. Generally, separator for HCs shall meet the following requirements: [20] (i) nonconductive and ease of wetting; (ii) minimum ionic resistance; (iii) introduction of intrinsic resistance between electrolytes and electrode materials; and (iv) sufficient mechanical resistance.

Various materials have been used as separators, such as glass, paper, and other polymer-based materials. Owing to the low cost and good mechanical and chemical resistance, the polymer-based separators have been widely used in the HC, such as polypropylene, polyethylene, and polyvinylidene difluoride (PVDF) [21, 22]. Among these polymer materials, polypropylene is commonly used due to its easy wettability. However, due to the composition variation of HC, it is difficult and time-consuming to select and design the proper separators so as to attain good overall performance. Researchers are working on developing high-quality separators to enable accelerated ion and mass transfer. As depicted earlier, a good separator shall have good surface properties to enable surface cleaning and wettability.

10.2.1.4 Current Collectors

Current collectors have been used for transporting current from a current source to electrode and from electrodes to external loads. The commercial current collectors include steel, nickel, aluminum, iron, and alloys. The active materials have been uniformly coated onto current collectors, and the close contact between them can minimize the interface resistance, thus improving the overall performance of cells. In addition, 3D porous current collector had been designed for capacitor fabrication [23]. To enhance the contact between electrode materials and current collector, polymeric binding agents have been introduced, such as Nafion, PVDF, and polytetrafluoroethylene (PTFE). Abbas et al. investigated the difference in binding agents affecting the performance of capacitors [24]. The research results showed that PTFE-based electrodes were able to extend the thermodynamic potential limit of di-oxygen evolution (in 1 mol l^{-1} $NaNO_3$, $E_{ox} = 0.834$ V vs. SHE) to 1.4 V. It is important to select suitable binding agents, which can improve the performance of electrode materials. To enhance the contact without the binding agents, one strategy is that the electrode materials can be *in situ* grown on the surface of the collector. It can be a new research direction and perspective for development of future hybrid capacitor devices. Huang et al. proposed a flexible supercapacitor electrode by growing 3D TiO_2 nanoflowers on activated Ti foils and then coated with Au film and MnO_2 nanowires [25]. Au film can greatly improve the electrical conductivity of Ti/TiO_2 substrate and adhesion of MnO_2 nanowires. Pei et al. designed 3D Co_3O_4/CoS nanoneedle arrays to fabricate asymmetric capacitors [26]. Specifically, Co_3O_4 nanoneedle arrays were grown on nickel foam, followed by electrodeposition of ultrathin CoS nanosheets. Apart from the aforementioned, there is growing demand for the design and fabrication of free-standing electrodes without binder or collector for fabricating flexible solid-state HCs.

10.2.1.5 Sealants

Sealant plays a role in blocking HCs to prevent cells from contaminants (air, water, and chemicals). To avoid performance decay of HCs, it is essential to choose sealants of high quality. Owing to excellent moisture resistance and flexibility, polymer-based materials are widely used as sealants.

10.2.2 Energy Storage Principles of Hybrid Capacitors

The energy storage principle of HCs has combined EDL capacitance with pseudocapacitance. The synergistic contribution of EDLC and pseudocapacitance storage mechanism enables HCs to achieve high capacitance, high-energy density/power density, and cycle stability.

EDLC and pseudocapacitor are assembled in a battery to form a hybrid capacitor whose capacitance is controlled by the EDLC and pseudocapacitor electrode materials. According to the difference in the electrode materials, hybrid electrodes can be divided into composites: battery-type and asymmetric electrodes. Based on composite electrodes, HCs can be fabricated by using carbon/metal oxide or conducting polymer composites, followed by assembling two composite electrodes in one cell.

Carbon materials play the part of EDL capacitance while metal oxides or conducting polymers deliver pseudocapacitance. The capacitance of EDL capacitive materials is owned by the adsorption/desorption at the electrode surface. Therefore, the capacitance of EDLC materials has been restricted by their surface areas. Metal oxides and conducting polymers are pseudocapacitive materials, which undergo redox reaction under the operation voltage and deliver high specific capacitance and high energy density. However, these pseudocapacitive electrode materials show poor cyclic life. The combination of two electrode materials aims to overcome the drawback of the individual type. Figure 10.1a shows the charge/discharge mechanism of a symmetric composite electrode-based HC. The electrons generated by metal oxides transfer with the assistance of carbon materials, which minimizes charge transfer resistance of the electrode and provides ease in movement of ions and electrons.

Asymmetric hybrid capacitor has been designed by using one EDLC electrode and one pseudocapacitive electrode [27]. This design shows both absorption/desorption and redox reaction of energy storage mechanism. In principle, EDLC materials have been used as negative electrodes while pseudocapacitive materials act as positive electrodes. The charge storage mechanism of asymmetric HC can be depicted in Figure 10.1b. During the charge and discharge process, asymmetric HC can take full advantage of different operation potential window of negative electrode and positive electrode, which allows the asymmetric HC to work in the greater potential window. The HC with redox-asymmetric electrode demonstrates better cycling stability than comparable symmetric pseudocapacitors [28]. To achieve high-performance asymmetric HCs, it is essential to design electrode materials with high specific surface area, controllable pore size, and good electrical conductivity.

The battery-type hybrid capacitor is composed of battery-type electrode and capacitor-type electrode. The battery-type electrode provides high energy density and high capacitance by intercalating electrolytic ions to store charge. The supercapacitor-type electrode stores charge by adsorption of ions on the electrode surface during charging and discharging. Amatucci et al. first fabricated lithium-ion hybrid capacitors (LIHCs) by using $Li_4Ti_5O_{12}$ (LTO) as the anode and AC as the cathode [29]. Thereafter, a large quantity of research emphasizes on the development of other MIHCs, such as sodium-, potassium-, and zinc-ion HCs. During charge and discharge process, metal ions shuttle back and forth between positive and negative electrodes. The energy storage mechanism of MIHCs can be depicted in Figure 10.1c. During charging, metal ions are intercalated into the battery-type anode while anions in the electrolyte are absorbed to the fast surface capacitor-type cathode surface. During discharging, metal ions return to the electrolyte from anode; meanwhile, anions desorb from cathode to achieve charge balance.

10.2.3 Performance Evaluation of Hybrid Capacitor

The electrochemical performance of capacitor can be evaluated by DC test procedures [30]. However, there is a large inconsistency between evaluation criteria and testing procedures, which makes it difficult to compare different testing techniques. Moreover, due to the diversity of electrode materials, it is unreasonable to evaluate

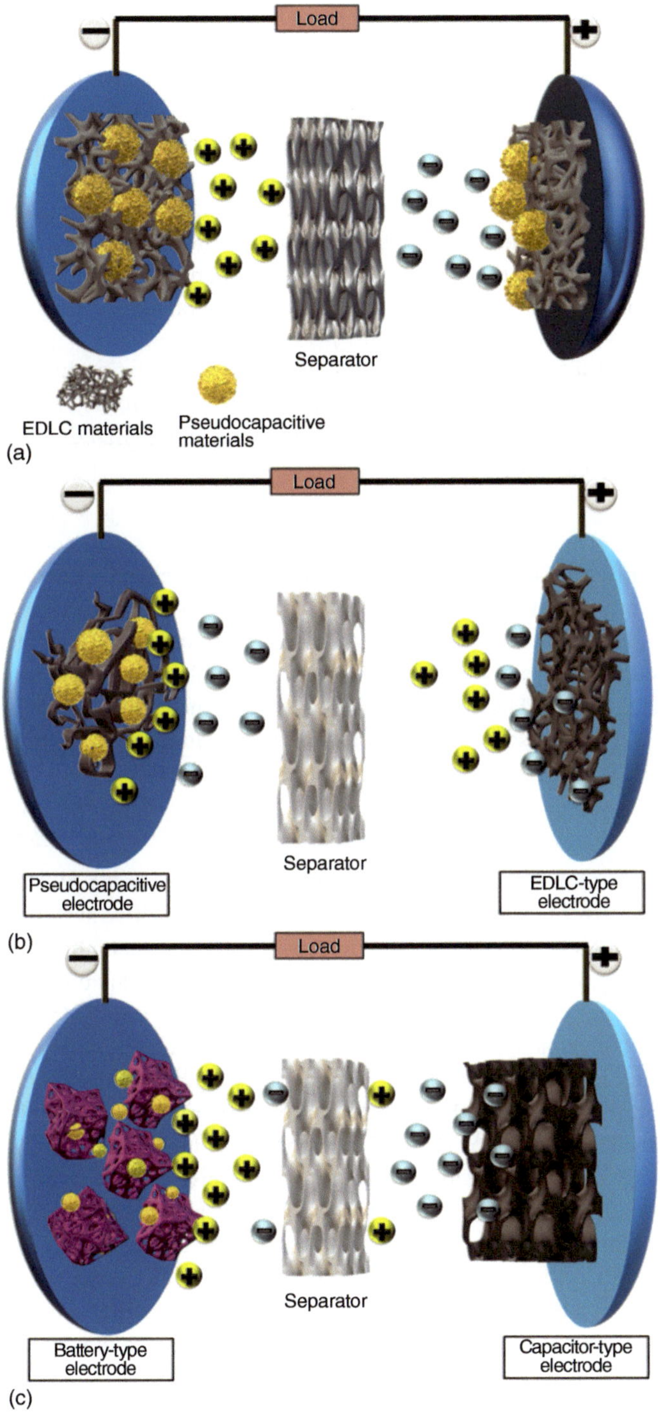

Figure 10.1 Schematic diagrams of energy storage mechanism for HCs: (a) composite-type HC, (b) asymmetric HC, and (c) battery-type HC.

the performance of HC by using simple test procedures. Hence, the proper testing procedure should be optimized on the basis of HC's energy storage mechanism. As mentioned in the previous section, HC has combined intercalation, electrostatic adsorption, capacitive Faradaic reaction, or noncapacitive Faradaic reaction in one or both electrodes. Therefore, evaluating the performance of HCs is more complicated than that of EDLCs. However, the current test indicators for evaluating the performance of HC are nearly the same as those of EDLC, including at least capacitance, operating voltage window, resistance, cycle life, energy density, and power density.

10.2.3.1 Capacitance

The evaluation of capacitance can be determined by the galvanostatic charge/discharge (GCD) test. In a typical charging–discharging curve (V–t profile), there is a linear variation for a constant current discharge process for EDLC. The capacitance of EDLC can be calculated by Eq. (10.1):

$$C = \frac{\Delta Q}{\Delta V} = \frac{i\Delta t}{\Delta V} \tag{10.1}$$

where "C" represents differential capacitance ($F\,g^{-1}$), "i" is the current density ($A\,g^{-1}$), "ΔV" is the operation voltage window, and "Δt" represents the discharging time (s) of the electrode. However, HC has combined the non-Faradaic and Faradaic charge storage. For pseudocapacitive materials, the operative voltage window can be fixed in certain electrolytes. The aforementioned capacitance calculation is not appropriate for HCs. It can also be observed that a highly nonlinear constant current discharge process in the V–t profile of HCs can be determined by Eq. (10.2):

$$C = \frac{i}{\mathrm{d}V/\mathrm{d}t} \tag{10.2}$$

In addition, the voltage drop (IR) at the initial discharge stage can be always observed in the constant current test for both EDLC and pseudocapacitive systems. Therefore, the resistance shall be taken into account for capacitance calculation. In this regard, more reliable parameters to evaluate the energy storage capability of HC devices are the performance characteristics in capacity (Ah) and energy (Wh). Therefore, the capacitance, capacity, and energy density of HCs are necessary to be tested in most studies related to capacitors. It is worth noting that the cutoff voltage is also an important parameter that needs to be considered as it is mainly dependent on the electrode materials of HCs. The discharge cannot be conducted below the cutoff shoulder voltage. In addition, some commercial HCs have been prepared by composite electrodes, which are composed of carbon materials and metal oxides. However, metal oxides only account for a small amount of the composition and their contributions to the overall capacitance are different. Metal oxides as pseudocapacitive materials often suffer from poor cycle life. Therefore, it is really challenging to give accurate assessment just on the basis of capacitance.

10.2.3.2 Steady Operating Voltage Window

The assessment of operating potential window shall take into account the resistance of electrode materials, electrochemical kinetics, and thermodynamic property of electrolytes. Cyclic voltammetry (CV) is usually used for determining the steady operative voltage window of HC. The determination is done by applying a potential while calculating the current in a three-electrode system. The operating voltage window can be determined by the transformation of CV shape and size according to the materials under investigation. However, it is not accurate for electrode materials involved in faradaic processes to identify potential window via the aforementioned method. Therefore, there have been different methods and procedures developed for determining operating voltage window [31–33]. One approach has been proposed to achieve maximum operating voltage, which is required to adjust the mass ratio of electrode materials and comply with charge balance principle according to Eq. (10.3) [31]:

$$\frac{m_+}{m_-} = \frac{C_-E_-}{C_+E_+} \tag{10.3}$$

where "m" is the mass of the anode or cathode material, "E" is the voltage window for charging/discharging, and "C" is the specific capacitance under the corresponding voltage window. Although this approach can be viable in the principle of mass balance, it is still required to do empirical correction according to practical applications. Xu et al. had proposed the coulombic efficiency as the stability criterion to assess the given potential window value according to Eq. (10.4) [32]:

$$S = \frac{Q_{\text{charge}}}{Q_{\text{discharge}}} - 1 \tag{10.4}$$

where "Q_{charge}" and "$Q_{\text{discharge}}$" are the stored and released charges, respectively. The maximum operating voltage window is identified where $S < 0.1$. This method was further re-examined and reformulated by Weingarth et al. [34]. Even though more and more methods have been developed for identifying operating voltage window, it is still difficult to guarantee the accuracy of final results. It is because the stability of HC devices has been affected by many factors, and the operating voltage window is only one of them. In addition, given the current procedures on the stability assessment, it is not sufficient to confirm the stability of HCs.

10.2.3.3 Resistance

Resistance is an important parameter of HCs, which is closely related to power performance. Electrochemical impedance spectroscopy (EIS) is regarded as an effective technique to measure the resistance of HCs by interpreting the Nyquist plots. The Nyquist plot includes three regions: a semicircle at higher frequency, a vertical line at high and medium frequency, and an imaginary line along a vertical line at low frequency. The semicircle in the plot is induced by the interfacial resistance. The vertical line in the plot is due to charge transfer resistance and an imaginary line along the vertical line represents the capacitive behavior [35]. A typical Nyquist plot provides information on the equilibrium differential capacitance (ESR) and resistance of electrode and electrolyte. However, there are multiple physical interpretations of the Nyquist plots of HCs.

In addition, GCD technique can also be used for determining the resistance of HCs. ESR is usually used to interpret the internal resistance of electrode active material, electrolyte, and contact resistance between the electrode and the current collectors. In addition, voltage drop has also been used to calculate the resistance. The obtained resistance (steady-state resistance, R_{ss}) value calculated by the voltage drop closes to the ESR after the micropores of electrode are fully engaged. The R_{ss} is the best performance parameter for evaluation of pulse power capability of HCs. In general, the resistances of the HCs are higher than that for typical EDLC cells of comparable capacitance.

10.2.3.4 Energy Density and Power Density

The Ragone plot has been used to evaluate the overall performance of HC. The energy density can be calculated by integrating the discharging curves. The stored energy (E_s) is related to the capacitance C, rated voltage V_r and lower limit of the discharge voltage (V_{min}). It can be expressed in Eq. (10.5):

$$E_s = \frac{C_{eff}\left(V_r^2 - V_{min}^2\right)}{2} \tag{10.5}$$

where C_{eff} is assumed to be constant independent of discharge power.

The energy density of HCs can be calculated according to Eq. (10.6):

$$E_s = \frac{CV_r^2}{2} \tag{10.6}$$

The energy density of hybrid capacitors should be determined by testing them over a wide range of power densities. However, the energy calculation as the aforementioned equation could overestimate or underestimate the energy content of HCs.

The correlation between energy and power density is depicted in Eq. (10.7):

$$P = \frac{E}{t} \tag{10.7}$$

where "P" represents power density (W kg^{-1}), "E" denotes energy density (Wh kg^{-1}), and "t" is time. Maximum power density (P_{max}) depends on the square of the maximum voltage (V_{max}) and equivalent series resistance denoted by R_{ss} as shown in Eq. (10.8).

$$P_{max} = \frac{V_{max}^2}{4R_{ss}} \tag{10.8}$$

The energy density of HC decreases along with the increase in power/current density due to the resistance of electrode. The ESR is the total resistance, including resistance of electrode, electrolyte, and the diffusion resistances of ions in electrode pores [36].

As mentioned earlier, power density and energy density are both related to the mass loading of active materials in the electrode. It is crucial to optimize the mass ratio of electrode materials so as to achieve good performance.

10.2.3.5 Cycle Life

HC's electrode contains pseudocapacitive materials, which have limited cycling stability due to the Faradaic reaction during the charge and discharge process. With the increase in cycling numbers, these pseudocapacitive materials are gradually decomposed, thus leading to mechanical failure and performance degradation of HC. The only available life cycle data for hybrid capacitors is from the manufacturers, which show good cycle life. There have been many factors that affect cyclic life, such as operating voltage, temperature, and humidity. The reliability assessment can be conducted by determining the lifetime of various devices at different specified voltage and temperature conditions.

10.3 Recent Advances in Hybrid Capacitors

10.3.1 Hybrid Capacitors with Composite Electrodes

Composite electrodes are commonly composed of EDLC-type materials and pseudocapacitive materials (metal oxides or conducting polymers). The EDL capacitive materials with high surface area can serve as frameworks for bearing metal oxides or conducting polymers. Carbon-based materials, as typical EDLC-type materials, can greatly improve the specific capacitance and operating potential windows. Owing to porous structure and high surface area, carbon materials can shorten ion diffusion path and reduce the ion transport resistance of HCs.

The application of carbon-based materials in capacitors has been extensively studied, such as AC, graphene, carbon nanofibers (CF), carbon nanotubes, and other derivatives. To further improve capacitance and energy density, many researchers mainly focus on the investigation of surface modification and morphology control for carbon materials. Hence, heteroatom doping has been introduced for increasing active sites for carbon materials. In this regard, the most impressive is biomass-derived carbon material, which has the advantage of abundant source, direct heteroatom-doped elements, and multiple structures. For composite electrode design, carbon materials play the part of EDL capacitance and also provide the anchor site for other pseudocapacitive materials. The overall performance improvement of HC depends on the composition ratio and interconnected porous structures of composite electrodes.

Transitional metal oxides have been widely used in the fabrication of capacitor due to their pseudocapacitive characteristics. Among these transitional metal oxides, manganese oxides and niobium pentoxides have been extensively designed to be combined with carbon materials to form composite electrodes [37–40]. Besides, other transition oxides incorporated with carbon materials have also been employed as composite electrodes, such as tungsten oxides and vanadium oxides [41, 42]. Most noteworthy, binary metal oxides have also been proposed and used as electrode materials, which could enhance the pseudocapacitance compared with the use of single metal oxides [43]. Park et al. had proposed the manganese/vanadium (Mn/V) oxide coupling with multiwalled carbon nanotube (MWCNT) to fabricate composite electrode (Mn/V oxide@MWCNT) [44]. The flexible HC was assembled using Mn/V oxide@MWCNT electrode and sulfone-based electrolytes. The as-fabricated HC behaved excellent areal capacitance (11.8 $mF\,cm^{-2}$), high areal energy density, and power density (6.58 μWh cm^{-2} at 200 μW cm^{-2}). Furthermore, the flexible HC had been integrated with a strain sensor and a gecko-inspired adhesive into a single device, which had been used for detecting strains generated by bio-signals and skin/body movements.

Transition metal sulfides have also been developed in the application of capacitors, due to their higher conductivity and more flexible structures compared to metal oxide counterparts. In particular, molybdenum disulfides are preferable to be utilized for electrode materials. Wu and coworkers had designed molybdenum disulfide-carbon nanotubes (MoS_2-CNTs) composite, which demonstrated good conductivity, fast ion diffusion, and abundant electrochemical activity [45]. Two as-prepared MoS_2-CNTs composite electrodes were further used for fabricating a flexible solid-state HC through direct-write printing construction (shown in Figure 10.2a). The as-fabricated flexible solid-state HC was integrated with a smart thermometer, which can be used for monitoring the human skin surface temperature in real time. The as-prepared integrated devices delivered high capacitance (723 F g^{-1}) and energy density (226 mWh g^{-1}), which could also attain good stability at high/low temperature (shown in Figure 10.2b–g). As listed in Figure 10.2h, the as-assembled flexible HC had the capability of powering the wearable temperature-monitoring device.

Binary metal sulfides show better pseudocapacitance in comparison to single metal sulfides, due to the presence of multiple redox reactions induced by two transition metal ions. $NiCo_2S_4$ has been investigated in the fabrication of capacitor due to its abundant electroactive sites, low cost, and high theoretical capacitance [46]. Wang et al. prepared hierarchically wood-derived carbon (WC)@Ag@$NiCo_2S_4$ composite electrodes for the HC [47]. The asymmetric HC was assembled by using WC@Ag composites as the anode and WC@Ag@$NiCo_2S_4$ composites as the cathode. The as-prepared asymmetric HC delivered a high areal and volumetric energy density of 0.59 mWh cm^{-2} and 3.93 mWh cm^{-3}, respectively.

Transition metal phosphate has also been used in the field of HC. The cobalt phosphate grown on a flexible woven carbon (CoPi/C) was prepared with the assistance of phytic acid (PA) [48]. The as-obtained flexible composite electrode delivered high energy density and power density (31.1 Wh kg^{-1} at 476.0 W kg^{-1}) for solid-state symmetric HC.

Apart from the aforementioned metal oxides, conducting polymers as typical pseudocapacitive materials are good candidates for capacitor fabrication, especially for flexible solid-state capacitors. The most commonly used conducting polymers are polypyrrole, polythiophene, polyaniline (PANI), and PEDOT. Conducting polymers with tunable morphologies and fast charge–discharge capability have been regarded as good candidate materials for supercapacitor devices [49]. Conducting polymers deliver pseudocapacitance by doping/dedoping anions or cations during redox reaction. However, the conducting polymer suffers from slow ion diffusion and low charge–discharge rate. Therefore, combining with carbon materials can be a good strategy to broaden the application of conducting polymers in the application of capacitors. Conducting polymers coupled with carbon materials can greatly improve capacitance, energy density, and power density of capacitor device.

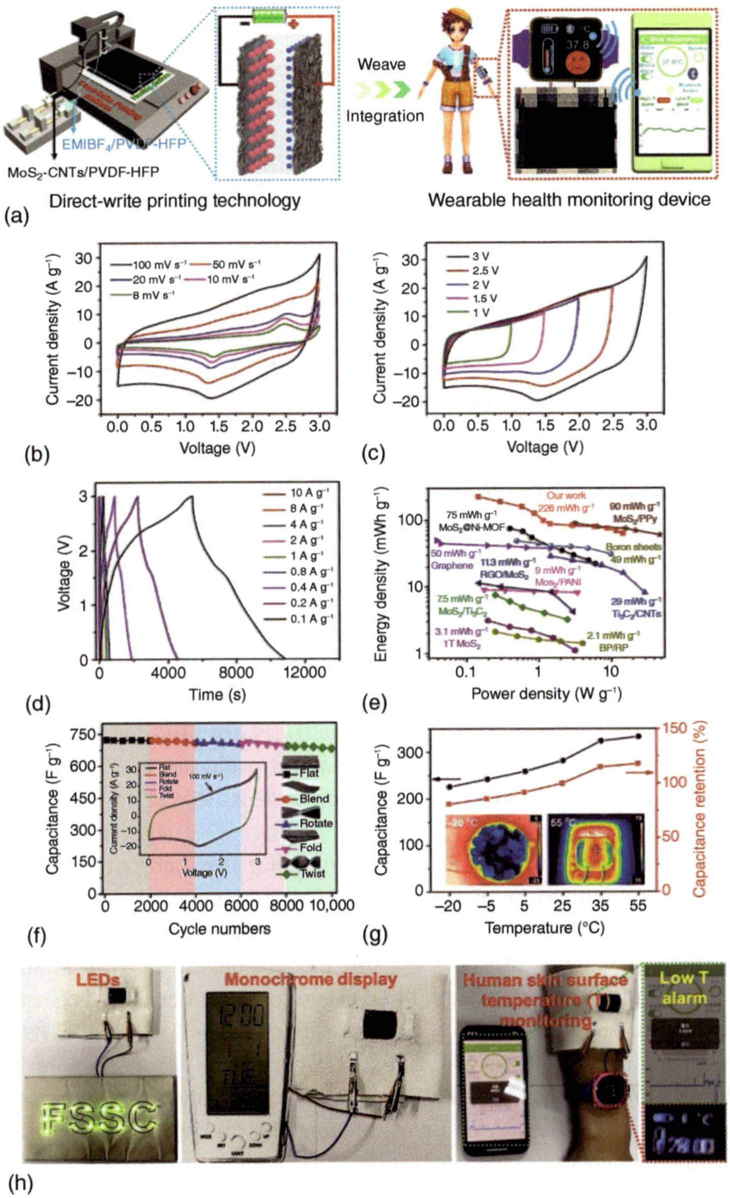

Figure 10.2 (a) The construction of flexible HC integration into a wearable health-monitoring device. (b) CV curves of the as-prepared flexible HC. (c) CV curves of as-prepared flexible HC under different voltage windows at 100 mV s^{-1}. (d) Galvanostatic charge/discharge curves of flexible HC at current densities from 0.1 to 10 A g^{-1}. (e) Energy density and power density comparison. (f) Cycling stability under various deformations of flexible HC. (g) The relationship between operating temperature and capacitances at 1 A g^{-1}. (h) Photographs of flexible HC powering LEDs, monochrome display, and the wearable temperature-monitoring device. Source: Reproduced with permission of Wu et al. [45]. © 2021/John Wiley & Sons.

Muthulakshmi et al. fabricated a symmetric HC by AC/polypyrrole composites, which displayed a specific capacitance of around 354 F g^{-1} [50]. Moyseowicz and Gryglewicz designed HC based on porous polypyrrole/reduced graphene oxide composites in a redox-active electrolyte. The assembled symmetric device achieved high energy density and excellent long-term stability [51]. In addition, a freestanding and thick polypyrrole/polydopamine@carbon foam electrode was rationally designed and used as electrodes for symmetric HCs [52]. The assembled symmetric HC exhibited a high capacitance (996 mF cm^{-2}) and a high energy density (0.12 mWh cm^{-2}) with excellent cyclic stability.

PANI is also a good electrode material for flexible solid-state HCs. Hu et al. developed a flexible solid-state symmetric HC by using polyvinyl alcohol (PVA)/carbon nanomaterials (CNMs)/PANI composite as electrodes [53]. The fabricated symmetric HC delivered high specific capacitance (284.6 mF cm^{-2}) and areal capacitance (216.2 mF cm^{-2} at 0.5 mA cm^{-2}) along with excellent cyclic stability.

Mxenes can be integrated with conducting polymers to form composite electrodes. Luo et al. prepared flexible $Ti_3C_2T_x$ (MXene)/PANI films for fabricating all-solid-state symmetric HC [54]. The assembled device exhibited a high specific capacitance (272.5 F g^{-1} at 1 A g^{-1}), high energy density, and power density (31.18 Wh kg^{-1} at 1079.3 W kg^{-1}).

In addition, redox-active metal–organic fragment can be introduced to conducting polymer for preparing conducting metallopolymer. Chen and Wong introduced a redox-active ferrocenyl moiety onto a polythiophene substrate through electropolymerization [55]. The proposed composite electrodes are used for a flexible all-solid-state symmetric HC, which exhibited ultrahigh capacitance (1.35 F cm^{-2}), high energy densities (over 0.37 mWh cm^{-2}), and power densities (22.4 mW cm^{-2}) with good cycling performance.

Conducting polymer can not only combine with carbon but also incorporate metal oxides/sulfides to prepare composite electrodes. Poly(3,4-ethylenedioxythiophene) (PEDOT) as a derivative of polythiophene inherits high electrical conductivity and good chemical stability. Wang et al. designed the PEDOT-glued V_2O_5/graphene (VP-G) films for fabricating solid-state symmetric HC, which exhibited a high energy density (0.18 μWh cm^{-2} at 11 μW cm^{-2}) [56]. He et al. prepared CNTs/PEDOT sponge electrodes for fabricating an all-solid-state symmetric HC with a high voltage range of 1.4 V [57]. The assembled HC showed the highest energy density of 12.6 Wh kg^{-1} under the power density of 1 kW kg^{-1} and the highest power density of 10.2 kW kg^{-1} with an energy density of 8 Wh kg^{-1}.

Das et al. designed binder-free CuS@PEDOT composite electrode (cathode) on carbon cloth (CC) substrate (CC/CuS@PEDOT) and the CC/Co-V-Se (CC/CVS) composite electrode as the anode for fabricating quasi-solid-state HC [58]. The as-assembled HC demonstrated a high volumetric energy density of 2.21 mWh cm^{-3} and good cyclic stability (96.7% capacity retention after 10,000 cycles). Yeasmin et al. prepared an all-solid-state symmetric HC based on graphene-polyaniline nanofiber electrode [59]. The as-prepared HC demonstrated a specific capacitance of 28.37 F g^{-1}, high energy density (8.32 μWh cm^{-2}), and power density (39.97 μW cm^{-2}). Liu and coworkers used PEDOT: polystyrene sulfonate/CNT

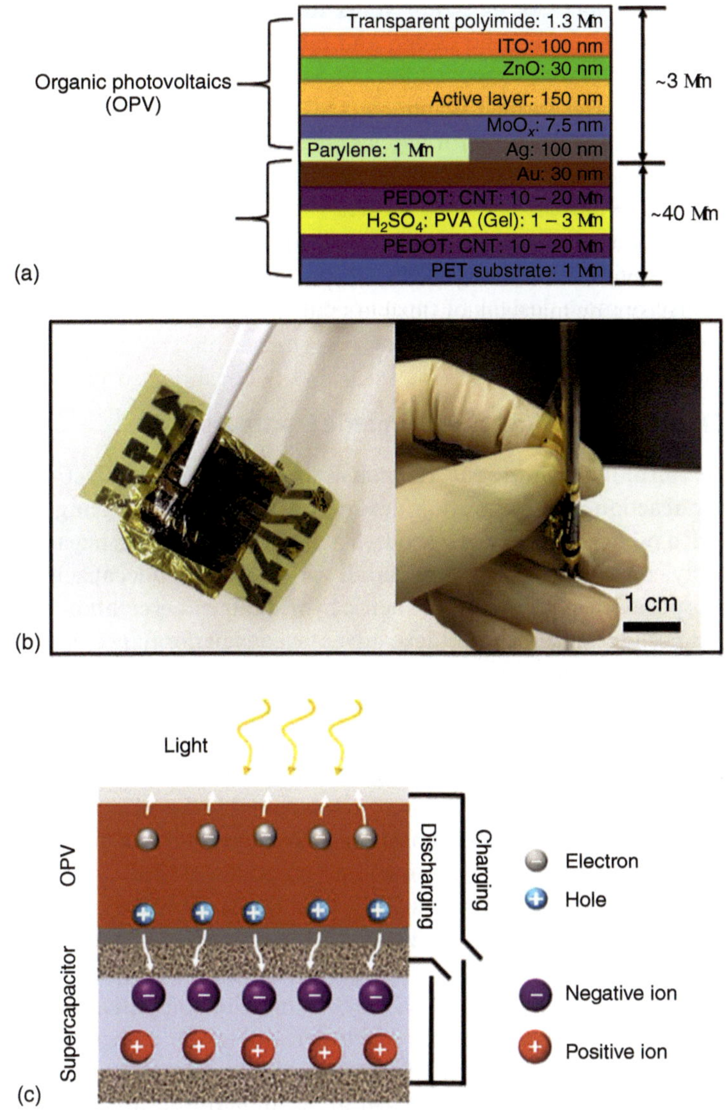

Figure 10.3 (a) Schematic diagram of the integrated photocharging capacitor. (b) Photographs of the as-fabricated ultrathin device (left) and the device wrapped onto a rod (diameter: 2 mm). (c) Illustration of the working mechanism of the photocharging process. Source: Reproduced with permission of Liu et al. [60]. © 2020/John Wiley & Sons.

composites to fabricate a flexible photocharging symmetric HC [60]. As depicted in Figure 10.3a,b, the photocharging HC was composed of organic photovoltaics (OPVs) and HC part. The photoactive layer of top electrode generated electron–hole pairs under the light illumination, followed by transferring to electrodes. The electrodes of HC then received the electrons and the other electrode would take the holes (shown in Figure 10.3c).

Currently, the research's focus of capacitor shifts from design and fabrication to its application in different fields. The design and fabrication of capacitor aims to support and broaden its practical application. Khodabandehlo et al. proposed a symmetric HC by using laser-scribed graphene (LSG) coupled with PANI [61]. The as-prepared HC-based LSG-PANI with wireless charging capability had been integrated with thin-film humidity sensor, a resistor, and a near-field communication (NFC) antenna in one device. The as-fabricated flexible HC exhibited a high energy density/power density (0.407 mWh cm^{-3}/196 mW cm^{-3}) with good rate capability. This study provides new thoughts for fabricating the flexible integrated system, which can inspire more designs of capacitor-integrated devices for diverse applications.

10.3.2 Hybrid Capacitors with Redox-Asymmetric Electrodes

As listed in Figure 10.1b, the asymmetric HC consists of two different electrode materials for redox reaction and absorption/desorption process by coupling an EDLC electrode with a pseudocapacitive electrode. Theoretically speaking, negative electrodes are usually composed of carbon material, while some pseudocapacitive materials are used as positive electrodes. Asymmetric hybrid capacitors combine the merits of EDLC and pseudocapacitor to achieve advantageous performance. Many pseudocapacitive materials, such as metal oxides, MOFs, metal sulfides, conductive polymers, and their composites, have been extensively utilized as positive electrodes in asymmetric supercapacitors.

Transition metal oxides including but not limited to manganese oxides, vanadium oxides, cobalt oxides, and zinc oxides (ZnO), have been reported in the fabrication of asymmetric HC. Alagar et al. proposed mesoporous Mn_2O_3 nanospheres/nanocubes as positive electrode and reduced graphene oxide as negative electrode to fabricate asymmetric solid-state HC [62]. The as-fabricated HC delivered an energy density of 46 Wh kg^{-1} at a power density of 247 W kg^{-1} and excellent capacitance retention of 95% after 5000 cycles in PVA-$LiNO_3$ gel solid-state electrolyte. Chen et al. assembled an asymmetric HC by a CNT/V_2O_5 nanowire composite as the anode and AC as the cathode in an organic electrolyte [63]. The as-assembled asymmetric HC showed a specific energy of 40 Wh kg^{-1} at a power density of 210 W kg^{-1} at 0.5 mA cm^{-2}. Zhou et al. fabricated an asymmetric flexible HC by using V_2O_5/polyindole@activated carbon cloth (V_2O_5/PIn@ACC) as the anode and a reduced graphene oxide@activated carbon cloth (rGO@ACC) as the cathode [64]. The as-prepared HC demonstrated superior electrochemical performance including high energy density and power density (38.7 Wh kg^{-1} at 900 W kg^{-1}) along with good cyclic stability. Ndiaye et al. fabricated an asymmetric HC based on the vanadium dioxide/activated expanded graphite (VO_2/AEG) composites (positive electrode) and porous carbon-vanadium oxynitride (C-V_2NO) (negative electrode) operating in a 6 mol l^{-1} KOH electrolyte [65]. The prepared asymmetric HC showed a specific energy of 41.6 Wh kg^{-1} with a corresponding specific power of 904 W kg^{-1} at a 1 A g^{-1} in a large operating voltage of 1.8 V. Devarayapalli et al. fabricated V_2O_5 nanobelt/cetyltrimethylammonium bromide (CTAB)-modified g-C_3N_4 nanosheet

(CCN/VO) composites to assemble an asymmetric HC [66]. The as-fabricated HC exhibited a high energy density (96.6 Wh kg^{-1} at 811.0 W kg^{-1}) in the voltage window of 1.5 V. Nickel oxide (NiO) and hydroxide can also be prepared for asymmetric HCs. The NiO/reduced graphene oxide (rGO) composites had been designed with tunable microstructures and used as the anode materials for fabricating solid-state asymmetric HC [67]. The as-fabricated HC in aqueous 6 M KOH solution, organic $LiPF_6$ solution and ionogel polymer showed energy densities of 31.6, 49, and 146 Wh kg^{-1}, respectively. In addition, the Co_3O_4@$Ni(OH)_2$ electrode materials had been proposed for fabricating asymmetric HC [68]. The as-assembled HC delivered an energy density of 112.5 Wh kg^{-1} at a power density of 1350 W kg^{-1}. Throughout the aforementioned studies, it can be concluded that the combination utilization of one metal oxide with carbon materials and conducting polymers to fabricate HCs is an effective way to improve the overall performance as the combination use of two or more materials can show a synergic effect. Additionally, this further corroborates that the HC's performance is closely related to its fabrication and design.

ZnO with the tunable structures and morphologies are often used for the fabrication of triboelectric nanogenerators (TENGs). However, TENG is required to be integrated with energy storage devices for practical applications [69]. Jayababu and Kim had proposed ZnO nanorods (NRs)@conductive carbon black (CB) nanocomposite (ZnO NRs@CB) for fabricating flexible triboelectric nanogenerator (FZCT) and hybrid capacitor (FZCS) [70]. FZCS was fabricated by using ZnO NRs@CB composite-coated nickel foam as the anode and AC-coated nickel foam as the cathode. The as-fabricated FZCS had been integrated with the as-fabricated triboelectric nanogenerator (FZCT), which can efficiently convert and store the energy generated by flexible TENGs.

Apart from the aforementioned metal oxides, lead dioxide (PbO_2) with high theoretical energy density is another material that is commonly used as the anode for fabricating lead-carbon hybrid capacitors (LCHCs). The cathode of LCHC is usually made of carbon-based materials, which stores energy by adsorption–desorption process. However, the effective utilization of the positive active materials is just 12.5%, which limits the development of LCHCs [71]. Therefore, two main strategies have been employed for improving the performance of LCHCs: (i) accelerating fast double-layer absorption/desorption of carbon materials (negative electrode) and (ii) balancing the kinetics of oxygen evolution reaction and positive charge reaction. It is of great significance to design PbO_2 electrode with high specific surface area, which can improve the utilization ratio of positive active materials. Doping strategy has been applied to functionalize PbO_2 electrodes, such as Mn-PbO_2 and F-PbO_2 [72, 73]. Muduli et al. proposed nitrogen- and phosphorous-doped carbon materials as the cathode active materials and *in situ* synthesis of PbO_2 as the positive electrode to fabricate a LCHC. The as-fabricated LCHC showed good cycling stability [74].

Binary metal oxides have been employed in the fabrication of asymmetric HCs due to their redox capability and electrochemical activity. $NiCo_2O_4$ arrays had been designed and grown on Ni foam [75]. The asymmetric HC was fabricated by using $NiCo_2O_4$ arrays as the positive electrode and nitrogen-doped porous carbon (NPC) as the negative electrode. The $NiCo_2O_4$//NPC-assembled asymmetric HC exhibited a maximum energy density of 32.08 Wh kg^{-1} at a power density of 700.43 W kg^{-1} in a 2 mol l^{-1} KOH electrolyte. Acharya et al. proposed $CoFe_2O_4$/MWCNT hybrids for fabricating an asymmetric HC [76]. The as-fabricated HC delivered a maximum specific capacitance of 390 F g^{-1} at a current density of 1 mA cm^{-2} and good rate capability. $ZnCo_2O_4$ as electrode material has also been extensively investigated in the application of capacitor owing to its unique structure characteristics and high capacitance. Liu et al. prepared $ZnCo_2O_4$/$CoMoO_4$ heterostructure electrode as the cathode and active carbon as anode for assembling an asymmetric HC [77]. The assembled asymmetric HC demonstrated an energy density of 135.6 Wh kg^{-1} at 2704.1 W kg^{-1} and capacity of 95.1% after 8000 cycles.

Transition metal layered double hydroxides (TMLDHs) have been employed in fabricating capacitor due to their redox activity. Tsai et al. had prepared SWCNTs/Ni-Co-Mn layered double hydroxide (NCMH) nanocomposites to fabricate asymmetric hybrid capacitors [78]. The SWCNT-NCMH composites served as positive electrodes while SWCNTs acted as negative electrodes. The fabricated asymmetric HCs delivered a high energy density of 11.17 Wh kg^{-1} with the corresponding power density of 66.2 W kg^{-1} in 1 mol l^{-1} KOH. Acharya et al. designed 3D hierarchical zinc–nickel–cobalt (ZNCO)@Co-Ni-LDH core–shell nanostructured arrays as the pseudocapacitive electrode (positive electrode) and N-doped graphene hydrogel (NGH) as the negative electrode for assembling asymmetric HC [79]. The as-fabricated HC demonstrated specific capacitance (178 F g^{-1} at 1 A g^{-1}) and very high energy density (63.28 Wh kg^{-1} at a power density of 796.53 W kg^{-1}). The electrochemical method had been used for Ni-Al layered double hydroxide (Ni-Al LDH) that was intercalated with electrochemically reduced graphene oxide (ERGO) to form LDH/ERGO composites [80]. The asymmetric HC was fabricated by using LDH/ERGO as the positive electrode and AC as the negative electrode, which exhibited capacity retention of 88% and the charge/discharge efficiency of 95%.

Transition metal hexacyanoferrates (MHFs) have also been developed for fabricating capacitors due to their unique archetypal hexacyanometalate structure. The unique ability to accommodate metal cations endows MHFs with good electrochemical property, extraordinary reversibility, and excellent stability. Nickel hexacyanoferrate (NiHCF) had been integrated with diallyldimethylammonium chloride (PDDA)-induced reduced graphene oxide (PRGO) to form PRGO-NiHCF composites [81]. The asymmetric HC had been fabricated using the as-prepared PRGO-NiHCF as positive electrode and interconnected sandwiched graphene carbon (SGC) as negative electrode, which delivered a wide operating voltage of 1.8 V and a high energy density of 25.4 Wh kg^{-1}. In addition, other Prussian blue analogs (PBAs) have been explored for fabricating electrodes of HC as well. Choi et al. prepared CoFe PBA/nitrogen-doped carbon composites as the positive electrode and nitrogen-doped carbon as the negative electrode for fabricating a quasi-solid-state HC [82]. The as-fabricated HC demonstrated a high energy density and power density (42.9 Wh kg^{-1}/14.6 kW kg^{-1}), which had been used for powering LED paper watch.

Metallic sulfides and binary metallic sulfides have been reported to be used in the application of HC. Wang and coworkers constructed a hierarchical Ni_3S_2/CoFe

LDH/Ni as the anode and AC as cathode for fabricating asymmetric HCs [83]. The as-constructed HC delivered a high power density of 986 W kg^{-1} and a high energy density of 47.31 Wh kg^{-1} with good cycling stability. $NiCo_2S_4$ nanostructures are widely reported as pseudocapacitive materials. Liu et al. prepared hybrid $NiCo_2S_4$/carbon materials as the positive electrode and porous carbon electrode as the negative electrode to assemble an asymmetric HC [84]. The as-obtained HC delivered a high energy density of 53.7 Wh kg^{-1} at 184.4 W kg^{-1} and a good cyclic stability. Khalafallah et al. designed hierarchical yolk-shelled nickel cobalt sulfide hollow cages that were modified with carbon ($NiCo_2S_4$/C) [85]. The as-obtained $NiCo_2S_4$/C composites as the positive electrodes were used for fabricating asymmetric HC, which achieved a maximum energy density of 60.2 Wh kg^{-1} and a power density of 375 W kg^{-1} with a good cycling stability. Cu_2MoS_4 has also been studied as pseudocapacitive materials due to the presence of the various oxidation states of Mo and Cu ions. The hetero-network-based MoS_2@Cu_2MoS_4-210 (MS@CMS-210) and Co-containing MS@CMS-210 had been prepared by hydrothermal synthesis method and then were used for constructing asymmetric hybrid capacitor [86]. The pouch-type HC (Co-MS@CMS-210//AC) had provided maximum energy and power density values of 41.6 Wh kg^{-1} and 6240 W kg^{-1}, respectively.

Conducting polymers have been employed in the fabrication of asymmetric HCs, which are served as pseudocapacitive electrode materials. He et al. designed monolithic nickel sulfide@PEDOT arrays binder-free cathode for asymmetric HC [87]. The assembled Ni_3S_2@PEDOT//active carbon asymmetric HC delivered capacity of 243.6 F g^{-1} at 0.5 A g^{-1} and outstanding cycling durability. It has been noted that the common synthesis process for conducting polymer is electropolymerization, which may affect the stability of device due to the solvent used in the electrochemical deposition. Chemical vapor deposition (CVD) has been introduced to synthesize conducting polymers. Zhou and coworkers had proposed horizontally aligned CNT arrays (HACNTs) coated with poly(3-methylthiophene) (P3MT) by oxidative CVD to obtain P3MT/HACNT flexible electrode [88]. The as-obtained electrode was further used as the positive electrode and HACNTs as the negative electrode to assemble a flexible asymmetric HC. The assembled HC delivered maximum energy and power densities of 1.08 mWh cm^{-2} and 1.75 W cm^{-2}, respectively.

Transition metal pyrophosphates have been investigated and used as a novel electrode material for the HC because of their multidimensional pathway for ionic conduction. Nickel–cobalt pyrophosphate nanoparticles coupled with graphitic carbon nitride ($NiCoP_2O_7$/g-C_3N_4) had been proposed and used as an anode for assembling an asymmetric HC [89]. The as-fabricated HC delivered high-energy density and exceptional cyclability.

MXene-based materials have also been proposed for fabricating capacitors. Xie et al. fabricated a high-voltage asymmetric on-chip microsupercapacitor (MSC) based on MXene (listed in Figure 10.4a–c), which can deliver an operating voltage window up to 1.6 V in neutral gel electrolyte (PVA/Na_2SO_4) and a high areal capacitance of 7.8 mF cm^{-2} [90].

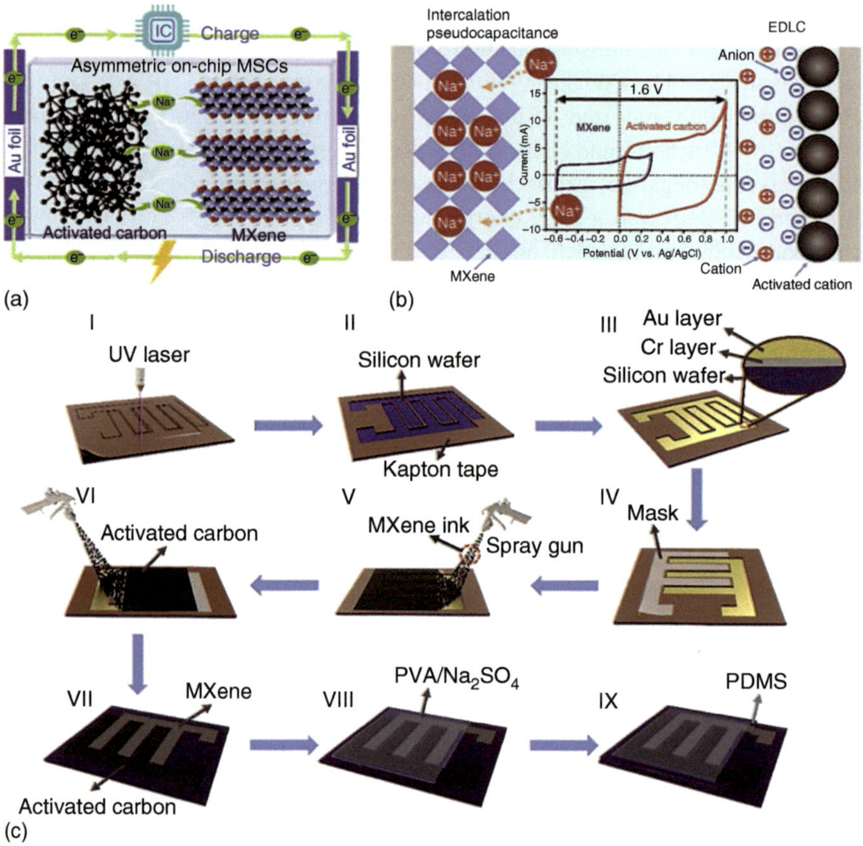

Figure 10.4 (a) Schematically illustrating the internal structure and work principle of on-chip MSCs. (b) CV curves of MXene-negative and activated carbon (AC)-positive electrodes at a scan rate of 10 $mV\,s^{-1}$ in a three-electrode configuration, and the realization of high voltage for on-chip MSCs is clarified based on analyzing the charge storage mechanism. (c) The cutting-spraying manufacture process of high-voltage MXene-based on-chip MSCs on silicon wafer. Source: Reproduced with permission Xie et al. [90]. Copyright 2020, Elsevier.

10.3.3 Hybrid Capacitors with Battery-Type Electrodes

The HC with battery-type electrodes is fabricated by the combination of a capacitor electrode with a battery electrode. This configuration replicates the demand for higher energy capacitors and higher power batteries, combining the energy characteristics of the batteries with the power, cycle life, and recharging times of capacitors. MIHC is a typical battery-type HC. According to the configuration, both the negative and positive electrodes can play the role of intercalating metal ions [91]. If the positive electrode intercalates metal ions and negative electrode for EDLC storage, it can be called "rocking chair" mechanism; if the negative electrode intercalates metal ions and positive electrode for EDLC storage, it can be classified as "accordion" type. "Hybrid"-type MIHCs demonstrate that the positive or/and negative electrode contains both battery-type and capacitor-type materials. With regard to the design and fabrication of MIHC, the battery-type material plays a key role in accelerating electrochemical reaction and enhancing performance.

According to the different shuttle metal ions, metal ions in HCs can be divided into Li ion, Na ion, Mg ion, Al ion, Ca ion, Zn ion, etc. The energy storage mechanism of MIHCs is dependent on the reversible anions adsorption/desorption and the surface of the cathode and reversible intercalation/deintercalation of metal ions in the anode. There have been several good literatures reviewed the research advances of MIHC [91]. The research progress of MIHCs mainly focuses on the design of metal-ion intercalation electrode materials. In particular, pore size and interlayer spacing of electrode materials for certain MIHCs shall be taken into consideration on account of metal-ion size. Hence, the materials with large interlayer space and interconnected network can freely intercalate/deintercalate metal ions and tackle with volume expansion during the charge–discharge process.

As far as research concerning MIHC, univalent metal ion (Li^+, Na^+, K^+)-based HCs have been widely developed and designed, whereas only a small number of studies have been conducted on multivalent metal ion (Zn^{2+}, Mg^{2+}, Ca^{2+}, Al^{3+})-based HC, especially Ca-ion and Al-ion HC. In the view of Ca^{2+} diameter, 3D tunnels and layered structures can be suitable for hosting Ca^{2+}. Wu et al. proposed Ca-ion hybrid capacitors (CIHCs) by using AC-positive electrode, Sn foil negative electrode, and 0.8 M $Ca(PF_6)_2$ electrolyte solution in mixed carbonate solvent [92]. The as-fabricated CIHCs demonstrated reversible capacities of 92 mAh g^{-1} (at 0.1 A g^{-1}) and 82 mAh g^{-1} (at 0.4 A g^{-1}) with the operating voltage window of 1.5–4.8 V. Al-ion hybrid capacitor (AIHC) is also multivalent ion-based device and involved in the three-electron redox system. TiO_2 has been used as anode materials for Al ion intercalation/deintercalation. Zhou et al. had prepared porous TiO_2 microrods as anode and AC as the cathode for fabricating AIHC [93]. The as-assembled HC delivered a high energy density of 26.3 Wh kg^{-1} at 627.3 W kg^{-1} as well as an outstanding high power of 5454.5 W kg^{-1} at 10.9 Wh kg^{-1}.

Theoretically, the multivalent metal-ion HC demonstrates faster charge transfer compared to that of univalent metal ion-based HC. According to current research on MIHC, various materials have been explored to be used in the electrode materials, such as transition metal oxides, binary metal oxides, metal sulfides, MXenes, conducting polymers, and carbon materials. The battery-type materials are required to have good chemical and mechanical properties and thus can host shuttle metal ions and deal with volume expansion. Generally, the battery-type materials have combined the utilization of two or more types of materials. The capacitor-type electrode shall have large specific area and good conductivity. Among them, AC has been widely used as cathode material in the assembling of MIHCs.

Carbon material is an important component of MIHC electrode materials. To achieve good capacitance, it is necessary to create pores and increase active sites for carbon materials. Chen et al. proposed onion-like carbon materials as battery-type anode and AC as capacitor-type cathode to fabricate potassium-ion hybrid capacitor, which delivered a high energy density of 142 Wh kg^{-1} and a high power output of 21 kW kg^{-1} [94]. The main research efforts on carbon-based materials are devoted to designing the porous structure and doping with elements to improve their electroactivity. Shang et al. had prepared hierarchical porous carbon spheres co-doped with N, P, and S, which enabled fast mass transfer [95]. The as-prepared carbon spheres were used as cathode materials to fabricate Zn-ion hybrid capacitor, which showed a specific capacity of 104.7 mAh g^{-1} and an energy/power density of 90.17 Wh kg^{-1}/81.2 W kg^{-1} in the operating voltage range of 0.1–0.7 V. In addition, the source of carbon materials is also closely associated with their properties and structures. Some carbon-based materials have been synthesized by using wastes and biowastes [96]. These biomass-derived carbon materials demonstrate various morphologies, such as sheets, fakes, and 3D porous structure, and also have rich inherent functional groups and heteroatoms. Li et al. used pencil shavings to prepare porous carbon, which served as electrode materials for fabricating a Zn-ion HC [97]. The as-prepared HC could achieve a high energy density of 147.0 Wh kg^{-1} at 136.1 W kg^{-1} and a maximum power density of 15.7 kW kg^{-1} at 65.4 Wh kg^{-1}. Xu et al. synthesized the scalable carbon nanosheets with high contents of inbuilt N/S sites using waste plastic as precursors [98]. The prepared carbon nanosheets as battery-type anode materials demonstrated good capability of K^+ storage, which had been utilized for fabrication of a K-ion HC. The as-assembled K-ion HC delivered large energy/power density (61 Wh kg^{-1} at 36,576 W kg^{-1}) and good cycling stability. Qian et al. proposed oriented carbon microspheres (OCMSs) and 3D porous carbon materials (3DPC) through high-temperature hydrothermal "disproportionation" of biomass [99]. The potassium-ion hybrid capacitor was fabricated by using OCMS (anode materials) and 3DPC (cathode materials), which exhibited an energy of 140.7 Wh kg^{-1} at 643.8 W kg^{-1}, with a long cycle life over 8500 cycles. Even though biomass and waste-derived carbon materials have been widely developed and used in energy storage, it is difficult to control the morphologies and attain uniform structures. Therefore, it is important for selection of biomass precursors and controlling the synthesis process. Cai et al. proposed directional selection strategy of biomass precursors through an advanced prediagnosis method [100]. The derived carbon materials with fewer aliphatic chains and more aromatic carbons enabled more K ions insertion. The fabricated K-ion HC based on the carbon-based materials had delivered a high energy density of 151 Wh kg^{-1} and an ultrahigh power output of 10 kW kg^{-1}. All-cellulose-based quasi-solid-state sodium-ion HC had been designed by using nitrogen-doped carbon cathode derived from cellulose microfibrils and nonporous hard carbon anode derived from cellulose nanocrystals [101]. The assembled HC delivered 181 Wh kg^{-1} at 250 W kg^{-1} in cellulose-based gel electrolytes. Leng et al. designed N, P, O tri-doped carbon nanocage as the cathode and Zn anode coated with ion modulation layer for fabrication of Zn-ion hybrid capacitors (ZIHC) [102]. The as-assembled ZIHC showed a specific capacitance of 310 F g^{-1} at 0.5 A g^{-1}, and high energy density/power density (43 Wh kg^{-1} at 137.9 kW kg^{-1}).

Metal oxides coupling with carbon materials can be good candidates for MIHCs. Chen et al. fabricated binder-free 3D electrodes based on MnO/graphene composites (MnO/3DGS) as the anode for LIHCs [103]. MnO nanoparticles were uniformly distributed onto 3D graphene scroll through *in situ* synthesis. The 3D structure of MnO/3DGS offered abundant interconnected channels for ion transfer and good capability for handling with volume expansion. The as-fabricated HC (MnO/3DGS//AC) could deliver a maximum energy density of 179.3 Wh kg^{-1} at a power density of 139.2 W kg^{-1}. Lang et al. prepared an anode by using nanoporous gold and nanocrystalline MnO_2, where nanoporous gold ameliorated the conductivity of MnO_2 and accelerated the ion diffusion between the MnO_2 and the electrolyte [104]. Huang et al. designed a moderate amount of oxygen vacancies in manganese oxide nanosheets/reduced graphene oxide (OV-MnO_2/rGO) composites, which was then used as electrode materials for lithium-ion capacitors [105]. The assembled lithium-ion capacitor was fabricated by using OV-MnO_2/rGO negative electrode and N-doped carbon nanosheet-positive electrode, which delivered an energy density of 206.2 Wh kg^{-1} at 250 W kg^{-1} and a power density of 25,000 W kg^{-1} at 41.7 Wh kg^{-1}. Wang et al. proposed MnO_2 nanowires coated with rGO, which were used as electrode materials for both K- and Na-ion HCs and demonstrated good electrochemical performance [106]. Yang et al. designed 3D hierarchical ternary aerogel of TiO_2 nanoparticles@porous carbon nanofiber-reduced graphene oxide (TiO_2@PCNF-GA) as an anode for lithium-ion capacitor [107]. The fabricated lithium-ion capacitor consists of TiO_2@PCNF-GAs anode and an AC cathode, which demonstrated the energy density vs. power density (79.7 Wh kg^{-1} vs. 75 W kg^{-1}, 15.0 Wh kg^{-1} vs. 15 kW kg^{-1}). Sennu et al. prepared mesoporous Co_3O_4 nanosheets as battery-type electrode materials and Jack fruit-derived AC as capacitor-type electrode materials, which were used to fabricate Li-ion capacitors [108]. The as-fabricated Li-ion capacitor delivered a maximum energy density of around 118 Wh kg^{-1} at a high temperature of 50 °C, and good cycle stability at high rate of 2.6 kW kg^{-1} at 50 °C.

Binary metal oxides can be used as anode materials in the fabrication of MIHCs. Cheng et al. prepared carbonized $NiCo_2O_4$ by annealing NiCo MOF, which served as anode and carbon nanoflakes as the cathode to fabricate a LIHC [109]. The as-fabricated device demonstrated a high energy density (136.9 Wh kg^{-1} at 200 W kg^{-1}) in the operative voltage window of 1–4.2 V. The binder-free CuO@FeO_x@TiO_2 composite electrode was proposed to fabricate a full Li-ion capacitor device, which exhibited high energy/powder densities (154.8 Wh kg^{-1} at 200 W kg^{-1}; 66.2 Wh kg^{-1} at 30 kW kg^{-1}) [110]. Molybdenum oxide (MoO_2) has been investigated in the application of HC. Ock et al. designed MoO_2 coupled with rGO shells as the anode and PANI chain-integrated rGO structures (PANI@rGO) as the cathode to assemble a lithium-ion HC [111]. The as-fabricated HC demonstrated high energy density (up to 242 Wh kg^{-1}), ultrafast chargeable power density (up to 28,750 W kg^{-1}), and long-life stability over 10,000 cycles.

Tian et al. fabricated a printable magnesium ion quasi-solid-state asymmetric HC [14]. The flexible HC was assembled with nanoflower-like MnO_2 as the positive electrode, nanowire-shaped vanadium nitride (VN) as the negative electrode, and $MgSO_4$-PAM gel as the electrolyte. The as-fabricated Mg ion HC demonstrated a wide operative voltage window (up to 2.2 V) and high energy density up to 13.1 mWh cm^{-3}. In addition, the fabricated HC had combined with a flexible solar-charging integrated unit, which enabled the photocharging to provide power for electronic devices. Li and coworkers had fabricated a photo-charging LIHC by using lithium titanium oxide/reduced graphene oxide ($Li_4Ti_5O_{12}$/rGO) as the anode and AC as the cathode [112]. The as-prepared LIHC had been combined with a flexible perovskite solar cell (PSC), which demonstrated a high energy/power density, high output voltage of 3 V, and good overall efficiency. The flexible PSC-LIHC had been further integrated with strain sensors, which had been used for monitoring on-skin physiological signals induced by body motions. This PSC-LIHC-sensor unit had integrated energy conversion, storage, and sensor unit in one device, which could have good prospect in the application of wearable electronics.

In addition to flexible planar HCs, fiber electrode-based HCs have great potential in the application of flexible electronics. The binder-free diamond fibers had been developed by using carbon fibers coated with boron-doped diamond films [113]. The morphologies of diamond fibers are shown in Figure 10.5a–c. The formed diamond fibers were then used as the positive electrode along with diamond fibers coated onto zinc nanosheet as the negative electrode for flexible ZIHCs. The as-fabricated ZIHCs demonstrated a high energy density (70.7 Wh kg^{-1} at 709.0 W kg^{-1}) and good cyclic stability (Figure 10.5d–i).

In addition, metal sulfides have gained much research interest in MIHCs. Xu et al. designed bowl-like VS_2 nanosheet arrays anchored on carbon nanofibers, which was served as anode materials and graphene sheet-wrapped carbon nanofibers as cathode for Na-ion HC [114]. The fabricated Na-ion HC displayed prominent energy power densities in the voltage range of 0.0–4.0 V (116 Wh kg^{-1} at 0.4 kW kg^{-1}; 66 Wh kg^{-1} at 40 kW kg^{-1} based on the total mass of both electrodes). Cui and co-workers prepared N/S-doped carbon combined with MoS_2 bilayers for K-ion HC [115]. The bilayer of MoS_2 was grafted on 3D N/S-doped carbon skeleton, which can expose more S-Mo-S to adsorb K-ions and avoid structure collapse of electrode, thus improving the stability of the as-prepared K-ion HC. Wang et al. proposed freestanding porous carbon fiber membrane, which was acted as 3D substrate for the growth of MoS_2@ PEDOT [116]. The resultant PCNF@MoS_2@PEDOT double core/shell nanofiber had been used as anode materials for fabricating quasi-solid-state flexible Na-ion HC, which exhibited outstanding capacity retention during high-current charge/discharges cycling after 5000 cycles.

Metal selenides show great promise as electrode materials because of their high capacity and suitable voltage platform. Lu and coworkers studied the electrochemical performance of N-doping carbon-coated $FeSe_2$ clusters as the anode for a potassium-ion hybrid capacitor, which performed competitive special capacity and high-rate capability via the evolution from $FeSe_2$ to Fe_3Se_4 [117]. Chen et al. designed a nonaqueous K-ion hybrid capacitor (KIHC) by using a binder-free anode made of $NbSe_2$ and N, Se co-doped carbon nanofibers (NSeCNFs) [118]. The $NbSe_2$/NSeCNF composites with porous structure demonstrated good capability for hosting K ions. The as-prepared KIHCs demonstrated good electrochemical performances including a high energy density (145 Wh kg^{-1} at 50 mA g^{-1}) and long cycle life.

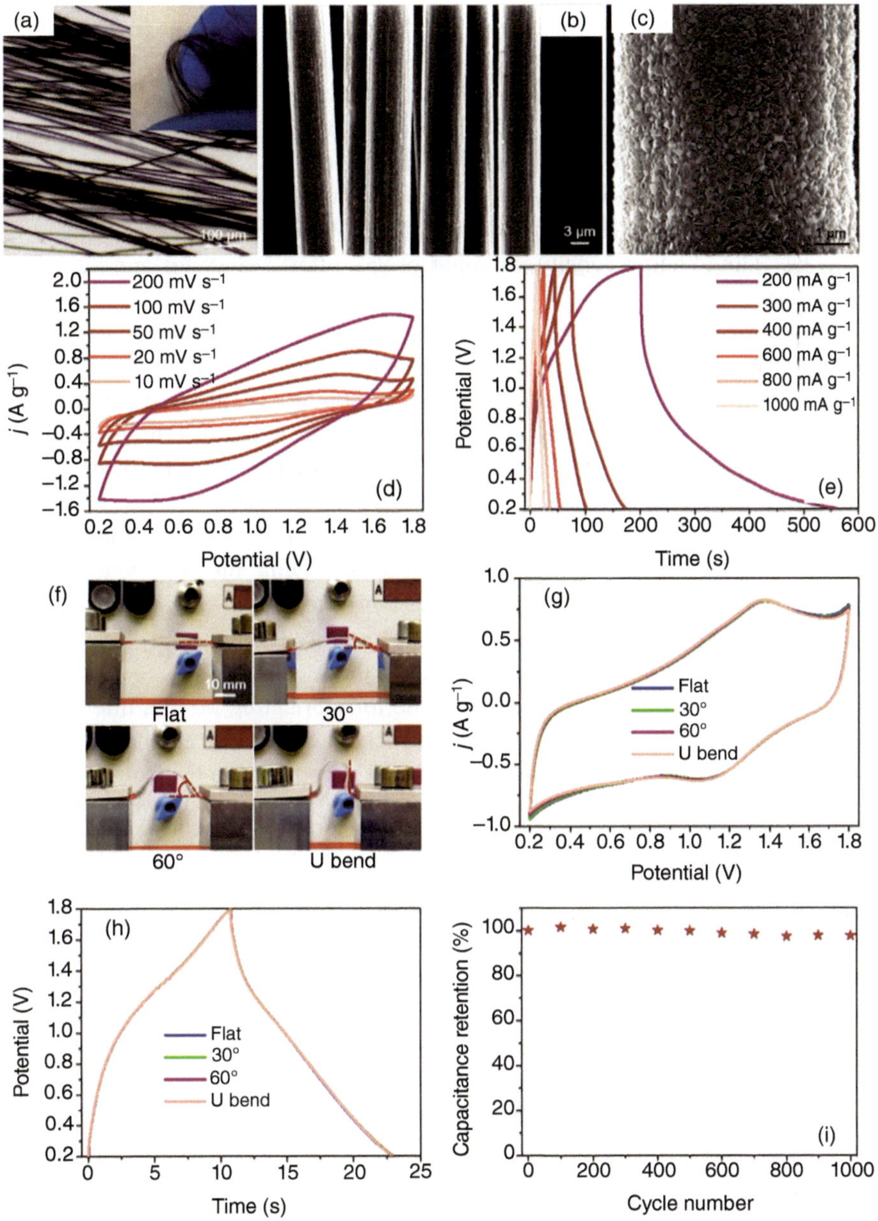

Figure 10.5 (a) Optical micrograph with an inset digital photograph of diamond fibers. (b,c) field emission scanning electron microscopy (FESEM) images of diamond fibers. (d) CVs of flexible Zn-ion HC at different scan rates in 1.0 mol l^{-1} $ZnSO_4$. (e) GCD curves at different current densities. (f) Photographs of diamond zinc-ion HC at different bending stages. (g) CVs and (h) GCD curves of diamond zinc-ion HC bent with different bending angles. (i) Capacitance retention during the periodic bending test (up to 60°) for 1000 times. Source: Reproduced with permission Jian et al. [113] © 2020/with permission of John Wiley & Sons.

MXenes, as well-known 2D materials, with hydrophilicity, high metallic conductivity, and good mechanical properties have been used as pseudocapacitance materials. Porous, crumpled, and 3D MXenes with enhanced surface area are able to accommodate more ions and mitigate volume expansion. MXenes have been demonstrated to expand interlayer spacing and expose more active sites. Li et al. designed a zinc-ion HC based on amorphous $Ti_3C_2T_x/BiCuS_{2.5}$ electrode material, which was prepared by *in situ* deposition strategy [119]. The fabricated Zn-ion HC exhibited an energy density of 298.4 Wh kg^{-1} at a power density of 7200 W kg^{-1}, and energy density retention of 95% after 7000 cycles and 82% after 10,000 cycles.

Sodium super ion conductor structure (NASICON) compounds also have been developed and applied in hybrid ion capacitors, which demonstrated excellent metal-ion storage performance, owing to their robust 3D frameworks, fast ion transportation, and controllable volume change upon insertion/extraction [120]. Han et al. prepared $K_3V_2(PO_4)_3/C$ nanocomposites to fabricate K-ion HC, which delivered a high-potential plateau at 3.6–3.9 V and a reversible capacity of 50 mAh g^{-1} over 100 cycles [121]. Zhang et al. proposed hierarchical $Ca_{0.5}Ti_2(PO_4)_3$@C microspheres by the electro-spraying method, which were employed to fabricate a K-ion HC [122]. The fabricated K-ion HC demonstrated a good reversible capacity (239 mAh g^{-1}), and high energy/power density (80 Wh kg^{-1} and 5144 W kg^{-1}) in a potential window of 1.0–4.0 V.

Metal phosphides have been used as anode for fabricating MIHCs due to their excellent electronic conductivity, superb electrochemical activity, and high capacity. However, capacity decay is a challenge for metal phosphides to overcome. By coupling with carbon-based materials is an effective way to suppress the capacity decay of metal phosphide-based anodes. Gómez-Cámer et al. proposed carbon-coated tin phosphide composites (Sn_4P_3@C) as the negative electrode and olive pit-derived AC as the positive electrode for fabricating a Li-ion HC, which can achieve high energy density and power density [123].

Apart from the aforementioned MIHCs, dual-ion HC has been developed by integrating energy storage principle of dual-ion battery and capacitor. Xiong et al. proposed N-doped carbon as the electrode materials for a symmetric dual-ion HC. The assembled HC utilized the electrolyte of LiCl in dual-salt ionic liquid and realized a high mass capacitance (374 F g^{-1}) and high energy and power density (208 Wh kg^{-1} at 1144 W kg^{-1}) [124]. This work implies the significance of research on dual-ion or multi-ion HCs. As electrode materials may have various pore sizes, different metal ions can take full advantage of these pore structures and improve the utilization ratio of electrode materials. In addition, dual ions or multiple ions can further accelerate ionic diffusion and chemical reaction, thus improving the performance of MIHCs.

Currently, more and more flexible and microsized metal-ion HCs have been developed, which has attracted extensive attention of researchers. These flexible and wearable HCs can be further integrated with charging units, sensors, and other function unit to prepare wearable and integrated device, which can greatly promote their applications in portal devices.

10.4 Conclusion

This chapter summarizes the current status and advancement of hybrid capacitors and gives an insight into development prospects of capacitors. The energy storage principle and electrochemical performance assessment are clearly explained in the chapter. The advancement of HC device underlines its leading position in the field of energy storage. Consequently, it is of great significance to develop more creative and diversified HCs, thus enabling to meet the growing demand for specific application devices. High-performance HCs can be realized by designing electrode materials with tunable structures and optimizing the design and fabrication of HCs. Apart from this, flexible and microsized HCs can be the promising energy storage devices, which can be integrated with functional units to fabricate multifunctional integrated devices. Microsized HC has paved a new avenue in the application of microelectronics. Its novel design and facile fabrication could promote the commercialization of HC and broaden their applications.

References

1 Simon, P. and Gogotsi, Y. (2008). Materials for electrochemical capacitors. *Nature Materials* 7 (11): 845–854.

2 Bai, Y., Wang, F., Wu, F. et al. (2008). Influence of composite LiCl–KCl molten salt on microstructure and electrochemical performance of spinel $Li_4Ti_5O_{12}$. *Electrochimica Acta* 54 (2): 322–327.

3 Aravindan, V., Gnanaraj, J., Lee, Y.-S., and Madhavi, S. (2014). Insertion-type electrodes for nonaqueous Li-ion capacitors. *Chemical Reviews* 114 (23): 11619–11635.

4 Naskar, P., Kundu, D., Maiti, A. et al. (2021). Frontiers in hybrid ion capacitors: a review on advanced materials and emerging devices. *ChemElectroChem* 8 (8): 1393–1429.

5 Liu, X., Sun, Y., Tong, Y. et al. (2021). Exploration in materials, electrolytes and performance towards metal ion (Li, Na, K, Zn and Mg)-based hybrid capacitors: a review. *Nano Energy* 86: 106070.

6 Aslam, M.K., Niu, Y., and Xu, M. (2021). MXenes for non-lithium-ion (Na, K, Ca, Mg, and Al) batteries and supercapacitors. *Advanced Energy Materials* 11 (2): 2000681.

7 Dubey, P., Shrivastav, V., Maheshwari, P.H., and Sundriyal, S. (2020). Recent advances in biomass derived activated carbon electrodes for hybrid electrochemical capacitor applications: challenges and opportunities. *Carbon* 170: 1–29.

8 Huang, J., Sumpter, B.G., and Meunier, V. (2008). Theoretical model for nanoporous carbon supercapacitors. *Angewandte Chemie International Edition* 47 (3): 520–524.

9 Vivekchand, S.R.C., Rout, C.S., Subrahmanyam, K.S. et al. (2008). Graphene-based electrochemical supercapacitors. *Journal of Chemical Sciences* 120 (1): 9–13.

10 Jiang, J., Zhang, Y., Nie, P. et al. (2018). Progress of nanostructured electrode materials for supercapacitors. *Advanced Sustainable Systems* 2 (1): 1700110.

11 Ibrahim, I., Zheng, S., Foo, C.Y. et al. (2021). Hierarchical nickel-based metal-organic framework/graphene oxide incorporated graphene nanoplatelet electrode with exceptional cycling stability for coin cell and pouch cell supercapacitors. *Journal of Energy Storage* 43: 103304.

12 Bhattacharjya, D., Carriazo, D., Ajuria, J., and Villaverde, A. (2019). Study of electrode processing and cell assembly for the optimized performance of supercapacitor in pouch cell configuration. *Journal of Power Sources* 439: 227106.

13 Li, Z., Chang, S., Khuje, S., and Ren, S. (2021). Recent advancement of emerging nano copper-based printable flexible hybrid electronics. *ACS Nano* 15 (4): 6211–6232.

14 Tian, Z., Tong, X., Sheng, G. et al. (2019). Printable magnesium ion quasi-solid-state asymmetric supercapacitors for flexible solar-charging integrated units. *Nature Communications* 10 (1): 4913.

15 Tian, X., Ma, H., Gao, Y., and Xu, B. (2022). Flexible in-plane zinc-ion hybrid capacitors with synergistic electrochemical behaviors for self-powered energy systems. *Journal of Materials Chemistry A* 10 (26): 14011–14019.

16 Aleksenskii, A.E., Brunkov, P.N., Dideikin, A.T. et al. (2013). Single-layer graphene oxide films on a silicon surface. *Technical Physics* 58 (11): 1614–1618.

17 Verma, M.L., Minakshi, M., and Singh, N.K. (2014). Synthesis and characterization of solid polymer electrolyte based on activated carbon for solid state capacitor. *Electrochimica Acta* 137: 497–503.

18 Wang, J., Chen, H., Xiao, Y. et al. (2019). PAMPS/PVA/MMT semi-interpenetrating polymer network hydrogel electrolyte for solid-state supercapacitors. *International Journal of Electrochemical Science* 14 (2): 1817–1829.

19 Peng, H., Lv, Y., Wei, G. et al. (2019). A flexible and self-healing hydrogel electrolyte for smart supercapacitor. *Journal of Power Sources* 431: 210–219.

20 Zhang, L., Huang, D., Hu, N. et al. (2017). Three-dimensional structures of graphene/polyaniline hybrid films constructed by steamed water for high-performance supercapacitors. *Journal of Power Sources* 342: 1–8.

21 Liu, X., Men, C., Zhang, X., and Li, Q. (2016). An extraordinary sulfonated-graphenal-polymer-based electrolyte separator for all-solid-state supercapacitors. *Small* 12 (36): 4973–4979.

22 Feng, W., Zhang, J., Yusuf, A. et al. (2022). Quasi-solid-state sodium-ion hybrid capacitors enabled by UiO-66@PVDF-HFP multifunctional separators: selective charge transfer and high fire safety. *Chemical Engineering Journal* 427: 130919.

23 Kuratani, K., Okuno, K., Iwaki, T. et al. (2011). Converting rice husk activated carbon into active material for capacitor using three-dimensional porous current collector. *Journal of Power Sources* 196 (24): 10788–10790.

24 Abbas, Q., Pajak, D., Frąckowiak, E., and Béguin, F. (2014). Effect of binder on the performance of carbon/carbon symmetric capacitors in salt aqueous electrolyte. *Electrochimica Acta* 140: 132–138.

25 Huang, D., Lu, Z., Xu, Q. et al. (2022). TiO_2 nanoflowers@Au@MnO_2 core-shell composite based on modified Ti foil for flexible supercapacitor electrode. *Electrochimica Acta* 407: 139866.

26 Pei, D., Bao, J., Li, Y. et al. (2022). Three-dimensional Co_3O_4/CoS hierarchical nanoneedle arrays electrode grown on nickel foam for high-performance asymmetric capacitors. *Journal of Energy Storage* 51: 104483.

27 Kötz, R. and Carlen, M. (2000). Principles and applications of electrochemical capacitors. *Electrochimica Acta* 45 (15): 2483–2498.

28 Shao, Y., El-Kady, M.F., Sun, J. et al. (2018). Design and mechanisms of asymmetric supercapacitors. *Chemical Reviews* 118 (18): 9233–9280.

29 Amatucci, G.G., Badway, F., Du Pasquier, A., and Zheng, T. (2001). An asymmetric hybrid nonaqueous energy storage cell. *Journal of the Electrochemical Society* 148 (8): A930.

30 Burke, A. and Miller, M. (2010). Testing of electrochemical capacitors: capacitance, resistance, energy density, and power capability. *Electrochimica Acta* 55 (25): 7538–7548.

31 Zheng, J.P. (2003). The limitations of energy density of battery/double-layer capacitor asymmetric cells. *Journal of the Electrochemical Society* 150 (4): A484.

32 Xu, K., Ding, S.P., and Jow, T.R. (1999). Toward reliable values of electrochemical stability limits for electrolytes. *Journal of the Electrochemical Society* 146 (11): 4172.

33 Bahdanchyk, M., Hashempour, M., and Vicenzo, A. (2020). Evaluation of the operating potential window of electrochemical capacitors. *Electrochimica Acta* 332: 135503.

34 Weingarth, D., Noh, H., Foelske-Schmitz, A. et al. (2013). A reliable determination method of stability limits for electrochemical double layer capacitors. *Electrochimica Acta* 103: 119–124.

35 Bruce, P.G., Freunberger, S.A., Hardwick, L.J., and Tarascon, J.-M. (2012). Li–O_2 and Li–S batteries with high energy storage. *Nature Materials* 11: 19–29.

36 Wang, Y., Song, Y., and Xia, Y. (2016). Electrochemical capacitors: mechanism, materials, systems, characterization and applications. *Chemical Society Reviews* 45 (21): 5925–5950.

37 Lim, E., Kim, H., Jo, C. et al. (2014). Advanced hybrid supercapacitor based on a mesoporous niobium pentoxide/carbon as high-performance anode. *ACS Nano* 8 (9): 8968–8978.

38 Vicentini, R., Nunes, W., Freitas, B.G.A. et al. (2019). Niobium pentoxide nanoparticles@multi-walled carbon nanotubes and activated carbon composite material as electrodes for electrochemical capacitors. *Energy Storage Materials* 22: 311–322.

39 Jyothibasu, J.P., Wang, R.-H., Ong, K. et al. (2021). Cellulose/carbon nanotube/MnO_2 composite electrodes with high mass loadings for symmetric supercapacitors. *Cellulose* 28 (6): 3549–3567.

40 Zhao, N., Deng, L., Luo, D., and Zhang, P. (2020). One-step fabrication of biomass-derived hierarchically porous carbon/MnO nanosheets composites for symmetric hybrid supercapacitor. *Applied Surface Science* 526: 146696.

41 Pang, S., Gong, L., Du, N. et al. (2019). Formation of high-performance Cu-WO_x@C tribasic composite electrode for aqueous symmetric supercapacitor. *Materials Today Energy* 13: 239–248.

42 Tiwari, B., Joshi, A., Munjal, M. et al. (2020). A heterojunction of $VO(OH)_2$ nanorods onto hemp stem derived carbon for high voltage (1.5 V) symmetric supercapacitors. *Sustainable Energy & Fuels* 4 (10): 5102–5113.

43 Ahuja, P., Ujjain, S.K., and Kanojia, R. (2018). Electrochemical behaviour of manganese & ruthenium mixed oxide@ reduced graphene oxide nanoribbon composite in symmetric and asymmetric supercapacitor. *Applied Surface Science* 427: 102–111.

44 Park, H., Song, C., Jin, S.W. et al. (2021). High performance flexible micro-supercapacitor for powering a vertically integrated skin-attachable strain sensor on a bio-inspired adhesive. *Nano Energy* 83: 105837.

45 Wu, X., Zhu, X., Tao, H. et al. (2021). Covalently aligned molybdenum disulfide–carbon nanotubes heteroarchitecture for high-performance electrochemical capacitors. *Angewandte Chemie International Edition* 60 (39): 21295–21303.

46 Chen, W., Xia, C., and Alshareef, H.N. (2014). One-step electrodeposited nickel cobalt sulfide nanosheet arrays for high-performance asymmetric supercapacitors. *ACS Nano* 8 (9): 9531–9541.

47 Wang, F., Liu, X., Duan, G. et al. (2021). Wood-derived, conductivity and hierarchical pore integrated thick electrode enabling high areal/volumetric energy density for hybrid capacitors. *Small* 17 (35): 2102532.

48 Kim, T., Tiwari, A.P., Chhetri, K. et al. (2020). Phytic acid controlled in situ synthesis of amorphous cobalt phosphate/carbon composite as anode materials with a high mass loading for symmetrical supercapacitor: amorphization of the electrode to boost the energy density. *Nanoscale Advances* 2 (10): 4918–4929.

49 Naskar, P., Maiti, A., Chakraborty, P. et al. (2021). Chemical supercapacitors: a review focusing on metallic compounds and conducting polymers. *Journal of Materials Chemistry A* 9 (4): 1970–2017.

50 Muthulakshmi, B., Kalpana, D., Pitchumani, S., and Renganathan, N.G. (2006). Electrochemical deposition of polypyrrole for symmetric supercapacitors. *Journal of Power Sources* 158 (2): 1533–1537.

51 Moyseowicz, A. and Gryglewicz, G. (2020). High-performance hybrid capacitor based on a porous polypyrrole/reduced graphene oxide composite and a redox-active electrolyte. *Electrochimica Acta* 354: 136661.

52 Liu, J., Wang, Z., Liu, Q. et al. (2022). Rational design of freestanding and high-performance thick electrode from carbon foam modified with polypyrrole/polydopamine for supercapacitors. *Chemical Engineering Journal* 447: 137562.

53 Hu, R., Wang, Y., Zhao, J. et al. (2019). Fabrication of stretchable multi-element composite for flexible solid-state electrochemical capacitor application. *Chemical Engineering Journal* 361: 109–116.

54 Luo, W., Wei, Y., Zhuang, Z. et al. (2022). Fabrication of $Ti_3C_2T_x$ MXene/polyaniline composite films with adjustable thickness for high-performance flexible all-solid-state symmetric supercapacitors. *Electrochimica Acta* 406: 139871.

55 Chen, B. and Wong, W.-Y. (2022). Introducing a redox-active ferrocenyl moiety onto a polythiophene derivative towards high-performance flexible all-solid-state symmetric supercapacitors. *Journal of Materials Chemistry A* 10 (14): 7968–7977.

56 Wang, L., Shu, T., Guo, S. et al. (2020). Fabricating strongly coupled V_2O_5@PEDOT nanobelts/graphene hybrid films with high areal capacitance and facile transferability for transparent solid-state supercapacitors. *Energy Storage Materials* 27: 150–158.

57 He, X., Yang, W., Mao, X. et al. (2018). All-solid state symmetric supercapacitors based on compressible and flexible free-standing 3D carbon nanotubes (CNTs)/poly(3,4-ethylenedioxythiophene) (PEDOT) sponge electrodes. *Journal of Power Sources* 376: 138–146.

58 Kumar Das, A., Ramulu, B., Girija Shankar, E., and Su Yu, J. (2022). Binder-free CuS@PEDOT and Co–V–Se electrodes for flexible quasi-solid-state asymmetric supercapacitor. *Chemical Engineering Journal* 429: 132486.

59 Yeasmin, S., Talukdar, S., and Mahanta, D. (2021). Paper based pencil drawn multilayer graphene-polyaniline nanofiber electrodes for all-solid-state symmetric supercapacitors with enhanced cyclic stabilities. *Electrochimica Acta* 389: 138660.

60 Liu, R., Takakuwa, M., Li, A. et al. (2020). An efficient ultra-flexible photo-charging system integrating organic photovoltaics and supercapacitors. *Advanced Energy Materials* 10 (20): 2000523.

61 Khodabandehlo, A., Noori, A., Rahmanifar, M.S. et al. (2022). Laser-scribed graphene–polyaniline microsupercapacitor for internet-of-things applications. *Advanced Functional Materials* 32 (39): 2204555.

62 Alagar, S., Madhuvilakku, R., Mariappan, R., and Piraman, S. (2019). Nano-architectured porous Mn_2O_3 spheres/cubes vs rGO for asymmetric supercapacitors applications in novel solid-state electrolyte. *Journal of Power Sources* 441: 227181.

63 Chen, Z., Augustyn, V., Wen, J. et al. (2011). High-performance supercapacitors based on intertwined CNT/V_2O_5 nanowire nanocomposites. *Advanced Materials* 23 (6): 791–795.

64 Zhou, X., Chen, Q., Wang, A. et al. (2016). Bamboo-like composites of V_2O_5/polyindole and activated carbon cloth as electrodes for all-solid-state flexible asymmetric supercapacitors. *ACS Applied Materials & Interfaces* 8 (6): 3776–3783.

65 Ndiaye, N.M., Sylla, N.F., Ngom, B.D. et al. (2019). High-performance asymmetric supercapacitor based on vanadium dioxide/activated expanded graphite composite and carbon-vanadium oxynitride nanostructures. *Electrochimica Acta* 316: 19–32.

66 Devarayapalli, K.C., Lee, K., Do, H.B. et al. (2021). Mesostructured g-C_3N_4 nanosheets interconnected with V_2O_5 anobelts as electrode for coin-cell-type-asymmetric supercapacitor device. *Materials Today Energy* 21: 100699.

67 Liu, A., Zhang, H., Wang, G. et al. (2018). Sandwich-like NiO/rGO nanoarchitectures for 4 V solid-state asymmetric-supercapacitors with high energy density. *Electrochimica Acta* 283: 1401–1410.

68 Hu, P., Liu, Y., Song, J. et al. (2019). Transition metal oxide@hydroxide assemblies as electrode materials for asymmetric hybrid capacitors with excellent cycling stabilities. *RSC Advances* 9 (56): 32510–32516.

69 Xia, K., Tang, H., Fu, J. et al. (2020). A high strength triboelectric nanogenerator based on rigid-flexible coupling design for energy storage system. *Nano Energy* 67: 104259.

70 Jayababu, N. and Kim, D. (2021). ZnO nanorods@conductive carbon black nanocomposite based flexible integrated system for energy conversion and storage through triboelectric nanogenerator and supercapacitor. *Nano Energy* 82: 105726.

71 Bao, J., Lin, N., Guo, J. et al. (2020). Effects of nano-SiO_2 doped PbO_2 as the positive electrode on the performance of lead-carbon hybrid capacitor. *Journal of Colloid and Interface Science* 574: 377–384.

72 Dan, Y., Lin, H., Liu, X. et al. (2012). Porous quasi three-dimensional nano-Mn_3O_4+PbO_2 composite as supercapacitor electrode material. *Electrochimica Acta* 83: 175–182.

73 Pei, D., Bao, J., Liang, C. et al. (2020). Facile fabricate of F-doped lead dioxide electrode coating onto porous 3D titanium plate for advanced lead-carbon hybrid capacitor. *Materials Letters* 280: 128553.

74 Muduli, S., Rotte, N.K., Naresh, V., and Martha, S.K. (2020). Nitrogen phosphorous derived carbons from *Peltophorum pterocarpum* leaves as anodes for lead–carbon hybrid ultracapacitors. *Journal of Energy Storage* 29: 101330.

75 Parveen, N., Al-Jaafari, A.I., and Han, J.I. (2019). Robust cyclic stability and high-rate asymmetric supercapacitor based on orange peel-derived nitrogen-doped porous carbon and intercrossed interlinked urchin-like $NiCo_2O_4$@3DNF framework. *Electrochimica Acta* 293: 84–96.

76 Acharya, J., Raj, B.G.S., Ko, T.H. et al. (2020). Facile one pot sonochemical synthesis of $CoFe_2O_4$/MWCNTs hybrids with well-dispersed MWCNTs for asymmetric hybrid supercapacitor applications. *International Journal of Hydrogen Energy* 45 (4): 3073–3085.

77 Liu, H., Dai, M., Zhao, D. et al. (2020). Realizing superior electrochemical performance of asymmetric capacitors through tailoring electrode architectures. *ACS Applied Energy Materials* 3 (7): 7004–7010.

78 Tsai, K.-J., Ni, C.-S., Chen, H.-Y., and Huang, J.-H. (2020). Single-walled carbon nanotubes/Ni–Co–Mn layered double hydroxide nanohybrids as electrode materials for high-performance hybrid energy storage devices. *Journal of Power Sources* 454: 227912.

79 Acharya, J., Ko, T.H., Seo, M.-K. et al. (2020). Engineering the hierarchical heterostructures of Zn–Ni–Co nanoneedles arrays@Co–Ni-LDH nanosheets core–sheath electrodes for a hybrid asymmetric supercapacitor with high energy density and excellent cyclic stability. *ACS Applied Energy Materials* 3 (8): 7383–7396.

80 Musella, E., Gualandi, I., Ferrari, G. et al. (2021). Electrosynthesis of Ni/Al layered double hydroxide and reduced graphene oxide composites for the development of hybrid capacitors. *Electrochimica Acta* 365: 137294.

81 Zhou, Q., Wei, T., Liu, Z. et al. (2019). Nickel hexacyanoferrate on graphene sheets for high-performance asymmetric supercapacitors in neutral aqueous electrolyte. *Electrochimica Acta* 303: 40–48.

82 Choi, J., Lim, J., Kim, D. et al. (2022). CoFe Prussian blue analogues on 3D porous N-doped carbon nanosheets boost the intercalation kinetics for a high-performance quasi-solid-state hybrid capacitor. *Journal of Materials Chemistry A* 10 (27): 14501–14512.

83 Wang, Y., Zhang, W., Guo, X. et al. (2021). Construction of high-performance asymmetric supercapacitor based on the hierarchical Ni_3S_2/CoFe LDH/Ni foam hybrid. *Applied Surface Science* 561: 150049.

84 Liu, C., Wang, H., Zhao, X. et al. (2020). Cellulose-derived carbon-based electrodes with high capacitance for advanced asymmetric supercapacitors. *Journal of Power Sources* 457: 228056.

85 Khalafallah, D., Zou, Q., Zhi, M., and Hong, Z. (2020). Tailoring hierarchical yolk-shelled nickel cobalt sulfide hollow cages with carbon tuning for asymmetric supercapacitors and efficient urea electrocatalysis. *Electrochimica Acta* 350: 136399.

86 Khaja Hussain, S., Vamsi Krishna, B.N., Nagaraju, G. et al. (2021). Porous Co-MoS_2@Cu_2MoS_4 three-dimensional nanoflowers via in situ sulfurization of Cu_2O nanospheres for electrochemical hybrid capacitors. *Chemical Engineering Journal* 403: 126319.

87 He, Y., Zhang, P., Huang, H. et al. (2020). Rationally design nickel sulfide@ PEDOT arrays as binder-free cathode for durable asymmetric supercapacitor and aqueous Ni–Zn battery. *Electrochimica Acta* 343: 136140.

88 Zhou, Y., Wang, X., Acauan, L. et al. (2019). Ultrahigh-areal-capacitance flexible supercapacitor electrodes enabled by conformal P3MT on horizontally aligned carbon-nanotube arrays. *Advanced Materials* 31 (30): 1901916.

89 Matheswaran, P., Karuppiah, P., Chen, S.-M. et al. (2018). Fabrication of g-C_3N_4 nanomesh-anchored amorphous $NiCoP_2O_7$: tuned cycling life and the dynamic behavior of a hybrid capacitor. *ACS Omega* 3 (12): 18694–18704.

90 Xie, Y., Zhang, H., Huang, H. et al. (2020). High-voltage asymmetric MXene-based on-chip micro-supercapacitors. *Nano Energy* 74: 104928.

91 Dong, S., Lv, N., Wu, Y. et al. (2021). Lithium-ion and sodium-ion hybrid capacitors: from insertion-type materials design to devices construction. *Advanced Functional Materials* 31 (21): 2100455.

92 Wu, N., Yao, W., Song, X. et al. (2019). A calcium-ion hybrid energy storage device with high capacity and long cycling life under room temperature. *Advanced Energy Materials* 9 (16): 1803865.

93 Zhou, X., Lu, Y., Zhen, Y. et al. (2020). Facile fabrication of highly porous TiO_2 microrods anode with enhanced Al-ion storage for hybrid capacitors. *Journal of Power Sources* 453: 227857.

94 Chen, J., Yang, B., Li, H. et al. (2019). Candle soot: onion-like carbon, an advanced anode material for a potassium-ion hybrid capacitor. *Journal of Materials Chemistry A* 7 (15): 9247–9252.

95 Shang, K., Liu, Y., Cai, P. et al. (2022). N, P, and S co-doped 3D porous carbon-architectured cathode for high-performance Zn-ion hybrid capacitors. *Journal of Materials Chemistry A* 10 (12): 6489–6498.

96 Muduli, S., Pati, S.K., and Martha, S.K. (2022). Bio-waste derived carbon nano-onions as an efficient electrode material for symmetric and lead-carbon hybrid ultracapacitors. *International Journal of Energy Research* 46 (10): 14074–14087.

97 Li, Z., Chen, D., An, Y. et al. (2020). Flexible and anti-freezing quasi-solid-state zinc ion hybrid supercapacitors based on pencil shavings derived porous carbon. *Energy Storage Materials* 28: 307–314.

98 Xu, J., Dou, S., Zhou, W. et al. (2022). Scalable waste-plastic-derived carbon nanosheets with high contents of inbuilt nitrogen/sulfur sites for high performance potassium-ion hybrid capacitors. *Nano Energy* 95: 107015.

99 Qian, Y., Li, Y., Pan, Z. et al. (2021). Hydrothermal "disproportionation" of biomass into oriented carbon microsphere anode and 3D porous carbon cathode for potassium ion hybrid capacitor. *Advanced Functional Materials* 31 (30): 2103115.

100 Cai, P., Momen, R., Tian, Y. et al. (2022). Advanced pre-diagnosis method of biomass intermediates toward high energy dual-carbon potassium-ion capacitor. *Advanced Energy Materials* 12 (5): 2103221.

101 Xu, Z., Xie, F., Wang, J. et al. (2019). All-cellulose-based quasi-solid-state sodium-ion hybrid capacitors enabled by structural hierarchy. *Advanced Functional Materials* 29 (39): 1903895.

102 Leng, C., Fedoseeva, Y.V., Zhao, Z. et al. (2022). Rational-design heteroatom-doped cathode and ion modulation layer modified Zn anode for ultrafast zinc-ion hybrid capacitors with simultaneous high power and energy densities. *Journal of Power Sources* 536: 231484.

103 Chen, P., Zhou, W., Xiao, Z. et al. (2020). In situ anchoring MnO nanoparticles on self-supported 3D interconnected graphene scroll framework: a fast kinetics boosted ultrahigh-rate anode for Li-ion capacitor. *Energy Storage Materials* 33: 298–308.

104 Lang, X., Hirata, A., Fujita, T., and Chen, M. (2011). Nanoporous metal/oxide hybrid electrodes for electrochemical supercapacitors. *Nature Nanotechnology* 6 (4): 232–236.

105 Huang, Z., Liu, S., Li, J. et al. (2022). Pre-lithiation optimized voltage ranges and MnO_2/rGO negative electrodes with oxygen vacancies for enhanced performance of lithium-ion capacitors. *Electrochimica Acta* 421: 140406.

106 Wang, Z., Yan, X., Wang, F. et al. (2021). Reduced graphene oxide thin layer induced lattice distortion in high crystalline MnO_2 nanowires for high-performance sodium- and potassium-ion batteries and capacitors. *Carbon* 174: 556–566.

107 Yang, C., Lan, J.-L., Ding, C. et al. (2019). Three-dimensional hierarchical ternary aerogels of ultrafine TiO_2 nanoparticles@porous carbon nanofibers-reduced graphene oxide for high-performance lithium-ion capacitors. *Electrochimica Acta* 296: 790–798.

108 Sennu, P., Madhavi, S., Aravindan, V., and Lee, Y.-S. (2020). Co_3O_4 nanosheets as battery-type electrode for high-energy Li-ion capacitors: a sustained Li-storage via conversion pathway. *ACS Nano* 14 (8): 10648–10654.

109 Cheng, C.-F., Li, X., Liu, K. et al. (2019). A high-performance lithium-ion capacitor with carbonized $NiCo_2O_4$ anode and vertically-aligned carbon nanoflakes cathode. *Energy Storage Materials* 22: 265–274.

110 Li, Y., Liang, T., Wang, R. et al. (2019). Encapsulation of Fe_3O_4 between copper nanorod and thin TiO_2 film by ALD for lithium-ion capacitors. *ACS Applied Materials & Interfaces* 11 (21): 19115–19122.

111 Ock, I.W., Lee, J., and Kang, J.K. (2020). Metal–organic framework-derived anode and polyaniline chain networked cathode with mesoporous and conductive pathways for high energy density, ultrafast rechargeable, and long-life hybrid capacitors. *Advanced Energy Materials* 10 (48): 2001851.

112 Li, C., Cong, S., Tian, Z. et al. (2019). Flexible perovskite solar cell-driven photo-rechargeable lithium-ion capacitor for self-powered wearable strain sensors. *Nano Energy* 60: 247–256.

113 Jian, Z., Yang, N., Vogel, M. et al. (2020). Flexible diamond fibers for high-energy-density zinc-ion supercapacitors. *Advanced Energy Materials* 10 (44): 2002202.

114 Xu, D., Wang, H., Qiu, R. et al. (2020). Coupling of bowl-like VS_2 nanosheet arrays and carbon nanofiber enables ultrafast Na^+-storage and robust flexibility for sodium-ion hybrid capacitors. *Energy Storage Materials* 28: 91–100.

115 Cui, Y., Zhao, L., Li, B. et al. (2022). Tailored MoS_2 bilayer grafted onto N/S-doped carbon for ultra-stable potassium-ion capacitor. *Chemical Engineering Journal* 450: 137815.

116 Wang, H., Xu, D., Jia, G. et al. (2020). Integration of flexibility, cyclability and high-capacity into one electrode for sodium-ion hybrid capacitors with low self-discharge rate. *Energy Storage Materials* 25: 114–123.

117 Ge, J., Wang, B., Wang, J. et al. (2020). Nature of $FeSe_2$/N-C anode for high performance potassium ion hybrid capacitor. *Advanced Energy Materials* 10 (4): 1903277.

118 Chen, M., Wang, L., Sheng, X. et al. (2020). An ultrastable nonaqueous potassium-ion hybrid capacitor. *Advanced Functional Materials* 30 (40): 2004247.

119 Li, Y., Zhang, W., Yang, X. et al. (2021). A high-voltage and high-capacity $Ti_3C_2T_x/BiCuS_{2.5}$ heterostructure to boost up the energy density and recyclability of zinc-ion-hybrid capacitors. *Nano Energy* 87: 106136.

120 Chen, S., Wu, C., Shen, L. et al. (2017). Challenges and perspectives for NASICON-type electrode materials for advanced sodium-ion batteries. *Advanced Materials* 29 (48): 1700431.

121 Han, J., Li, G.-N., Liu, F. et al. (2017). Investigation of $K_3V_2(PO_4)_3$/C nanocomposites as high-potential cathode materials for potassium-ion batteries. *Chemical Communications* 53 (11): 1805–1808.

122 Zhang, Z., Li, M., Gao, Y. et al. (2018). Fast potassium storage in hierarchical $Ca_{0.5}Ti_2(PO_4)_3$@C microspheres enabling high-performance potassium-ion capacitors. *Advanced Functional Materials* 28 (36): 1802684.

123 Gómez-Cámer, J.L., Arnaiz, M., Rojo, T., and Ajuria, J. (2019). Novel lithium-ion capacitor based on tin phosphide and olive pit derived activated carbon. *Journal of Power Sources* 434: 226695.

124 Xiong, Z., Guo, P., Yang, Y. et al. (2022). A high-performance dual-ion battery-supercapacitor hybrid device based on LiCl in ion liquid dual-salt electrolyte. *Advanced Energy Materials* 12 (7): 2103226.

11

Flexible Energy Storage Devices

Chuan Xie[1] *and Zijian Zheng*[1,2,3,4]

[1] *The Hong Kong Polytechnic University, Faculty of Science, Department of Applied Biology and Chemical Technology (ABCT), 11 Yuk Choi Rd, Hung Hom, Hong Kong, SAR 999077, China*
[2] *The Hong Kong Polytechnic University, Faculty of Science, School of Fashion and Textiles, 11 Yuk Choi Rd, Hung Hom, Hong Kong, SAR 999077, China*
[3] *The Hong Kong Polytechnic University, Research Institute for Intelligent Wearable Systems (RI-IWEAR), 11 Yuk Choi Rd, Hung Hom, Hong Kong, SAR 999077, China*
[4] *The Hong Kong Polytechnic University, Research Institute for Smart Energy (RISE), 11 Yuk Choi Rd, Hung Hom, Hong Kong, SAR 999077, China*

The widespread use of wireless and portable devices has greatly shaped our daily lives and influenced our behaviors in all areas. As a result, there has been a significant push toward developing next-generation flexible and wearable electronics that can be seamlessly integrated into our clothing or even onto our bodies [1–5]. These advanced electronics, including roll-up displays, electronic textiles (E-textiles), smart medical sensors and monitors, and the Internet of Things (IoT), require reliable energy storage devices (ESDs) to power them. However, the currently used rechargeable ESDs, such as lithium-ion batteries (LIBs) and supercapacitors (SCs), lack mechanical flexibility despite their high energy and power densities. Thus, considerable research efforts have been devoted to designing and fabricating flexible energy storage devices (f-ESDs) over the past two decades [6–9].

Over the years, numerous studies and research efforts have been dedicated to achieving mechanical flexibility in ESDs, resulting in significant progress both theoretically and practically [10–19]. This chapter will introduce f-ESDs, with a focus on novel materials and strategies for flexible lithium batteries (FLBs). Firstly, the development and pathway of FLBs will be briefly summarized, along with a brief overview of general principles and fundamental knowledge. Next, the mechanisms behind various strategies for achieving high-performance FLBs will be elucidated, with representative examples. The strengths and limitations of each strategy will be discussed and compared. Lastly, the remaining challenges in the field of f-ESDs will be analyzed, along with potential solutions.

Electrochemical Energy Storage Devices: Non-Conventional Technologies and Materials,
First Edition. Edited by Yongbing Tang and Luojiang Zhang.

11.1 Introduction of FLBs

11.1.1 Mechanical Foundation of ESDs with a Multilayer-Stacking Configuration

As the electrochemical mechanisms of ESDs are extensively covered in other reviews and literature [20–28], this aspect will not be further discussed here. Instead, the focus will be on mechanical considerations. The majority of rechargeable ESDs share a similar multilayer-stacking configuration, where several pairs of metal foil-supported electrodes (MFEs) are separated by a thin, porous film and encapsulated with two packaging layers. In the case of LIBs, the polymer separators and packaging materials are inherently flexible. However, the Cu and Al foil-supported electrodes exhibit limited flexibility, which can hinder the overall mechanical properties of the device. The failure of the device during bending is typically due to the cracking and delamination of the active materials, as well as the fracture and breakage of the metal foils [19, 29, 30].

Suo et al. have provided a mechanical explanation for this "film-on-substrate" model (Figure 11.1a,b), where the porous electrode can be viewed as a uniform film [31]. The thickness and Young's modulus of the film and metal substrate are represented by d_f, d_s and Y_f, Y_s, respectively. The tensile strain on the outer surface of the stacked electrode can be calculated using the following Eq. (11.1):

$$\varepsilon = \left(\frac{d_f + d_s}{2R}\right)\frac{(1 + 2\eta + \chi\eta^2)}{(1 + \eta)(1 + \chi\eta)} \tag{11.1}$$

where $\eta = d_f/d_s$ and $\chi = Y_f/Y_s$. Given that χ is usually lower than 1 due to the high Young's modulus of metal foils (e.g. 69 GPa for Al foil), the MFEs can experience significant strain upon bending.

This equation also highlights the importance of substituting metal foils with more pliable materials in the design of FLBs. On the one hand, Cu and Al metal foils exhibit very low yield strains (ε_y) of less than 1%, making them unable to withstand the repeated tension and compression experienced during bending. On the other hand, due to differences in mechanical properties (e.g. elastic modulus and Poisson's ratio) and thicknesses of the metal foil and electrode, MFEs may easily delaminate during bending due to stress concentrations on the film or at the interface between layers. As a result, electrode structures without current collectors or monolithic electrodes, where electrode materials are interpenetrated into conductive networks, are preferred for improved flexibility.

Structurally, incorporating pores into materials can reduce bending stiffness and enhance flexibility. The relative flexibility (α) of porous materials increases exponentially with higher porosity (ρ), as shown by Eq. (11.2) [17, 32]:

$$\alpha = \frac{Y_p}{Y_s} = \frac{2(1 + \sqrt{2})^2}{15}\frac{5\pi + 3(1 - \rho)(1 + \sqrt{2})^3}{25\pi + 7(1 - \rho)(1 + \sqrt{2})^3}(1 - \rho) \tag{11.2}$$

where Y_p and Y_s represent Young's modulus of porous and solid materials, respectively. For conventional electrodes used in current LIBs, a porosity of ~25% is typically adopted to achieve satisfactory energy density. Increasing the porosity to

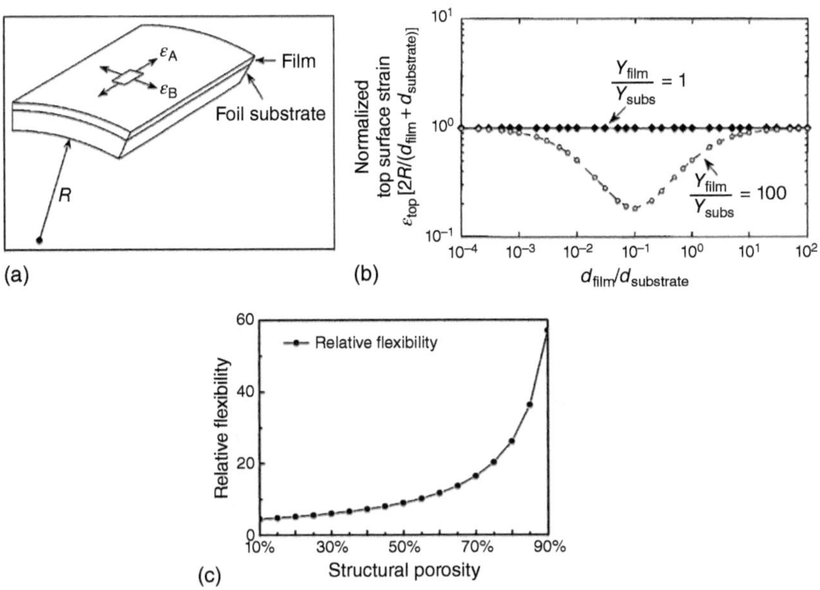

Figure 11.1 (a) A film-on-foil structure bends into a cylindrical roll. (b) Normalized strain in the film as a function of film/substrate thickness ratio. Source: Reproduced with permission [31]. Copyright 1999, AIP. (c) Relative flexibility of the porous structure as a function of porosity with respect to the solid material. Source: Reproduced with permission [17]. Copyright 2019, American Chemical Society.

50% can result in an 88% reduction in bending stiffness (Figure 11.1c). However, the porous electrodes, composed of polymeric binders, nanosized carbon additives, and microsized active particles, may be prone to cracking due to defects and particle agglomerations. In recent years, freestanding and flexible electrodes have been developed through modifications in manufacturing techniques (e.g. dry electrode technology) and the use of different polymeric binders (e.g. polytetrafluoroethylene [PTFE] and polyurethane [PU]) [33, 34]. Further optimization of electrode structure and composition holds promise for achieving high mechanical deformability in FLBs without compromising energy density.

11.1.2 Pathway and Research Strategies of the FLBs

Chang et al. have provided a summary of the development pathway of FLBs (Figure 11.2) [19]. After years of development, two main strategies have emerged for achieving good flexibility in lithium batteries: the soft materials strategy and the soft structure strategy. The former involves the use of unconventional components, such as flexible current collectors, to improve yield strain. The latter involves reconfiguring the basic structure of traditional LIBs. Both of these methods have shown promising results in achieving good electrochemical and mechanical performance in FLBs.

11.1.2.1 FLBs with Soft Structure

For the soft structure strategy, materials can remain mostly unchanged, with flexibility being achieved by reducing the overall thickness of the device. The maximum strain (ε) of a multilayer-stacking FLB can be estimated using the equation $\varepsilon = t/2r$, where t and r represent the thickness and bending radius, respectively [35]. For example, when the thickness is reduced to 0.5 mm, the ε is only 1.25% when bent to a radius of 20 mm. Koo et al. fabricated a FLB by sandwiching a super-slim $LiCoO_2$/LiPON/Li cell (~10 μm) between two soft polydimethylsiloxane (PDMS) packaging films. This device showed excellent durability, withstanding 20,000 bending cycles at a low radius of 3.1 mm with negligible voltage fluctuation. However, the total energy density was relatively low at only ~2.2 Wh l^{-1} due to the use of PDMS packaging [36]. Alternatively, structured metal foils can become more flexible and be used to fabricate FLBs as demonstrated by Tomohiro Ueda from Panasonic [37]. The as-made FLB showed more than 1000 times of bending at a bending radius of 25 mm with a capacity retention more than 80%.

Alternatively, the ability to deform mechanically can be achieved by separating the functions of energy storage and mechanical flexibility. In 2000, Hikmat and Feil proposed a segmented lithium battery made up of individual subcell units connected through internal interconnects [38]. This design allowed for the pliability of the battery while the subcell units remained stable during bending since the interconnects bear strains. However, the estimated energy density of this segmented battery was relatively low (~40 Wh l^{-1}) compared to rigid pouch cells. Based on a similar principle of decoupling flexibility and energy storage, FLBs with serpentine structures, origami/kirigami structures, human spine-like structures, human joint-like structures, and 3D interlocked structures have been developed and demonstrated [17].

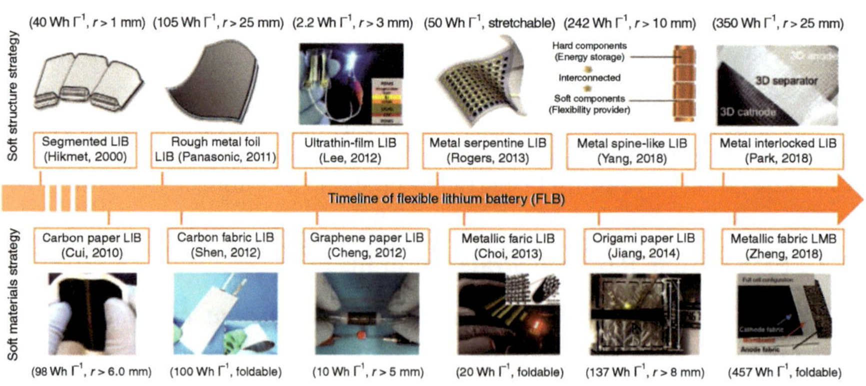

Figure 11.2 Timeline of developments in FLBs. Source: Reproduced with permission of Chang et al. [19]. © 2021/John Wiley & Sons. These representative proof-of-concept examples of FLBs are achieved by two major strategies: (1) The engineering of the soft structure; (2) the use of soft material electrodes.

One such example is the FLB with an energy density of 350 Wh l^{-1} and the ability to withstand 5000 bending cycles at a radius of 25 mm through a simple one-step patterning of a preassembled cell [39]. However, these structured FLBs have limited flexibility, with only uniaxial flexibility achievable and restrictions on the minimum bending radius (r) due to the low strain of metal interconnects. Further improvements in flexibility and energy density are expected through the use of innovative materials.

11.1.2.2 FLBs with Soft Materials

One example of the soft materials strategy is the use of flexible current collectors, pioneered by Hu et al. [40, 41]. To overcome the low yield strain of metal foils, thin, lightweight, conductive, and flexible carbon nanotube (CNT) films were used in FLB fabrication. The 2 μm CNT film served as a flexible current collector, with active materials coated on its surface using the doctor-blading technique. The resulting paper-like thin film battery had a thickness of ~300 μm and delivered a volumetric energy density of ~98 Wh l^{-1}, with a bending radius down to 6 mm without failure, although the prototype cell only showed a short cycle number of ~20 times. Following this strategy, various flexible and conductive films, papers, textiles, and foams have been investigated as potential replacements for metal foils as current collectors [42–48]. Among these soft current collectors, FLBs based on fabric materials have shown excellent flexibility and even foldability. For example, Liu et al. developed a hierarchical 3D $ZnCo_2O_4$ nanowire array/carbon cloth composite as a binder-free anode for FLBs [49]. The device exhibited preserved discharge capacities even after 120 folding cycles. The 3D structure of the carbon fabric enabled high reversible capacities of the novel anode nanomaterials, contributing to an energy density of over 100 Wh l^{-1}. When high-specific-capacity anode (e.g. Li metal) and cathode (e.g. S) materials were used with metal-coated carbon fabrics as current collectors, the energy density of FLBs could be further increased to over 450 Wh l^{-1} [50, 51].

Several commercial FLB products are currently available, such as the slim and lightweight flexible LIB exhibited by Panasonic in 2016, with a thickness of approximately 0.55 mm. This battery can maintain 99% of its capacity after 1000 bends at a radius of 25 mm and 1000 twists at an angle of 25°, making it suitable for use in devices such as smart cards and smart clothing [52]. The Korean company Jenax also offers a flexible lithium polymer battery that can be bent or folded into different shapes while maintaining safe and normal charge/discharge functions. The charge/discharge curves under dynamic bending at 20 mm match well with the nonbending curves, and the battery can withstand over 10,000 bending cycles [53]. The next-generation electronics, such as smartwatches, fitness bands, smart apparel, and military helmets, could potentially benefit from the use of FLBs. However, despite their excellent flexibility, the energy densities of these batteries are still relatively low (<200 Wh l^{-1}) compared to nonflexible commercial LIBs (>600 Wh l^{-1}). The issue of energy density will be discussed in Section 11.3.2.

11.2 Materials and Structures for Achieving High-Performance FLBs

11.2.1 Flexible Current Collectors

Replacing brittle metal foils with soft materials is a straightforward and effective method for achieving flexibility in batteries. The desired candidates should be thin and lightweight, comparable to currently used metal foils to maintain high energy densities. The electrical conductivity of the current collector should also be high ($>10^7$ S m^{-1}) to ensure optimal electrochemical performance, as low conductivity can result in large overpotential and low charge capacity. The yield strain (ε_y) is also a crucial factor, with a recommended yield strain of 5% for FLB components as suggested by Zheng and coworkers [54]. Additionally, the use of industrial roll-to-roll fabrication techniques requires that Young's modulus of the substrates should not be too low. Several potential current collectors, other than metal foils, are discussed in the following sections.

11.2.1.1 Conductive Nanomaterials

Lightweight and highly conductive carbon-based conductors have attracted significant attention for their potential use not only in flexible batteries (FBs) but also in electric powertrains, next-generation communications, and military applications. Carbon conductors, composed of nanomaterials such as 1D CNTs, 2D graphene, and carbon nanofibers, can be produced using techniques such as chemical vapor deposition (CVD), slurry coating, vacuum filtration, and electrospinning. These films have the advantage of being extremely thin (<5 μm) and lightweight (<2 mg cm^{-2}), making them suitable for use in FBs with high energy densities. However, their main disadvantage is their inferior conductivity compared to metal foils. To address this issue, chemical modification and posttreatment techniques can be used to enhance both electrical conductivity and mechanical properties. For example, Wan et al. reported ultrathin, densified CNT films with "metal-like" properties after high-temperature annealing and treatment with H_2SO_4 and $HClSO_3$ (Figure 11.3a,b) [55]. These films exhibited a conductivity of approximately 9.02×10^5 S m^{-1}, a thickness of only 1.85 μm, and a density of only 0.139 mg cm^{-3}. Additionally, they demonstrated a high Young's modulus (36.94 ± 6.59 GPa), yield strain ($7.59 \pm 0.27\%$), and tensile strength (822 ± 21 MPa).

The 2D graphene materials, with their sheet-like morphology, have even better theoretical conductivity than metals ($>10^7$ S m^{-1}). Highly conductive graphene films can be fabricated through physical and chemical cross-linking between the nanoflakes. Wan et al. reported the preparation of highly conductive and extremely tough graphene films through sequential cross-linking of filtrated graphene oxide (GO) sheets using π–π and covalent cross-linking agents (Figure 11.3c,d) [56]. The resulting film exhibited a tensile strength of approximately 944.5 MPa, toughness of approximately 20.6 MJ m^{-3}, and yield strain of approximately 5.5%. The electrical conductivity exceeded 5×10^4 S m^{-1} and could be further improved to 1.2×10^5 S m^{-1} through optimization of the cross-linking agents and procedures

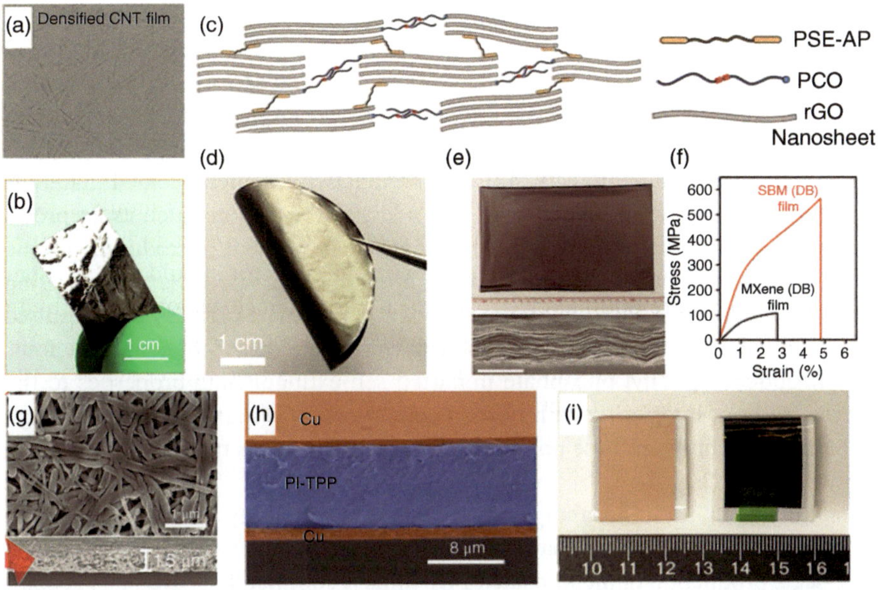

Figure 11.3 Flexible current collectors. (a,b) scanning electron microscope (SEM) and digital images of a thin and flexible densified CNT film. Source: Reproduced with permission of Wan et al. [55]. © 2020/American Chemical Society. (c,d) Structure scheme and the digital image of a thin and flexible graphene film. Source: Reproduced with permission of Wan et al. [56]. © 2018/National Academy of Sciences. (e,f) Images and tensile stress–strain curve of a cross-linked conductive MXene film. Source: Reproduced with permission of Wan et al. [57]. © 2021/American Association for the Advancement of Science - AAAS. (g) SEM images of a flexible Cu nanowire foil. Source: Reproduced with permission of Chu et al. [58]. © 2017/Elsevier. (h) Cross section of the TPP embedded composited Cu-PI film. (i) Digital photo of the cell made from Cu foil and TPP-embedded composited Cu-PI film. Source: Reproduced with permission of Ye et al. [59]. © 2020/Springer Nature.

[60]. Similar properties have also been achieved using other 2D materials, such as Mxene, where a 3 μm film exhibited a high electrical conductivity of 6×10^5 S m^{-1} and maintained excellent mechanical toughness (Figure 11.3e,f) [57]. These properties make these conductive films promising candidates for use as novel current collectors [61]. However, scalability and cost may be limiting factors when producing them on a large scale.

The assembly of nanomaterials can be expanded to the metals. The Cu nanowires (CuNWs), with good electrical conductivity, were used for the preparation of ultrathin, lightweight, and flexible CuNW foils through solvent evaporation and roll pressing procedure by Chu and Tuan (Figure 11.3g). The densified 1.5 μm CuNWs foil showed a weight of only 1.2 mg cm^{-2} and very low sheet resistance of ~24 mΩ sq^{-1}, close to that of the Cu foil (~10 mΩ sq^{-1}), and lower than the CNT and graphene films. The CuNW foils showed good foldability and interfacial adhesion to the graphite anode. Due to the high surface area of the CuNWs, the rate performance and the energy density of the full cells were increased compared to the Cu foil-based batteries [58].

11.2.1.2 Metal Composites

Metal/polymer composite materials, which combine the good conductivity of metal with the excellent flexibility of polymers, are also potential options for novel current collectors. Metal-coated polyethylene terephthalate (PET) films have been extensively studied as substrates for a wide range of applications, including flexible solar cells, displays, and interconnects [62]. The conductivity of metal-coated materials can be comparable to pure metal foils, and additional functions such as fireproofing can be achieved through additive engineering. Ye et al. introduced unique fire-extinguishing current collectors for high-energy and safe LIBs by depositing metal on the surface of flame-retardant-embedded polyimide (PI) [59]. This resulted in an energy density increase of 16–26% with thin and lightweight composite foils. The embedded triphenyl phosphate imparted self-extinguishing properties to the batteries without interfering with normal cell operation (Figure 11.3h,i). The high mechanical flexibility of these polymer substrates also makes them suitable for use in future safe FLBs.

In addition to their advantage of excellent conductivity and flexibility, the fabrication methods of metal-coated films/papers/fabrics are versatile and scalable. Large-scale production of metal-coated PET films is commercially available at comparable costs to pure Cu foil, and Ni/Cu-coated PET fabrics can be fabricated using a mature electroless metal deposition technique [47]. To enhance the interfacial adhesion between the metal layer and the substrate, Zheng and coworkers have developed a solution-processed polymer-assisted metal deposition (PAMD) method [63, 64]. This method allows for the facile coating of different kinds of metals (e.g. Cu, Ni, Ag, and Au) on various substrates, including 1D fiber materials, 2D planar substrates, and 3D porous fabrics. The solution-based PAMD method enables the production of enhanced flexing durable metal-coated conductive materials, which have been applied in the areas of optoelectronics, sensors, energy harvesting, and energy storage. For example, a 14 μm-thick Cu-coated PET fabric with an areal density of $3\,mg\,cm^{-2}$ and conductivity exceeding $10^7\,S\,m^{-1}$ can be obtained using the PAMD method. Additionally, roll-to-roll processed Cu-PET can withstand the current fabrication process of LIBs due to the favorable mechanical properties of PET, such as yield strain, elastic modulus, and tensile strength.

A comparison of different types of flexible current collectors with currently used Cu and Al foil is summarized in Table 11.1. Metal-coated composites show promising potential as an alternative to pure metal foils, with better mechanical properties, thickness, weight, and comparable conductivity.

11.2.2 Flexible Electrodes

When fabricating FLBs, simply replacing the current collectors is not enough to ensure the integrity of the electrodes during flexing and bending. One common issue when bending the battery is the cracking and detachment of active materials from the current collectors. To address this problem, various strategies have been proposed to redesign traditional electrodes, including the current collectors, active

Table 11.1 Representative flexible conductive films compared with Cu and Al foil.

	Thickness (μm)/ weight ($mg\,cm^{-2}$)	Conductivity ($S\,m^{-1}$)	Elastic modulus (GPa)	Yield strain (%)	Tensile strength (MPa)
Cu foil [65]	10/8.9	5.96×10^{7}	70–102	0.5–1.0	100–300
Al foil	15/4.05	3.77×10^{7}	69	0.2	100
CNT film [55]	1.85/0.35	9.03×10^{5}	36.94 ± 6.59	7.59 ± 0.27	822 ± 21
rGO film [56]	3.3/0.68	5.12×10^{4}	15.6 ± 1	5.9 ± 0.4	944.5 ± 46.6
MXene film [57]	3.0/0.90	6.11×10^{5}	26.8 ± 1.3	4.62 ± 0.21	559 ± 10
Cu nanowire [58]	1.5/1.20	2.48×10^{7} a) ($24\,m\Omega\,sq^{-1}$)	N/A b)	N/A b)	N/A b)
Metal-coated PI foil [59]	9/1.54	5.71×10^{7}	2.01	~30	~60
Metal-coated PET fabric [63, 64]	14/3.00	$>5.00\times10^{7}$	~2	~30	120–150

a) Estimated from the sheet resistance.
b) Not provided from the reference.

materials, and even the separator. Examples of these strategies can be seen in Figure 11.4.

11.2.2.1 Freestanding Electrodes Based on Polymeric Binders

The type and ratio of polymeric binders used in electrodes directly affect their mechanical properties. Exploring novel polymeric binder materials with stronger adhesion than commonly used materials like polyvinylidene fluoride (PVDF) and carboxymethyl cellulose (CMC) can improve the mechanical flexibility of batteries. For example, PU was investigated as a binder for flexible electrode fabrication [70]. It was found that ether-based PU can provide 35% higher adhesion strength compared to PVDF, leading to improved cohesion between active materials and preventing disintegration during flexing while maintaining electrochemical performance. Similar results can be achieved by increasing the ratio of polymeric binders. By optimizing the ratio of binders and conductive additives, freestanding, bendable, and foldable electrodes can be obtained (Figure 11.4a) [33, 66]. However, a high ratio of binder may decrease the electrochemical performance and energy density of the batteries due to blocked electrical transfer and lower active material content.

Leveraging the good mechanical properties of polymer materials, freestanding and flexible electrodes can be directly constructed using the solvent-induced phase inversion method. Instead of drying the slurry through heating, a nonsolvent can be introduced to precipitate the polymer [67, 71–74]. During the solvent exchange process, polymer-rich and polymer-lean phases form, leading to the formation of a flexible microporous structure. Commonly used polymer materials for this method

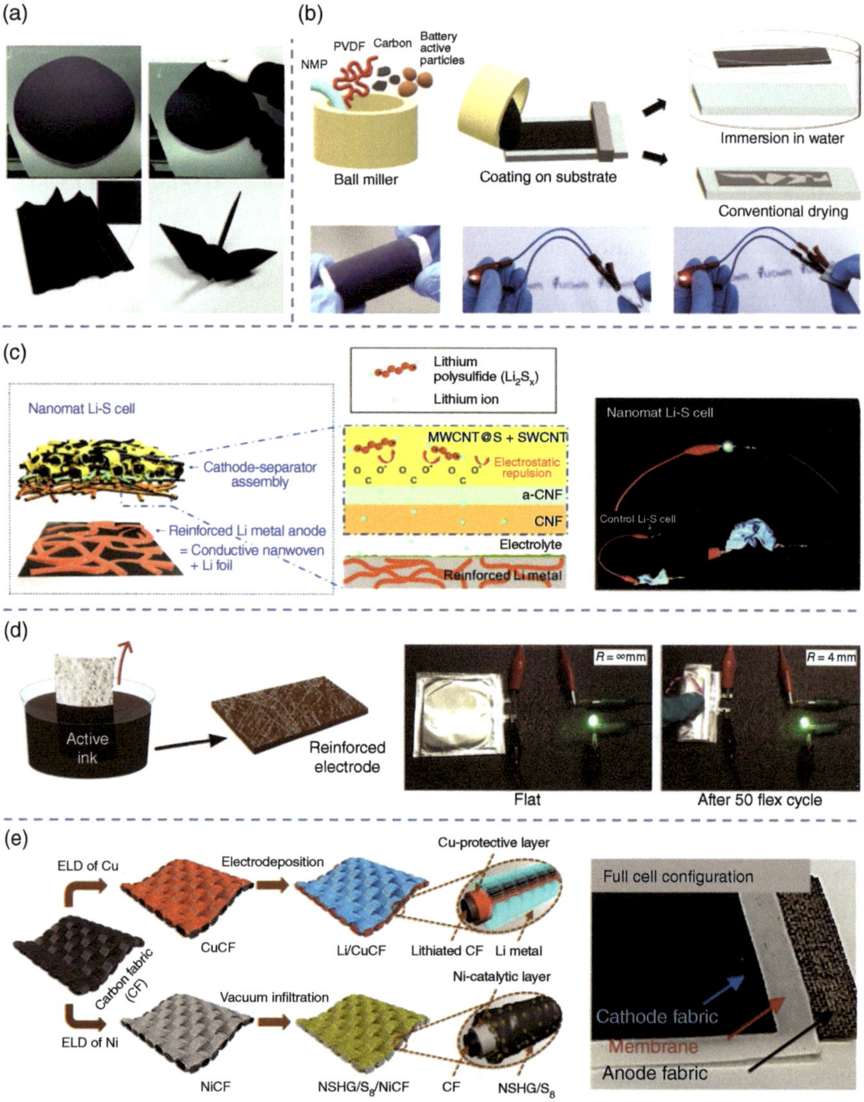

Figure 11.4 Strategies of fabricating flexible electrodes and the as-made FLBs. (a) Flexible electrodes were prepared by using thermoplastic polyurethane as the binder. Source: Reproduced with permission of Park et al. [66]. © 2017/Royal Society of Chemistry. (b) Scheme of preparing flexible electrodes by the solvent-induced phase separation method. Source: Reproduced with permission of Harks et al. [67]. © 2019/Elsevier. The as-made FLBs can light LED under different bending states. (c) Scheme of binder-free flexible S cathode and Li metal/polyethylene terephthalate composited anode. Source: Reproduced with permission of Kim et al. [68]. © 2019/Royal Society of Chemistry. A green LED lamp connected to the nanomat Li–S cell (vs. the control Li–S cell (inset)) in the severely crumpled state. (d) Scheme showing the preparation of TCEs by method of dip-coating. Source: Reproduced with permission of Gaikwad et al. [69]. © 2015/John Wiley & Sons. The battery was able to power the green-LED continuously even after flexing 50 times to a bend radius of 4 mm. (e) Schematic illustration of the preparation of fabric Li-S full batteries made from Cu- and Ni-coated carbon cloth and the digital picture of the inner cell configuration in full-cell Li-S batteries. Source: Chang et al. [50]/Springer Nature/CC BY 4.0.

include PU, PVDF, poly(vinylidene fluoride-*co*-hexafluoropropylene) (PVDF-HFP), and others. In addition to excellent flexibility, this method allows for the fabrication of high-loading and thick electrodes, increasing the energy density. Areal capacities of up to ~21 mAh cm^{-2} have been reported using this method (Figure 11.4b) [67].

11.2.2.2 Flexible Electrodes Based on Binder-Free Techniques

Paper-like, freestanding electrodes can also be fabricated using binder-free methods such as vacuum filtration, electrospinning, electrospray, and others [75–77]. 1D and 2D nanomaterials are commonly used as building blocks for these binder-free electrodes (Figure 11.4c) [68]. The advantage of these electrodes is their good electrochemical performance, as they allow for direct electrical transfer between active materials and conductive additives and smooth Li^+ transfer without the use of insulating polymer materials.

In terms of mechanical properties, 1D and 2D materials form a structural framework that supports the integrity of the electrode during repeated flexing. For example, Wang et al. synthesized 2D conductive nanosheets by modifying cellulose with 1D carboxylated CNTs, which act as a binder. Through a simple filtration method, self-supporting, high-strength electrodes with a mass loading of active materials up to ~90 mg cm^{-2} can be obtained. Due to strong van der Waals forces and hydrogen bonding, the conductive nanosheets physically wrap around and grasp the electrode particles, allowing for repeated folding and bending without electrochemical degradation [78]. A similar principle has been applied to other high-capacity active materials, such as Si and S, to further enhance the energy density [79].

11.2.2.3 Textile Composite Electrodes

Recent studies have shown that FLBs fabricated with textile composite electrodes (TCEs), electrodes coated onto or grown on conductive textile current collectors, offer both high energy density and good mechanical flexibility. These strain-compliant textiles can withstand stress during bending and provide a stable electrical path during charge/discharge cycles (Figure 11.4d) [69]. FLBs based on Ni-coated PET have shown excellent deformation tolerance, with over 5500 cycles of dynamic bending/unbending at a low radius of 0.65 mm [44]. The use of textiles as 3D hosts for active materials can also enhance electrochemical stability. For example, Chang et al. reported a Li-S full battery with a high energy density of 457 Wh l^{-1} and remarkable bending stability down to a radius of 1.0 mm. By rationally designing Cu-coated and Ni-coated carbon fabrics as hosts for Li metal and S, respectively, the dendrite problem with Li metal and the shuttle effect of polysulfides can be effectively inhibited (Figure 11.4e) [50].

The fabrication of TCEs can be achieved through various methods. Active materials can be coated onto porous textiles using techniques such as doctor-blading, dip-coating, and printing. Alternatively, active materials can be *in situ* grown onto the surface of fibers through hydrothermal reaction, electrochemical deposition, or chemical/physical vapor deposition. Additionally, incorporating active materials into fibers or yarns through electrospinning or dry/wet spinning, often followed by postpyrolysis treatment, is also reported [30, 80–82]. The use of paper-based and textile-based composite electrodes is estimated to significantly enhance the energy density of FLBs. Depending on the paired cathode and anode materials, an energy density of 800 Wh l^{-1} can be achieved, meeting the commercial standard for many applications, such as medical patches, watch belts, bendable phones, and roll-up displays. With the development of next-generation electrode materials, an energy density of 1000 Wh l^{-1} is also conceivable [19].

11.2.3 Flexible Solid-State Electrolytes (SSEs)

Safety is a crucial consideration when using wearable electronics on the human body. Liquid electrolyte-based LIBs pose potential risks such as leakage of corrosive chemicals (e.g. HF) and thermal runaway upon short circuits during flexing [83, 84]. SSEs with high ionic conductivity (>1 mS cm^{-1}), mechanical strength, soft texture, and thermal stability offer better alternatives to liquid electrolytes. Considering the requirement for flexibility, solid polymer electrolytes (SPEs) and their composites outperform inorganic electrolytes due to their higher mechanical toughness [85–88]. As noted by Liang et al., resilient SSEs can serve as filler and binder, tightly anchoring the entire structure and ensuring battery integrity during bending, twisting, and even stretching, avoiding slippage and delamination between the electrodes, electrolyte, and current collectors [89]. As a good example, Mackanic et al. demonstrated a conformable and stretchable FLB by using a supramolecular ion conductor to enable intimate bonding between the electrode and electrolyte through dynamic bonding [90]. This SPE-based FLB presents a promising route for future wearable and flexible electronics.

11.2.3.1 Solid Polymer Electrolytes (SPEs) and Gel Polymer Electrolytes (GPEs)

Although the chemistry of SPEs can vary depending on the polymers and salts used, their main limitation is the very low ionic conductivity (<0.1 mS cm^{-1} at RT) compared to other electrolytes, as shown in Table 11.2. This low ionic conductivity severely limits the practical applications of SPEs. However, the addition of inorganic nanoparticles and organic small molecules can enhance the ionic conductivity of SPEs [91–93]. The incorporation of a certain amount of liquid solvent can also significantly increase the ionic conductivity to a competitive level with liquid electrolytes. These GPEs have been successfully applied to various battery systems, including flexible LIBs, dual-ion batteries (DIBs), lithium-metal batteries (LMBs), and lithium-sulfur (Li-S) batteries, and others.

However, polymer-immobilized liquid solvents are still flammable [88]. To address this issue, nonflammable plasticizers such as ionic liquids, deep eutectic solvents, and fluorinated molecules have been explored to enhance the ionic conductivity of SPEs [94–97]. One of the most effective additives is the use of plastic crystals, such as succinonitrile (SN) [98]. The ionic conductivity of SN-doped GPEs can reach 1 mS cm^{-1} at room temperature. Ha and Choi et al. demonstrated the superior electrochemical performance and flexibility of UV-curable semi-interpenetrating polymer network and SN-based plastic crystal electrolyte

Table 11.2 Comparison of commonly reported electrolytes.

	Ionic conductivity (RT)	Mechanical flexibility	Thermal tolerance
Liquid electrolyte	10^{-3}–10^{-2} S cm^{-1}	—	Low (<100 °C)
Inorganic electrolyte	10^{-7}–10^{-3} S cm^{-1}	Brittle	Very high (>400 °C)
Polymer electrolyte	10^{-8}–10^{-4} S cm^{-1}	Soft	Medium (100–300 °C)
Composite electrolyte	10^{-5}–10^{-3} S cm^{-1}	Soft	High (200–400 °C)

[99, 100]. This resulted in soft, mechanically strong, and electrochemically stable GPEs with high ionic conductivity. The assembled FLBs exhibited good thermal stability, excellent deformability, and fast-charging capability and even successfully lit an LED under a crumbled state.

11.2.3.2 Aqueous Gel Polymer Electrolytes

A more desirable additive is H_2O, as aqueous electrolyte-based lithium batteries are considered very safe. However, the narrow electrochemical voltage window (0–1.23 V vs. standard hydrogen electrode [SHE] for pure water) due to the irreversible hydrogen evolution reaction (HER) limits the pairing with existing anode materials [101, 102]. One effective strategy is to decrease free water molecules by adding excessive salts and using water-chelating polymers [103–105].

Zhang et al. demonstrated a stable "water-in-salt" polymer electrolyte with tightly mobilized water molecules and a high salt content [106]. They presented a long-cycling, high energy density (151 Wh kg^{-1}), and 3 V-class $LiMn_2O_4//Li_4Ti_5O_{12}$ flexible aqueous battery with a practical cathode/anode ratio of 1.14. However, further exploration is needed to expand the voltage window down to that of the currently used anode such as graphite (0.2 V vs. Li^+/Li), which would significantly boost the energy density [101, 107].

11.2.3.3 Hybrid SSEs

Hybrid SSEs are promising electrolytes for all-solid-state lithium batteries, as they offer both flexibility and high ionic conductivity through the meticulous design of the composite structure. The common method for fabricating hybrid SSEs involves sequentially creating a porous framework (e.g. inorganic SSEs and porous membrane) and infiltrating a polymer electrolyte as a reinforcement matrix [93, 108]. In contrast to the use of inorganic fillers, the inorganic SSE framework provides continuous Li^+ pathways and maintains high ionic conductivity. The infiltrated polymer electrolyte, on the other hand, eliminates weak grain boundaries and pores, ensuring high mechanical properties and good contact between interfaces. For example, Fu et al. synthesized a flexible composite SSE consisting of electrospun garnet-type $Li_{6.4}La_3Zr_2Al_{0.2}O_{12}$ (LLZO) nanofibers and thermally infiltrated polyethylene oxide (PEO) polymer electrolyte for use in LMBs, demonstrating good flexibility and stable cycling of Li metal at room temperature [109]. Moreover, the thickness, mechanical properties, and ionic conductivity of these SSEs can be adjusted by selecting different frameworks and polymers [108–110]. Wan et al. found that the ionic conductivity of PEO electrolyte can be increased by five times to $2.3 \times 10^{-4}\,S\,cm^{-1}$ through the alignment of polymer chains. The vertically aligned nanochannel of the ultrathin PI film (8.6 μm) contributed to both rapid Li^+ diffusion and high energy density. A flexible and safe pouch cell successfully powered an LED under extreme mechanical deformation, such as folding, nail penetration, and cutting [110].

11.2.3.4 Interface Issues of the Solid-State Batteries

One noticeable issue when using SSEs is the poor contact at the interface. Unlike flowable liquid electrolytes that easily wet porous electrodes, the rigid–rigid contact of the cathode/electrolyte and anode/electrolyte leads to poor electrochemical performance [111–113]. Heat treatment and surface coating are effective methods for addressing interfacial problems [114]. Additionally, *in situ* fabrication methods can be used to eliminate the interface [115]. One such method is separator-assisted one-pot polymerization, where a liquid precursor containing polymerizable monomers, salts, and initiators can wet the electrodes well and transform into an SSE upon initiation [116, 117]. For example, Liu et al. converted the traditionally used 1,3-dioxolane (DOL)/1,2-dimethoxyethane (DME) electrolyte into GPE via $LiPF_6$-induced polymerization of DOL under ambient temperature, expanding the low electrochemical voltage window (<4 V) of the liquid ether-based electrolyte up to 4.6 V [118]. This allowed for multiple cathode material options, including S, $LiFePO_4$ (LFP), and $LiNi_{0.6}Co_{0.2}Mn_{0.2}O_2$ (NCM622), and exhibited excellent cycling stability and mechanical flexibility even under severe deformation. Acrylate-based monomers also provide more options as precursors for in-situ solidification. Lee et al. demonstrated a plastic-crystal-embedded elastomer electrolyte (PCEE) with poly(butyl acrylate) as the elastic matrix and a 3D interconnected plastic crystal phase as the ion-conducting pathway through polymerization-induced phase separation [119]. The *in situ* formed PCEE showed high ionic conductivity, mechanical robustness, and low interfacial resistance, allowing for the accommodation of large volume expansion of the Li metal anode during plating/stripping and achieved high reversibility. This resulted in a high-energy density LMB with high cathode loading and low N/P ratio at ambient temperature.

Another method for addressing the interface between the electrolyte and electrode is through sequential deposition of battery components layer-by-layer. Generally, the precursor cathode, electrolyte, and anode can be printed as viscous inks using techniques such as 3D printing, stencil printing, and screen-printing, followed by fast curing through heat or UV irradiation [120, 121]. In a study by Kim et al., solid-state flexible LIBs with arbitrary complex geometries were fabricated using a stencil printing and UV cross-linking process. The UV-curable anode slurry, composite electrolyte precursor, and cathode slurry were consecutively deposited and solidified on the Al current collector within 30 seconds using UV irradiation. In addition to exhibiting good flexibility and stable electrochemical performance, this method allows for the printing and seamless integration of customized batteries with different form factors onto various objects. The solvent-free and mold-free printing techniques greatly simplify the fabrication of complex-shaped batteries, including microbatteries and FBs.

11.2.4 Flexible Configuration of Batteries

11.2.4.1 Novel Configurations of FLBs

The concept of a soft structure, which revolutionizes the traditional multilayer-stacking configuration of batteries, has shown great potential for high-performance FLBs. Since 2000, unconventional FLBs with segmented, serpentine, spine-like, origami/kirigami, and 1D fiber-shaped structures have been proposed and demonstrated [122–125]. In these structured FLBs, there are two functional parts: subunit cells and interconnects. The interconnects bear large strains upon deformation, while the subunit cells maintain the original shape and function of energy storage. One example is the serpentine-structured stretchable LIB proposed by Xu et al., where an array of miniaturized batteries with a size of ~2 mm was constructed using photolithography and connected by "self-similar" interconnects [122]. The thin and soft silicon elastomer sheets supported and sealed the entire structure, enabling high stretchability (300%) without significant capacity loss. Additionally, it is possible to design FLBs with a high energy density comparable to that of commercial LIBs through structural reconfiguration. Qian et al. proposed a spine-like FLB by cutting and winding conventional electrodes. The wound thick stacks maintained their function of energy storage, while the unwound parts connected the electrodes and provided good flexibility to the batteries [125]. The calculated energy density of the unoptimized device was preserved at ~85% (242.5 Wh l^{-1}) of the standard prismatic cell with the same components, and the unique structure ensured a much smaller strain (~0.08%) than that of the prismatic cell (1.8%) and stacked cell (1.1%) when bent into a 9 mm circle.

11.2.4.2 1D Fiber-Shaped FLBs

Among the various novel configurations of FLBs, 1D fiber-shaped devices show great promise in future wearable E-textiles due to their omnidirectional flexibility and potential for seamless integration. The ideal fiber-shaped FLB should be slim, lightweight, have adequate energy density, be mechanically strong, and compatible with the existing textile industry [13, 82, 126]. However, most reported fiber-shaped devices have a limited length (<10 cm) due to the complexity of fabrication and increased internal resistance. He et al. discovered that the internal resistance of fiber batteries follows a hyperbolic cotangent function with increasing fiber length, contrary to previous expectations that a longer fiber would result in a linear increase in internal resistance and poor electrochemical performance [127]. Based on this theory, a 1 m long fiber-shaped FLB with a gravimetric energy density of 85.69 Wh kg^{-1} was fabricated using Al and Cu fibers as current collectors, exhibiting excellent flexibility with 100,000 bending cycles and 80% capacity retention. Additionally, these fiber-shaped FLBs were woven into textiles using an industrial rapier loom, demonstrating excellent safety, flexibility, washability, and seamless integration with textile electronics [127, 128]. It is expected that more applications can be explored in the future using fiber-shaped FLBs.

The capacity of a battery linearly increases with the lengthening of fiber devices. However, enhancing the energy density of fiber-shaped batteries is challenging due to the limited loading of active materials determined by the fiber diameter. High-specific capacity materials are desired for high-energy fiber batteries. Gao et al. reported the fabrication of a 10-m-long fiber anode using capillary filling of Li metal into Cu-coated carbon yarn at a high temperature. The resulting Li metal composite yarn (LMCY) exhibited excellent flexibility, high purity, and good cycling stability [129]. Full cells with controllable loading capacities were assembled using yarn-type LFP and S cathodes, delivering a high energy density of ~292 Wh l^{-1} and a long lifetime of over 800 cycles. The seamless integration of these fiber batteries into gloves was demonstrated, showcasing the potential of wire-type FLBs for various applications.

11.3 Challenges and Perspectives

As the development of FLBs continues to progress rapidly, several key challenges have emerged that need to be addressed for further commercialization. These challenges include the lack of standardized testing and evaluation methods, as well as the difficulty in achieving high energy density compared to mature LIBs on the market.

11.3.1 Lack of Standard Testing and Evaluation Method

11.3.1.1 FOM of FLBs

Despite the significant progress made in the field of FLBs, there are currently no standardized benchmarks for evaluating the electrochemical and mechanical performance of f-ESDs. Among the reported results, there is no single factor that can be used to assess the merit of FLBs. In most cases, the bending radius (r) is used to evaluate flexibility, with a smaller r indicating better flexibility. However, due to the lack of standardized testing procedures, the bending test evaluation methods are diverse and cannot be directly compared. As previously mentioned, the actual strain of a device can vary even under the same bending radius (r) when the bending angle (θ) and length of the device (L) differ [14, 130]. This is logical because the longer the device, the higher the unaffected area upon bending, resulting in better performance. It is necessary to fix the θ and L when bending devices to a fixed r. The accurate determination of r relies on stricter regulations and standards.

However, the bending radius (r) alone cannot be used to benchmark the performance of different devices, as it is the yield strain (ε_y) that determines the failure of materials and devices. The greatest tensile strain of a device occurs on the outer surface when bending, where $\varepsilon = h/2r$ and h is the thickness of the device. As h can vary for different devices, ε is a more important factor to consider for flexible devices. A figure of merit (FOM) was proposed by Chang et al. to fairly compare the

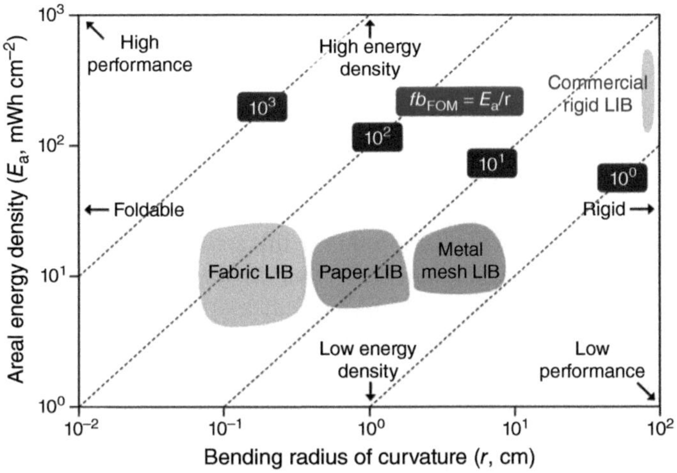

Figure 11.5 A figure of merit for flexible batteries. Source: Reproduced with permission [54]. Copyright 2020, Elsevier.

performance of different FBs [54]. It is assumed that the ε of cell components should be lower than 5%, as the *r* of flexible electronics should be lower than 10 times of their thickness, according to the Institute for Printed Circuits (IPC) (IPC-2292 standard). The areal energy density (E_A) is also introduced to evaluate the energy density of FBs. The FOM of FBs can be calculated using Equation 11.3:

$$fb_{FOM} = E_A/r \tag{11.3}$$

where the E_A can be calculated by areal capacity (C_a) and the discharge voltage (V).

Based on the FOM, a comparison of the currently reported FBs was made to evaluate the suitability of the reported results for commercialization as shown in Figure 11.5 [54]. It shows that among the numerous reports of FBs, paper-based and fabric-based batteries present attractive potential regarding energy density and flexibility.

11.3.1.2 Durability, Wearability, and Comfortableness of FLBs

Some research only tested devices under a fixed bending radius and angle without repeated bending cycles, which is insufficient for evaluating durability. The durability of flexible devices is typically determined by the number of bending cycles they can withstand. However, the required number of bending cycles varies for different flexible electronics. For example, a smartwatch belt needs to withstand at least 1000 bending cycles at a radius of 10 mm, while a roll-up display should withstand 2000 cycles at a larger radius of 25 mm. Furthermore, it is preferable to conduct *operando* testing during the bending test, meaning the device is charged and discharged while continuously bending. Other important factors, such as the frequency of bending, should also be considered when testing durability.

The wearability and comfortability of a flexible device can be evaluated using a commercially available leather and fabric softness tester, as proposed by Li et al.

[130]. This tester measures the distention height of a clamped fabric at a given applied force, with softer materials resulting in a higher displacement. Li et al. used this method to evaluate the softness of a series of current collectors and electrodes, including CNT paper, graphene paper, carbon cloth, and stainless-steel mesh, and compared them with commercial leather and nylon textiles. They found that while CNT paper and carbon cloth were promising flexible current collectors, their softness was significantly reduced when made into electrodes and devices. Furthermore, an increase in thickness compromised the softness of the device, making it even more uncomfortable as wearable devices, despite the enhancement in areal energy density.

11.3.2 Challenge of High Energy Density

11.3.2.1 The Heavy Packaging Materials

The main disadvantage of the current FBs is the trade-off between flexibility and energy density. In most thin-film FBs, the packaging materials account for a significant portion of the overall volume and weight, leading to a decrease in energy density. For example, the thickness of Al-laminated composite plastics used for flexible pouch cells is typically 115 μm, which is doubled after encapsulation and is close to the thickness of a pair of flexible electrodes (315 μm, including the anode, cathode, and separator). In contrast, commercial LIBs minimize this influence by stacking multiple pairs of electrodes in one cell. However, this is challenging to achieve in FBs due to the limit on total thickness, as shown in Figure 11.6. In some FLBs with soft structural configurations, such as human-joint-mimic FLBs, the influence of packaging materials can be overcame, and the volumetric energy density can reach 371.9 $Wh\,l^{-1}$, which is more than 90% of a conventional pouch cell [131]. On the other hand, fiber-shaped FLBs require more complex encapsulation and have a higher unnecessary volume and weight compared to soft pouch cells [82, 132].

In addition to packaging materials, increasing the loading of active materials and reducing the ratio of inactive components such as conductive additives, polymer binders, and electrolytes in electrodes can also improve energy density. However, the porosity of flexible electrodes must be carefully balanced to maintain flexibility, electrochemical performance, and energy density. Higher porosity can improve electrode flexibility and battery rate performance, but it also leads to lower energy density. Therefore, using high-surface-area 3D current collectors and high-porosity freestanding electrodes can result in a lower volumetric energy density for FLBs. A potential solution is the interpenetrating structure of TCEs, where conductive papers and fabrics act as "fillers" and the electrode materials are embedded as the "matrix." This structure has the potential to achieve high flexibility and energy density while maintaining good charge/discharge rates [19, 30].

11.3.2.2 Novel Electrode Materials with High Specific Capacity

Since the commercialization of the first rechargeable LIBs by Sony in the 1990s, there has been a continuous pursuit of electrode materials with high specific capacity. However, intercalation-type electrode materials, such as graphite (specific

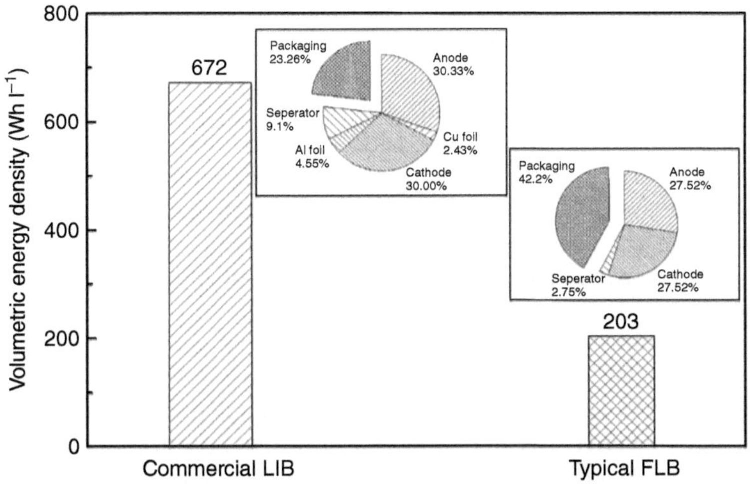

Figure 11.6 Comparison of volumetric energy density and volumetric ratio of the cell components for a commercial LIB and a typical FLB assembled from TCEs. For the commercial LIB, the assumed thickness of the anode, Cu foil, cathode, Al foil, separator, and packaging films are 50, 8, 50, 15, 15, and 115 μm, respectively. In addition, six pairs of electrodes are stacked to deliver an areal capacity of 18 mAh cm^{-2} with an output voltage of 3.7 V. For the TCE-based FLB, the assumed thicknesses of the anode, cathode, separator, and packaging films are 150, 150, 15, and 115 μm, respectively. Only one pair of electrodes is used to ensure high flexibility, with an areal capacity of 3 mAh cm^{-2} and a discharge voltage of 3.7 V.

capacity of 372 mAh g^{-1}) and $LiNi_xCo_yMn_zO_2$ (x+y+z=1) (NCM) (specific capacity of < 250 mAh g^{-1}), have relatively low specific capacities and would reach an energy density ceiling for FLBs [132]. Therefore, exploring innovative electrode materials and electrochemical systems is necessary for future high-energy FLBs. Conversion-type electrode materials have been extensively studied to increase battery energy density [22, 133–136]. For example, doping 5–10% of silicon (specific capacity of 4200 mAh g^{-1}) into the graphite anode can increase energy density by 5–20%. When paired with an NCM622 cathode, the estimated energy density of FLBs can reach ~900 and ~1000 Wh l^{-1} with 20% Si and 100% Si, respectively. On the cathode side, the use of sulfur (specific capacity of 1675 mAh g^{-1}) has shown promising results through the delicate structural design of the cathode and anode. For instance, Kim et al. developed a flexible, high-energy nanomat Li-S battery with a Ni-coated PET-embedded Li metal anode and S/CNTs cathode, along with a bi-layered cellulose separator. This battery demonstrated a high areal energy density of ~15.6 mWh cm^{-2}, excellent cycling stability of over 500 cycles, and outstanding bending durability of over 1000 cycles at a small radius of 2.5 mm [68].

In the long term, the ideal choice for FLBs would be an air cathode and metal anode, such as the Li/O_2 battery. This battery has a theoretical volumetric energy density of ~1100 Wh l^{-1}, comparable to that of petrol [137, 138]. However, due to the lack of deep understanding of these advanced electrochemical systems, stable operation of air batteries has only been achieved at a laboratory scale. Further theoretical and experimental work is needed at this early stage.

In addition to flexibility and energy density, other challenges must be addressed for practical applications and commercialization of FLBs, including cost, safety, and seamless integration with flexible electronics. To achieve cost-effective manufacturing, scalable industrial-compatible techniques such as continuous roll-to-roll fabrication must be used. Safety can be improved by using SSEs or aqueous electrolytes. Learning from textile techniques, various strategies such as "add-on," "build-in," and "direct fabrication" can be used to seamlessly integrate FLBs into flexible devices.

Due to space constraints, this chapter primarily focused on the introduction of flexible ESDs based on lithium battery technology. However, it is worth noting that the strategies explored for these devices can also be applied to other types of ESDs, such as SCs, sodium batteries, and zinc batteries. For further information on these systems, readers are encouraged to refer to additional literature and resources.

References

1 Yang, Y. and Gao, W. (2019). Wearable and flexible electronics for continuous molecular monitoring. *Chemical Society Reviews* 48 (6): 1465–1491.

2 Ray, T.R., Choi, J., Bandodkar, A.J. et al. (2019). Bio-integrated wearable systems: a comprehensive review. *Chemical Reviews* 119 (8): 5461–5533.

3 Kim, J., Campbell, A.S., de Ávila, B.E.-F., and Wang, J. (2019). Wearable biosensors for healthcare monitoring. *Nature Biotechnology* 37 (4): 389–406.

4 Koo, J.H., Kim, D.C., Shim, H.J. et al. (2018). Flexible and stretchable smart display: materials, fabrication, device design, and system integration. *Advanced Functional Materials* 28 (35): 1801834.

5 Shi, X., Zuo, Y., Zhai, P. et al. (2021). Large-area display textiles integrated with functional systems. *Nature* 591 (7849): 240–245.

6 Li, L., Wu, Z., Yuan, S., and Zhang, X.-B. (2014). Advances and challenges for flexible energy storage and conversion devices and systems. *Energy & Environmental Science* 7 (7): 2101–2122.

7 Zhou, G., Li, F., and Cheng, H.-M. (2014). Progress in flexible lithium batteries and future prospects. *Energy & Environmental Science* 7 (4): 1307–1338.

8 Huang, Q., Wang, D., and Zheng, Z. (2016). Textile-based electrochemical energy storage devices. *Advanced Energy Materials* 6 (22): 1600783.

9 Liu, W., Song, M.-S., Kong, B., and Cui, Y. (2017). Flexible and stretchable energy storage: recent advances and future perspectives. *Advanced Materials* 29 (1): 1603436.

10 Kaushik, V., Lee, J., Hong, J. et al. (2015). Textile-based electronic components for energy applications: principles, problems, and perspective. *Nanomaterials* 5 (3): 1493–1531.

11 Fu, K.K., Cheng, J., Li, T., and Hu, L. (2016). Flexible batteries: from mechanics to devices. *ACS Energy Letters* 1 (5): 1065–1079.

12 Zhai, S., Karahan, H.E., Wei, L. et al. (2016). Textile energy storage: structural design concepts, material selection and future perspectives. *Energy Storage Materials* 3: 123–139.

13 Zhang, Y., Zhao, Y., Ren, J. et al. (2016). Advances in wearable fiber-shaped lithium-ion batteries. *Advanced Materials* 28 (22): 4524–4531.

14 Mao, L., Meng, Q., Ahmad, A., and Wei, Z. (2017). Mechanical analyses and structural design requirements for flexible energy storage devices. *Advanced Energy Materials* 7 (23): 1700535.

15 Liu, Z., Mo, F., Li, H. et al. (2018). Advances in flexible and wearable energy-storage textiles. *Small Methods* 2 (11): 1800124.

16 Sumboja, A., Liu, J., Zheng, W.G. et al. (2018). Electrochemical energy storage devices for wearable technology: a rationale for materials selection and cell design. *Chemical Society Reviews* 47 (15): 5919–5945.

17 Qian, G., Liao, X., Zhu, Y. et al. (2019). Designing flexible lithium-ion batteries by structural engineering. *ACS Energy Letters* 4 (3): 690–701.

18 Zhao, Y. and Guo, J. (2020). Development of flexible Li-ion batteries for flexible electronics. *InfoMat* 2 (5): 866–878.

19 Chang, J., Huang, Q., Gao, Y., and Zheng, Z. (2021). Pathways of developing high-energy-density flexible lithium batteries. *Advanced Materials* 33 (46): 2004419.

20 Tarascon, J.M. and Armand, M. (2001). Issues and challenges facing rechargeable lithium batteries. *Nature* 414 (6861): 359–367.

21 Zhao, Y., Ding, Y., Li, Y. et al. (2015). A chemistry and material perspective on lithium redox flow batteries towards high-density electrical energy storage. *Chemical Society Reviews* 44 (22): 7968–7996.

22 Nitta, N., Wu, F., Lee, J.T., and Yushin, G. (2015). Li-ion battery materials: present and future. *Materials Today* 18 (5): 252–264.

23 Placke, T., Kloepsch, R., Dühnen, S., and Winter, M. (2017). Lithium ion, lithium metal, and alternative rechargeable battery technologies: the odyssey for high energy density. *Journal of Solid State Electrochemistry* 21 (7): 1939–1964.

24 Ding, J., Hu, W., Paek, E., and Mitlin, D. (2018). Review of hybrid ion capacitors: from aqueous to lithium to sodium. *Chemical Reviews* 118 (14): 6457–6498.

25 Li, M., Lu, J., Chen, Z., and Amine, K. (2018). 30 Years of lithium-ion batteries. *Advanced Materials* 30 (33): 1800561.

26 Winter, M., Barnett, B., and Xu, K. (2018). Before Li ion batteries. *Chemical Reviews* 118 (23): 11433–11456.

27 Zubi, G., Dufo-López, R., Carvalho, M., and Pasaoglu, G. (2018). The lithium-ion battery: state of the art and future perspectives. *Renewable and Sustainable Energy Reviews* 89: 292–308.

28 Xie, J. and Lu, Y.-C. (2020). A retrospective on lithium-ion batteries. *Nature Communications* 11 (1): 2499.

29 Vlad, A., Singh, N., Galande, C., and Ajayan, P.M. (2015). Design considerations for unconventional electrochemical energy storage architectures. *Advanced Energy Materials* 5 (19): 1402115.

30 Gao, Y., Xie, C., and Zheng, Z. (2021). Textile composite electrodes for flexible batteries and supercapacitors: opportunities and challenges. *Advanced Energy Materials* 11 (3): 2002838.

31 Suo, Z., Ma, E.Y., Gleskova, H., and Wagner, S. (1999). Mechanics of rollable and foldable film-on-foil electronics. *Applied Physics Letters* 74 (8): 1177–1179.

32 Hedayati, R., Sadighi, M., Mohammadi-Aghdam, M., and Zadpoor, A.A. (2016). Mechanical properties of regular porous biomaterials made from truncated cube repeating unit cells: analytical solutions and computational models. *Materials Science and Engineering: C* 60: 163–183.

33 Park, Y.-K., Park, G.-G., Park, J.-G., and Lee, J.-W. (2017). Robust free-standing electrodes for flexible lithium-ion batteries prepared by a conventional electrode fabrication process. *Electrochimica Acta* 247: 371–380.

34 Wahyudi, W., Cao, Z., Kumar, P. et al. (2018). Phase inversion strategy to flexible freestanding electrode: critical coupling of binders and electrolytes for high performance Li–S battery. *Advanced Functional Materials* 28 (34): 1802244.

35 Peng, J. and Snyder, G.J. (2019). A figure of merit for flexibility. *Science* 366 (6466): 690.

36 Koo, M., Park, K.-I., Lee, S.H. et al. (2012). Bendable inorganic thin-film battery for fully flexible electronic systems. *Nano Letters* 12 (9): 4810–4816.

37 Ueda, T. (2012). Thin flexible battery. US Patent 20120202101A1.

38 Hikmet, R.A.M. and Feil, H. (2003). Lithium secondary battery comprising individual cells with one another, as well as watches, computers and communication equipment provided with a battery. US Patent 6528204.

39 Cha, H., Lee, Y., Kim, J. et al. (2018). Flexible 3D interlocking lithium-ion batteries. *Advanced Energy Materials* 8 (30): 1801917.

40 Hu, L., Choi, J.W., Yang, Y. et al. (2009). Highly conductive paper for energy-storage devices. *Proceedings of the National Academy of Sciences* 106 (51): 21490–21494.

41 Hu, L., Wu, H., La Mantia, F. et al. (2010). Thin, flexible secondary Li-ion paper batteries. *ACS Nano* 4 (10): 5843–5848.

42 Gwon, H., Kim, H.-S., Lee, K.U. et al. (2011). Flexible energy storage devices based on graphene paper. *Energy & Environmental Science* 4 (4): 1277–1283.

43 Li, N., Chen, Z., Ren, W. et al. (2012). Flexible graphene-based lithium ion batteries with ultrafast charge and discharge rates. *Proceedings of the National Academy of Sciences* 109 (43): 17360–17365.

44 Lee, Y.-H., Kim, J.-S., Noh, J. et al. (2013). Wearable textile battery rechargeable by solar energy. *Nano Letters* 13 (11): 5753–5761.

45 Hu, J.W., Wu, Z.P., Zhong, S.W. et al. (2015). Folding insensitive, high energy density lithium-ion battery featuring carbon nanotube current collectors. *Carbon* 87: 292–298.

46 Ha, S.H., Shin, K.H., Park, H.W., and Lee, Y.J. (2018). Flexible lithium-ion batteries with high areal capacity enabled by smart conductive textiles. *Small* 14 (43): 1703418.

47 Lee, K., Choi, J.H., Lee, H.M. et al. (2018). Solution-processed metal coating to nonwoven fabrics for wearable rechargeable batteries. *Small* 14 (43): 1703028.

48 Shen, W., Li, K., Lv, Y. et al. (2020). Highly-safe and ultra-stable all-flexible gel polymer lithium ion batteries aiming for scalable applications. *Advanced Energy Materials* 10 (21): 1904281.

49 Liu, B., Zhang, J., Wang, X. et al. (2012). Hierarchical three-dimensional $ZnCo_2O_4$ nanowire arrays/carbon cloth anodes for a novel class of high-performance flexible lithium-ion batteries. *Nano Letters* 12 (6): 3005–3011.

50 Chang, J., Shang, J., Sun, Y. et al. (2018). Flexible and stable high-energy lithium-sulfur full batteries with only 100% oversized lithium. *Nature Communications* 9 (1): 4480.

51 Xie, C., Chang, J., Shang, J. et al. (2022). Hybrid lithium-ion/metal electrodes enable long cycle stability and high energy density of flexible batteries. *Advanced Functional Materials* 32 (34): 2203242.

52 Panasonic develops bendable, twistable, flexible lithium-ion battery. https://news.panasonic.com/global/press/data/2016/09/en160929-8/en160929-8.pdf.

53 Flexible lithium polymer battery J.FLEX. https://jenaxinc.com/products/batteries/.

54 Chang, J., Huang, Q., and Zheng, Z. (2020). A figure of merit for flexible batteries. *Joule* 4 (7): 1346–1349.

55 Wan, Y.-J., Wang, X.-Y., Li, X.-M. et al. (2020). Ultrathin densified carbon nanotube film with "Metal-like" conductivity, superior mechanical strength, and ultrahigh electromagnetic interference shielding effectiveness. *ACS Nano* 14 (10): 14134–14145.

56 Wan, S., Li, Y., Mu, J. et al. (2018). Sequentially bridged graphene sheets with high strength, toughness, and electrical conductivity. *Proceedings of the National Academy of Sciences* 115 (21): 5359–5364.

57 Wan, S., Li, X., Chen, Y. et al. (2021). High-strength scalable MXene films through bridging-induced densification. *Science* 374 (6563): 96–99.

58 Chu, H.-C. and Tuan, H.-Y. (2017). High-performance lithium-ion batteries with 1.5 μm thin copper nanowire foil as a current collector. *Journal of Power Sources* 346: 40–48.

59 Ye, Y., Chou, L.-Y., Liu, Y. et al. (2020). Ultralight and fire-extinguishing current collectors for high-energy and high-safety lithium-ion batteries. *Nature Energy* 5 (10): 786–793.

60 Wan, S., Chen, Y., Wang, Y. et al. (2019). Ultrastrong graphene films via long-chain π-bridging. *Matter* 1 (2): 389–401.

61 Blake, A.J., Kohlmeyer, R.R., Drummy, L.F. et al. (2016). Creasable batteries: understanding failure modes through dynamic electrochemical mechanical testing. *ACS Applied Materials & Interfaces* 8 (8): 5196–5204.

62 Wang, D., Zhang, Y., Lu, X. et al. (2018). Chemical formation of soft metal electrodes for flexible and wearable electronics. *Chemical Society Reviews* 47 (12): 4611–4641.

63 Yu, Y., Yan, C., and Zheng, Z. (2014). Polymer-assisted metal deposition (PAMD): a full-solution strategy for flexible, stretchable, compressible, and wearable metal conductors. *Advanced Materials* 26 (31): 5508–5516.

64 Li, P., Zhang, Y., and Zheng, Z. (2019). Polymer-assisted metal deposition (PAMD) for flexible and wearable electronics: principle, materials, printing, and devices. *Advanced Materials* 31 (37): 1902987.

65 Zhang, S., Sakane, M., Nagasawa, T., and Kobayashi, K. (2011). Mechanical properties of copper thin films used in electronic devices. *Procedia Engineering* 10: 1497–1502.

66 Park, G.-G., Park, Y.-K., Park, J.-k., and Lee, J.-W. (2017). Flexible and wrinkle-free electrode fabricated with polyurethane binder for lithium-ion batteries. *RSC Advances* 7 (26): 16244–16252.

67 Harks, P.-P.R.M.L., Robledo, C.B., George, C. et al. (2019). Immersion precipitation route towards high performance thick and flexible electrodes for Li-ion batteries. *Journal of Power Sources* 441: 227200.

68 Kim, J.-H., Lee, Y.-H., Cho, S.-J. et al. (2019). Nanomat Li–S batteries based on all-fibrous cathode/separator assemblies and reinforced Li metal anodes: towards ultrahigh energy density and flexibility. *Energy & Environmental Science* 12 (1): 177–186.

69 Gaikwad, A.M., Khau, B.V., Davies, G. et al. (2015). A high areal capacity flexible lithium-ion battery with a strain-compliant design. *Advanced Energy Materials* 5 (3): 1401389.

70 Jenkins, C.A., Coles, S.R., and Loveridge, M.J. (2020). Investigation into durable polymers with enhanced toughness and elasticity for application in flexible Li-ion batteries. *ACS Applied Energy Materials* 3 (12): 12494–12505.

71 Bao, J.-J., Zou, B.-K., Cheng, Q. et al. (2017). Flexible and free-standing $LiFePO_4$/TPU/SP cathode membrane prepared via phase separation process for lithium ion batteries. *Journal of Membrane Science* 541: 633–640.

72 Lv, Z., Yue, M., Ling, M. et al. (2021). Controllable design coupled with finite element analysis of low-tortuosity electrode architecture for advanced sodium-ion batteries with ultra-high mass loading. *Advanced Energy Materials* 11 (17): 2003725.

73 Wang, J., Wang, M., Ren, N. et al. (2021). High-areal-capacity thick cathode with vertically-aligned micro-channels for advanced lithium ion batteries. *Energy Storage Materials* 39: 287–293.

74 Wu, J., Ju, Z., Zhang, X. et al. (2021). Building efficient ion pathway in highly densified thick electrodes with high gravimetric and volumetric energy densities. *Nano Letters* 21 (21): 9339–9346.

75 Luo, S., Wang, K., Wang, J. et al. (2012). Binder-free $LiCoO_2$/carbon nanotube cathodes for high-performance lithium ion batteries. *Advanced Materials* 24 (17): 2294–2298.

76 Kim, J.-M., Kim, J.A., Kim, S.-H. et al. (2017). All-nanomat lithium-ion batteries: a new cell architecture platform for ultrahigh energy density and mechanical flexibility. *Advanced Energy Materials* 7 (22): 1701099.

77 Hwang, C., Song, W.-J., Han, J.-G. et al. (2018). Foldable electrode architectures based on silver-nanowire-wound or carbon-nanotube-webbed micrometer-scale fibers of polyethylene terephthalate mats for flexible lithium-ion batteries. *Advanced Materials* 30 (7): 1705445.

78 Wang, H., Fu, J., Wang, C. et al. (2021). A universal aqueous conductive binder for flexible electrodes. *Advanced Functional Materials* 31 (34): 2102284.

79 Wang, H., Fu, J., Wang, C. et al. (2020). A binder-free high silicon content flexible anode for Li-ion batteries. *Energy & Environmental Science* 13 (3): 848–858.

80 Lima, M.D., Fang, S., Lepró, X. et al. (2011). Biscrolling nanotube sheets and functional guests into yarns. *Science* 331 (6013): 51–55.

81 Kwon, Y.H., Woo, S.-W., Jung, H.-R. et al. (2012). Cable-type flexible lithium ion battery based on hollow multi-helix electrodes. *Advanced Materials* 24 (38): 5192–5197.

82 Mo, F., Liang, G., Huang, Z. et al. (2020). An overview of fiber-shaped batteries with a focus on multifunctionality, scalability, and technical difficulties. *Advanced Materials* 32 (5): 1902151.

83 Hu, Y.-S. (2016). Batteries: getting solid. *Nature Energy* 1 (4): 16042.

84 Janek, J. and Zeier, W.G. (2016). A solid future for battery development. *Nature Energy* 1 (9): 16141.

85 Yue, L., Ma, J., Zhang, J. et al. (2016). All solid-state polymer electrolytes for high-performance lithium ion batteries. *Energy Storage Materials* 5: 139–164.

86 Manthiram, A., Yu, X., and Wang, S. (2017). Lithium battery chemistries enabled by solid-state electrolytes. *Nature Reviews Materials* 2 (4): 16103.

87 Zhang, Q., Liu, K., Ding, F., and Liu, X. (2017). Recent advances in solid polymer electrolytes for lithium batteries. *Nano Research* 10 (12): 4139–4174.

88 Zhou, D., Shanmukaraj, D., Tkacheva, A. et al. (2019). Polymer electrolytes for lithium-based batteries: advances and prospects. *Chem* 5 (9): 2326–2352.

89 Liang, G., Zhu, J., Chen, A. et al. (2021). Adhesive and cohesive force matters in deformable batteries. *npj Flexible Electronics* 5 (1): 27.

90 Mackanic, D.G., Yan, X., Zhang, Q. et al. (2019). Decoupling of mechanical properties and ionic conductivity in supramolecular lithium ion conductors. *Nature Communications* 10 (1): 5384.

91 Zhang, Y., Chen, R., Wang, S. et al. (2020). Free-standing sulfide/polymer composite solid electrolyte membranes with high conductance for all-solid-state lithium batteries. *Energy Storage Materials* 25: 145–153.

92 Zhang, X., Liu, T., Zhang, S. et al. (2017). Synergistic coupling between $Li_{6.75}La_3Zr_{1.75}Ta_{0.25}O_{12}$ and poly(vinylidene fluoride) induces high ionic conductivity, mechanical strength, and thermal stability of solid composite electrolytes. *Journal of the American Chemical Society* 139 (39): 13779–13785.

93 Wan, J., Xie, J., Mackanic, D.G. et al. (2018). Status, promises, and challenges of nanocomposite solid-state electrolytes for safe and high performance lithium batteries. *Materials Today Nano* 4: 1–16.

94 Correia, D.M., Fernandes, L.C., Martins, P.M. et al. (2020). Ionic liquid–polymer composites: a new platform for multifunctional applications. *Advanced Functional Materials* 30 (24): 1909736.

95 Tan, S.-J., Yue, J., Tian, Y.-F. et al. (2021). In-situ encapsulating flame-retardant phosphate into robust polymer matrix for safe and stable quasi-solid-state lithium metal batteries. *Energy Storage Materials* 39: 186–193.

96 Jaumaux, P., Liu, Q., Zhou, D. et al. (2020). Deep-eutectic-solvent-based self-healing polymer electrolyte for safe and long-life lithium-metal batteries. *Angewandte Chemie International Edition* 59 (23): 9134–9142.
97 Xiang, J., Zhang, Y., Zhang, B. et al. (2021). A flame-retardant polymer electrolyte for high performance lithium metal batteries with an expanded operation temperature. *Energy & Environmental Science* 14 (6): 3510–3521.
98 Alarco, P.-J., Abu-Lebdeh, Y., Abouimrane, A., and Armand, M. (2004). The plastic-crystalline phase of succinonitrile as a universal matrix for solid-state ionic conductors. *Nature Materials* 3 (7): 476–481.
99 Ha, H.-J., Kil, E.-H., Kwon, Y.H. et al. (2012). UV-curable semi-interpenetrating polymer network-integrated, highly bendable plastic crystal composite electrolytes for shape-conformable all-solid-state lithium ion batteries. *Energy & Environmental Science* 5 (4): 6491–6499.
100 Choi, K.-H., Cho, S.-J., Kim, S.-H. et al. (2014). Thin, deformable, and safety-reinforced plastic crystal polymer electrolytes for high-performance flexible lithium-ion batteries. *Advanced Functional Materials* 24 (1): 44–52.
101 Chao, D., Zhou, W., Xie, F. et al. (2020). Roadmap for advanced aqueous batteries: from design of materials to applications. *Science Advances* 6 (21): eaba4098.
102 Liu, Z., Huang, Y., Huang, Y. et al. (2020). Voltage issue of aqueous rechargeable metal-ion batteries. *Chemical Society Reviews* 49 (1): 180–232.
103 Suo, L., Borodin, O., Gao, T. et al. (2015). "Water-in-salt" electrolyte enables high-voltage aqueous lithium-ion chemistries. *Science* 350 (6263): 938–943.
104 Xie, J., Liang, Z., and Lu, Y.-C. (2020). Molecular crowding electrolytes for high-voltage aqueous batteries. *Nature Materials* 19 (9): 1006–1011.
105 He, X., Yan, B., Zhang, X. et al. (2018). Fluorine-free water-in-ionomer electrolytes for sustainable lithium-ion batteries. *Nature Communications* 9 (1): 5320.
106 Zhang, J., Cui, C., Wang, P.-F. et al. (2020). "Water-in-Salt" polymer electrolyte for Li-ion batteries. *Energy & Environmental Science* 13 (9): 2878–2887.
107 Yang, C., Chen, J., Ji, X. et al. (2019). Aqueous Li-ion battery enabled by halogen conversion–intercalation chemistry in graphite. *Nature* 569 (7755): 245–250.
108 Zhao, N., Khokhar, W., Bi, Z. et al. (2019). Solid garnet batteries. *Joule* 3 (5): 1190–1199.
109 Fu, K., Gong, Y., Dai, J. et al. (2016). Flexible, solid-state, ion-conducting membrane with 3D garnet nanofiber networks for lithium batteries. *Proceedings of the National Academy of Sciences* 113 (26): 7094–7099.
110 Wan, J., Xie, J., Kong, X. et al. (2019). Ultrathin, flexible, solid polymer composite electrolyte enabled with aligned nanoporous host for lithium batteries. *Nature Nanotechnology* 14 (7): 705–711.
111 Luntz, A.C., Voss, J., and Reuter, K. (2015). Interfacial challenges in solid-state Li ion batteries. *The Journal of Physical Chemistry Letters* 6 (22): 4599–4604.
112 Wu, B., Wang, S., Evans Iv, W.J. et al. (2016). Interfacial behaviours between lithium ion conductors and electrode materials in various battery systems. *Journal of Materials Chemistry A* 4 (40): 15266–15280.

113 Tian, Y., Shi, T., Richards, W.D. et al. (2017). Compatibility issues between electrodes and electrolytes in solid-state batteries. *Energy & Environmental Science* 10 (5): 1150–1166.

114 Liu, Y., Lin, D., Jin, Y. et al. (2017). Transforming from planar to three-dimensional lithium with flowable interphase for solid lithium metal batteries. *Science Advances* 3 (10): eaao0713.

115 Vijayakumar, V., Anothumakkool, B., Kurungot, S. et al. (2021). In situ polymerization process: an essential design tool for lithium polymer batteries. *Energy & Environmental Science* 14 (5): 2708–2788.

116 Zhao, Q., Liu, X., Stalin, S. et al. (2019). Solid-state polymer electrolytes with in-built fast interfacial transport for secondary lithium batteries. *Nature Energy* 4 (5): 365–373.

117 Jiang, Z., Jin, L., Han, Z. et al. (2019). Facile generation of polymer–alloy hybrid layers for dendrite-free lithium-metal anodes with improved moisture stability. *Angewandte Chemie International Edition* 58 (33): 11374–11378.

118 Liu, F.-Q., Wang, W.-P., Yin, Y.-X. et al. (2018). Upgrading traditional liquid electrolyte via in situ gelation for future lithium metal batteries. *Science Advances* 4 (10): eaat5383.

119 Lee, M.J., Han, J., Lee, K. et al. (2022). Elastomeric electrolytes for high-energy solid-state lithium batteries. *Nature* 601 (7892): 217–222.

120 Kim, S.-H., Choi, K.-H., Cho, S.-J. et al. (2015). Printable solid-state lithium-ion batteries: a new route toward shape-conformable power sources with aesthetic versatility for flexible electronics. *Nano Letters* 15 (8): 5168–5177.

121 Kim, S.-H., Kim, J.-H., Cho, S.-J., and Lee, S.-Y. (2019). All-solid-state printed bipolar Li–S batteries. *Advanced Energy Materials* 9 (40): 1901841.

122 Xu, S., Zhang, Y., Cho, J. et al. (2013). Stretchable batteries with self-similar serpentine interconnects and integrated wireless recharging systems. *Nature Communications* 4 (1): 1543.

123 Song, Z., Ma, T., Tang, R. et al. (2014). Origami lithium-ion batteries. *Nature Communications* 5 (1): 3140.

124 Song, Z., Wang, X., Lv, C. et al. (2015). Kirigami-based stretchable lithium-ion batteries. *Scientific Reports* 5 (1): 10988.

125 Qian, G., Zhu, B., Liao, X. et al. (2018). Bioinspired, spine-like, flexible, rechargeable lithium-ion batteries with high energy density. *Advanced Materials* 30 (12): 1704947.

126 Tebyetekerwa, M., Marriam, I., Xu, Z. et al. (2019). Critical insight: challenges and requirements of fibre electrodes for wearable electrochemical energy storage. *Energy & Environmental Science* 12 (7): 2148–2160.

127 He, J., Lu, C., Jiang, H. et al. (2021). Scalable production of high-performing woven lithium-ion fibre batteries. *Nature* 597 (7874): 57–63.

128 Liao, M., Wang, C., Hong, Y. et al. (2022). Industrial scale production of fibre batteries by a solution-extrusion method. *Nature Nanotechnology* 17 (4): 372–377.

129 Gao, Y., Hu, H., Chang, J. et al. (2021). Realizing high-energy and stable wire-type batteries with flexible lithium–metal composite yarns. *Advanced Energy Materials* 11 (40): 2101809.

130 Li, H., Tang, Z., Liu, Z., and Zhi, C. (2019). Evaluating flexibility and wearability of flexible energy storage devices. *Joule* 3 (3): 613–619.

131 Chen, A., Guo, X., Yang, S. et al. (2021). Human joint-inspired structural design for a bendable/foldable/stretchable/twistable battery: achieving multiple deformabilities. *Energy & Environmental Science* 14 (6): 3599–3608.

132 Xie, C., Guo, Y., and Zheng, Z. (2023). Pushing the limit of flexible batteries. *CCS Chemistry* 5 (3): 531–543.

133 Choi, J.W. and Aurbach, D. (2016). Promise and reality of post-lithium-ion batteries with high energy densities. *Nature Reviews Materials* 1 (4): 16013.

134 Wu, F. and Yushin, G. (2017). Conversion cathodes for rechargeable lithium and lithium-ion batteries. *Energy & Environmental Science* 10 (2): 435–459.

135 Wu, F., Maier, J., and Yu, Y. (2020). Guidelines and trends for next-generation rechargeable lithium and lithium-ion batteries. *Chemical Society Reviews* 49 (5): 1569–1614.

136 Liu, B., Jia, Y., Yuan, C. et al. (2020). Safety issues and mechanisms of lithium-ion battery cell upon mechanical abusive loading: a review. *Energy Storage Materials* 24: 85–112.

137 Lin, D., Liu, Y., and Cui, Y. (2017). Reviving the lithium metal anode for high-energy batteries. *Nature Nanotechnology* 12 (3): 194–206.

138 Lu, J., Li, L., Park, J.-B. et al. (2014). Aprotic and aqueous Li–O_2 batteries. *Chemical Reviews* 114 (11): 5611–5640.

Index

Electrochemical Energy Storage Devices: Non-Conventional Technologies and Materials, First Edition. Edited by Yongbing Tang and Luojiang Zhang.

y

z